Seedborne Diseases and Their Control

Principles and Practice

This book is dedicated to
the memory of
Amanda Jill Maude
1962–1988

Seedborne Diseases and Their Control
Principles and Practice

R.B. Maude

Horticulture Research International
Wellesbourne
Warwick CV34 9EF, UK

CAB INTERNATIONAL

CAB INTERNATIONAL
Wallingford
Oxon OX10 8DE
UK

Tel: +44 (0)1491 832111
Fax: +44 (0)1491 833508
E-mail: cabi@cabi.org
Telex: 847964 (COMAGG G)

A catalogue record for this book is available from the British Library.

ISBN 0 85198 922 5

Typeset in 10/12 Stempel Garamond by AMA Graphics Ltd, Preston
Printed and bound in the UK at the University Press, Cambridge

Contents

Preface

I have worked as an applied plant pathologist investigating the seedborne diseases of temperate vegetables in a career spanning some 32 years. During that time I was a member of staff of the National Vegetable Research Station, Wellesbourne which later became Horticulture Research International, UK. In my research, which concerned fungal pathogens transmitted by seed, I have found it necessary to understand the biology of organisms before attempting to devise practical methods for their control.

Although my work applied mainly to field vegetables, many of the principles and practices are transferable to seedborne diseases of other crops. I took the opportunity to review the wider perspective in a series of lectures on the biology and control of seedborne diseases presented to final year students of the School of Biological Sciences of the University of Birmingham over a period of some 20 years. In those talks I discussed the seed-transmitted fungi, bacteria and viruses which cause economic diseases of many crops in different climates and examined the principles supporting the practices which have been developed for their control.

Those lectures and my own research experience form the basis of this book. It is written for those with an interest in seed pathology at student, graduate and research worker level. I believe that it is also suitable for the more general reader with an interest in the subject.

The first chapter briefly reviews historical aspects of seed pathology and seedborne organisms. The main thrust of the book is concerned with the processes of infection of seeds (Chapter 2), the location and the survival of inoculum (Chapter 3), the transmission (Chapter 4) and spread of seedborne pathogens (Chapter 5) and how, with a knowledge of the biology of organisms, strategies and methods are developed and employed at national and international levels to exclude and eradicate seed-transmitted diseases (Chapters 6–8). Effective methods for the detection of seedborne inoculum are important and the principles involved and progress made in this technology is considered in Chapter 9. Taken together, Chapters 2–5 describe the disease cycle of organisms. To preserve the continuity of this cycle I found it necessary to engage in some repetition between chapters.

In the preparation of this text I have made use of manual and modern abstracting methods to retrieve references. In addition, I read or consulted a number of books or chapters in books on seed pathology. I found that mistakes which occurred in some of the earlier texts were sometimes transferred and reproduced in the more recent accounts. For that reason I read the original papers, where I could, and any mistakes which occur are mine.

I have introduced such terms and definitions as appeared necessary to me at various appropriate points in the text.

The book is not a comprehensive account of seedborne organisms; rather it sets out to describe the main principles and practices involved in seed pathology.

Acknowledgements

The writing of this book has been greatly helped by the fact that after retirement in 1989 I continued at my workplace for two years as a member of a team in an Agriculture and Food Research Council-linked research programme between Horticulture Research International (HRI), Wellesbourne and the University of Birmingham working on the 'Process engineering of seeds'. This was followed by the award of an Emeritus Fellowship by HRI, Wellesbourne to complete this manuscript and to continue research on seed-borne diseases. I am extremely grateful to Professor C.C. Payne and the management of HRI for the support and the facilities they provided.

I am indebted to a number of scientists who read the complete manuscript or part of it and gave valued criticisms which have enabled me to strengthen and improve the text. In that respect I am particularly grateful to Dr Sarah F.L. Ball (Royal Horticultural Society's Gardens, Wisley, Surrey), Mr W.F. Rennie (Scottish Agricultural Science Agency, East Craigs, Edinburgh), Dr J.D. Taylor and Mr P. Hunter (HRI, Wellesbourne), Dr J.C. Reeves (National Institute of Agricultural Botany, Cambridge), Mr T.J. Martin (Bayer plc, Bury St Edmunds) and to Dr M. Holderness and Dr E. Punithalingham (International Mycological Institute, Egham, Surrey).

I wish to acknowledge with thanks the editors of journals and individuals who permitted me to use some of the illustrations which appear in this book. I also wish to thank the following HRI staff: Miss Lindsey Peach, who reproduced many of the figures and graphs, Miss Margaret Jones for the artwork and Mr Richard Sampson for photographic assistance.

Glossary

Abaxial – the side of a lateral organ away from the axis.
Acervulus plural acevuli – a compact cushion-like mass of hyphae having conidiophores, conidia and often setae (dark bristles), characteristic of the Melanconiales (Fungi Imperfecti).
Active ingredient – the active component in a formulated product.
Adaxial – the side of a lateral organ next to the axis.
Alarm pheromone – substance secreted by aphids in response to attack by predatory insects causing the aphids to disperse rapidly.
Anamorph – see Imperfect stage.
Anthesis – period of flowering of a crop.
Anti-feedants – plant-derived, insect behaviour-controlling chemicals.
Ascospores – the sexual spores of Ascomycetes, for example *Mycosphaerella pinodes, Pyrenophora teres* etc.
Asexual – without sex spores or sexual organs.
Autoradiograph – photograph of an object produced by radiation from radioactive material in the object.
Basidium – the structure on which the basidiospores (sex spores) in Basidiomycetes undergo development.
Break crop – crop of a different plant species (e.g. oilseed rape) grown to avoid the continual cropping of another species (e.g. cereals).
Bunt balls – teliospore masses of *Tilletia tritici* contained by the pericarps of cereal caryopses (fruits) whose ovaries have been replaced by that fungus.
Bunt kernels – cereal caryopses infected by the bunt fungus (*Tilletia tritici*).
Capsular fruit – a dry dehiscent fruit.
Chlamydospore – a thick-walled asexual resting spore which can be intercalary or terminal on the mycelium.

Coleoptile – protective sheath enclosing the germinating plumule of monocot plants.

Conidiophore – a simple or branched hypha on which conidia are produced.

Conidium (plural conidia) – any asexual spore which comes away from its conidiophore when mature.

Cultivar – a variety of a cultivated plant.

Dicotyledons (dicots) – flowering plants (Angiosperms) possessing embryos with two cotyledons, e.g. phaseolus beans, brassicas, etc.

Edaphic – produced or influenced by the soil.

Endophyte – an organism living within another organism.

Ergot – the sclerotium of *Claviceps* spp., especially *C. purpurea* (ergot).

False negative – the incorrect exclusion of an individual in a diagnostic test or another process, for example, the extraction of bacteria from seeds.

False positive – the incorrect inclusion of an individual in a diagnostic test or another process.

Filler – a diluent in powder form.

Formulation – the method and process of selecting the components of a mixture; the product of such a process.

Fruiting bodies – asexual or sexual spore-bearing structures, such as conidiophores, pycnidia, sporodochia, perithecia, etc.

Fumigation – use of chemical compounds in a volatile state to clear an area or a seed bulk of unwanted organisms (e.g. insects or nematodes).

Gamma globulin – any serum globulin (albumin-associated protein) having gamma electrophoretic mobility; broadly, any immunoglobulin.

Gel electrophoresis – protein separation that is carried out in a silica gel or acrylamide gel under the influence of an applied electric field.

Gene – an inherited factor that determines the constitution of an organism; a hereditary unit that is composed of a sequence of DNA and occupies a specific position on a chromosome.

Genome – complete gene complement of an organism contained in a set of chromosomes (animals, plants, fungi); in a single chromosome (bacteria); or in a DNA or RNA molecule (viruses).

Germplasm – the hereditary material contained in the gametes that is transmitted to the offspring; seeds carry the germplasm of their species.

Homologous reaction – a serological reaction in which an antiserum reacted against the antigen used in its preparation.

Honeydew – plant sap rich in sugars secreted as a yellowish fluid from *Claviceps purpurea*-infected cereal flowers.

Hypha (plural hyphae) – a thread of mycelium.

Imperfect stage – the state in which asexual spores (such as conidia) or no spores are produced; anamorph.

Infest – to be dispersed through soil or other substrate. When used in reference to fungi, etc., on the surface of seeds there is no implication that infection has occurred or will occur.

Lesion – a localized area of diseased or disordered tissue.
Mericarp – the one-seeded portion of a schizocarp.
Monocotyledons (monocots) – flowering plants (Angiosperms) possessing embryos with a single cotyledon, e.g. cereals, onions, etc.
Mycelium (plural mycelia) – a mass of hyphae; the thallus of a fungus.
Non-indigenous pathogen or disease – a pathogen that is not one of the naturally occurring organisms in a land or region.
Non-persistent virus – one which remains infective within its insect vector for only a short period.
Oomycetes – Phycomycetes having oospores.
Oospore – the fertilized resting spore of one of the lower fungi.
Palisade cells – elongated cells forming the outer epidermal layer beneath the cuticle of seeds.
Parenchyma – large thin-walled cellulose cells containing a living protoplast.
Pathovar – the subdivision of a species distinguished by common characters of pathogenicity, particularly in relation to host range.
Pedicel – the supporting axis of a single flower.
Perfect stage – the state of the fungal life cycle in which spores (such as ascospores and basidiospores) are formed after nuclear fusion or by parthenogenesis; teleomorph.
Perianth – protective whorls around the stamens and ovaries of flowers comprising the calyx (sepals) and corolla (petals).
Pericarp – formed from the wall of the fertilized ovary; this is the fruit wall and remains fleshy in some species or becomes hard or dry in others.
Perithecium – globose or flask-shaped sexual fruiting body of some Ascomycetes (Pyrenomycetes) containing asci which bear the sexual spores (ascospores).
Persistent virus – one which remains infective within its insect vector for a long period.
Phaseolus bean – of the genus *Phaseolus*, for example, *P. vulgaris* (common bean, French bean, dwarf bean), *P. coccineus* (runner bean), *P. lunatus* (lima bean) and *P. acutifolius* (tepary bean).
Phloem – large empty-looking conducting cells (sieve tubes) in the vascular system of plants.
Pistil – the female organ of a flower, comprising ovary, style and stigma.
Plasmodesmata – the cytoplasmic connections between cells.
Pseudomorph – stroma made up of fungal hyphae held together within a strong coat.
Pycnidiospore – a conidium in or from a pycnidium.
Pycnidium (plural pycnidia) – asexual fruiting body of the Sphaeropsidales (Fungi Imperfecti) and Ascomycetes; globose or flask-shaped, containing pycnidiospores.
Race – genetically, and usually geographically, distinct mating groups within a species.

Resistance, horizontal – resistance which is evenly spread against all races of the pathogen.

Resistance, vertical – resistance to some races of a pathogen but not to others.

Schizocarp – a pericarp which splits into one-seeded portions, i.e. mericarps.

Sclerenchyma – strong lignified protective cells with little living contents at maturity.

Sclerotium (plural sclerotia) – a resting body, usually with a dark hard rind, containing a compact mass of hyphae from which asexual or sexual fruiting bodies may develop; external on or internal in host tissues.

Sonication – the process of subjecting a cell, virus, etc. to high-frequency sound-wave energy, usually for purposes of dispersal or separation.

Spore – a general term for a reproductive structure in cryptogams, i.e. fungi, algae, mosses (Bryophyta) and ferns (Pteridophyta)

Sporodochium (plural sporodochia) – tightly packed conidiophores located on a stroma or on a mass of hyphae as in the Tuberculariaceae (Fungi Imperfecti).

Stigma – part of the pistil or style that receives the pollen.

Strain – many meanings, including a group of similar isolates; race; form; the descendants of a single isolation in pure culture; isolate; a culture of bacteria which corresponds to a cultivated variety (cultivar) of higher plants having some special significance; a group of plant viruses having most of its antigens in common with another group.

Stroma – a mass of vegetative hyphae, with or without tissue of the host or substrate, sometimes sclerotium-like, on or in which spores may be produced.

Style – the thin part of a pistil between the ovary and the stigma.

Stylodium – a style-like stigma, as in the grasses.

Symbiont – an organism living in symbiosis.

Symbiosis – the association of two different organisms living attached to each other or one within the other.

Synergism – the combined effects of two or more systems (for example, fungicides and organisms) producing a change which one only is not able to make.

Systemic – entering a plant via the roots or shoots and passing through the tissues.

Teleomorph – see Perfect stage.

Teliospore – a winter (resting) spore of certain Basidiomycetes, for example, the rusts (Uredinales) or the smuts (Ustilaginales), from which a basidium is produced.

Testa (plural testae) – the seed coat.

Thallus – the vegetative body of a fungus (mycelium) or alga.

Thermotherapeutic methods – use of heat treatments to control diseases, for example by treating the seedborne inoculum of viruses, bacteria and fungi.

Tiller – shoot of plant springing from the bottom of the original stalk.

Titre – the strength of a solution as determined by titration; the highest dilution of an antiserum that will react with its homologous virus.

Vascular system – comprises the principle water- and nutrient-conducting tissues (xylem and phloem) of plants.

Vector – an organism able to transmit a pathogen, especially an insect, nematode, etc. transmitting a virus.

Vicia bean – of the genus *Vicia*, for example, *V. faba* (broad bean, field bean).

Xylem – woody tissues in vascular plants which give support and conduct water and nutrients.

1 Seed Pathology

Introduction

Seeds are an essential component of world trade and are distributed nationally and globally, destined to be eaten as food or to be used to grow crops which themselves are consumed. About 90% of all food crops in the world are propagated by seeds (Schwinn, 1994). They are also the passive carriers of pathogens which are transmitted when the seed hosts are sown and emerge under suitable environmental conditions (Noble, 1957, 1971). Fungi, bacteria, viruses and nematodes are implicated and can be carried with, on and in seeds. Concern over the ease of distribution of seedborne pathogens by national and international trade in seeds has been expressed by pathologists down the years (Moore, 1946; Muskett, 1950; Noble, 1951; Neergaard, 1977). There is an awareness that the increasing movement of seed germplasm around the world also provides an avenue for the dispersion of all crop pathogens, and in particular viruses (Hampton *et al*., 1982). Neergaard (1981) identified the risks for the European and Mediterranean Plant Protection Organization (EPPO) area (see Smith, 1979) of the introduction of new seedborne pathogens and/or new pathotypes of existing diseases and of the redistribution of existing seedborne pathogens and their pathotypes within the region. Changes in seed health responsibilities may also affect the distribution of seedborne diseases. Such changes may follow the introduction of the single European market in June 1993 (Ebbels, 1993) where the onus for seed health in the European Community (EC) resides with exporting nations within the community (Anon., 1989; Rennie, 1993).

Seed pathology is concerned with the recognition and control of seedborne diseases. To proceed from one to the other, a complete understanding of the biology of pathogens is essential.

There is a prevailing image which associates seed pathology mainly with the development of laboratory practices for the detection of seedborne organisms (McGee, 1981) leaving an only partially fulfilled requirement to relate laboratory seed tests to the risk of subsequent field disease (Baker, 1972; McGee, 1981; Neergaard, 1986; Agarwal and Sinclair, 1987). Appreciation of 'risk' requires the discovery of life cycles and disease cycles of seedborne pathogens and an understanding of the interrelationships of these and of the many environmental factors which influence them. From such information are derived the principles and practices which are applied to minimize transmission of seedborne pathogens and to achieve control. These are the main areas treated in this book.

A Short History of Seed Pathology

It is thought that parasitic fungi and bacteria emerged with land plants from the Palaeozoic seas more than 330 million years ago (Baker, 1972). Buller (1950) suggested that fungi such as *Claviceps* spp. (ergots) had developed mechanisms for seed transmission 130 million years before the present time. This information has relevance to current seed pathology in so far as it indicates that, for certain, pathogens there has been considerable evolutionary time for the development of potentially sophisticated host–pathogen relationships which are intimately linked in the plant–disease cycle, and as such may prove difficult to separate when applications of disease control measures are considered. On such a time scale the science of seed pathology is extremely young. Its inadvertent beginnings may go back to the 18th century when Jethro Tull (1733) recorded the observation of farmers near Bristol, UK that seed wheat salvaged from the sea was free of bunt (*Tilletia tritici* syn. *T. caries*). It seems probable now that Tull's description may be based on a much earlier account (Buttress and Dennis, 1947) and 'brining' was a steep treatment in use in the UK by the middle of the 17th century for the treatment of wheat (Blith, 1652). Whatever the origins of brining, its discovery introduced the concept of disinfection by means of a seed treatment. Paradoxically it was not until 1755 that Tillet (1755) established unequivocally by experiment that the bunt fungus was seedborne. Many of the subsequent records of seedborne organisms related to the bunts and smuts of cereals and it was almost a century later before Frank (1883) demonstrated that the fungus causing anthracnose of phaseolus beans, *Colletotrichum lindemuthianum*, was seedborne. The chronological order in which the early discoveries of seedborne pathogens, including fungi, bacteria, viruses and nematodes, were made is listed by Baker (1972). Since then, this information has been updated in textbooks by Neergaard (1977) and more recently by Agarwal and Sinclair (1987), and all three publications describe the general progress made in the science of seed pathology.

Recognition and Detection of Seedborne Organisms

At the turn of the century, with the introduction of official seed testing stations (the first was in Lower Saxony, Germany in 1869 (Malone and Muskett, 1964)), seeds of a wide range of crops came under scrutiny for purity and quality. Quality was assessed by laboratory-designed germination tests done under moist conditions and sometimes it could be seen that seeds had microorganisms growing on them. Fungi were often easily recognizable and in some cases, for example *Gloeotinia granigena* (syn. *G. temulenta*) or blind seed disease of ryegrass, it was possible to see a relationship between the percentage of non-viable seeds with the percentage of those infected by the fungus. It was evident that the quality of seeds could be affected by the organisms they bore and that a statement concerning the health of seeds might be possible by identifying the organisms affecting them. However, it was also true that tests of germination, though excellent for assessing seed quality and for demonstrating the presence of some organisms, were not usually suitable techniques for the detection of other organisms.

Dorogin (1923) established a number of seed health testing methods for crop plants in Russia; later, crop seed testing was made obligatory for all of the former USSR (Budrina, 1935). In many cases the methods were simple and direct and could be used in tandem with the testing of seeds for purity and germination. They included, for example, observations for the occurrence of sclerotia mixed with the seed, the mummification of seeds by sclerotia of different organisms, surface-borne fructifications of fungi, surface-borne spores of fungi and bacteria, etc. Some of Dorogin's methods were recommended for cereal testing in Germany (Klemm, 1926) and were modified for use in the USA (Orton, 1931). Munn (1919) and Orton (1931) expressed the view that seed analysts should be trained to recognize seedborne organisms. At about that time, Doyer, in the Netherlands, advocated that the international seed testing rules developed for purity and quality testing of a wide range of seeds should be extended to include seed health testing. In fact, the First International Rules for Seed Testing published in 1928 drew special attention to seedborne pathogens and pests on peas, cereals, beans and flax (Wold, 1983). It was Doyer, however, working at the Official Seed Testing Station, Wageningen, who developed the basic methods of seed health testing (Wold, 1983). Doyer's *Manual for the Determination of Seed-borne Diseases* (1938) became a landmark publication in seed technology (Malone and Muskett, 1964), describing reproducible methods for the detection and identification of a considerable number of seedborne fungi, bacteria, nematodes and insects.

Porter (1949), in the USA, identified the 1930s as a period when interest in seed pathology expanded essentially because of: (i) a few very active research programmes concerned with the production of practical methods for determining the presence of seedborne organisms; and (ii) the compilation of lists of seedborne parasites. Twenty years later in the UK, Noble (1951) welcomed the report of a government committee which ruled in 1950 that seed testing practice should be extended to cover 'health' and that there should be further research into rapid methods for the detection of seedborne organisms. Although common objectives within seed pathology were recognized internationally, the rate of progress of officialdom varied.

There has been marked progress over the past 40 years in the development of accurate and rapid methods for the detection of seedborne pathogens. Nonetheless, for a considerable period 'traditional' methods prevailed, and for certain applications still do. These include the detection of seedborne fungi by the recognition of the cultural characteristics of fungi growing from seeds incubated on agar and of their fruiting bodies on seeds placed on moist blotter (filter or germination paper) with the aid of microscopes. Methods for the detection of seedborne bacteria and viruses have depended on growing seeds to produce disease symptoms, followed by pathogen isolation and biochemical tests for the identification of bacteria (Stead, 1992) and isolation and inoculation of indicator plants for the identification of viruses (Marrou and Messiaen, 1967; Kimble *et al.*, 1975). Technological advances have now provided selective media for the recognition of organisms, particularly bacteria, the use of viruses (i.e. phages) for the detection of bacteria and the development of an array of immunological (serological) techniques particularly for the detection of seedborne viruses but also bacteria (Lange, 1986; Franken and Van Vuurde, 1990; Ball and Reeves, 1992). Molecular biological techniques used in clinical microbiology (Tenover, 1988) are recent innovations which provide nucleic acid-based methods for the identification of organisms. These include deoxyribonucleic acid (DNA) probes and polymerase chain reaction (PCR) methods (Duncan and Torrance, 1992; Fox, 1993) which, when proven, may have considerable application for seed health testing technology in coming years (Ball and Reeves, 1992). Seed detection techniques are considered in greater detail in Chapter 9.

Occurrence of Seedborne Organisms

Information on seedborne organisms increased with the publication of records of their occurrence. Bessey in 1886, in the USA, issued bulletins on the results of seed germination tests with notes on the accompanying fungi (Noble, 1951). Specialized studies can also be found in the literature including, for example, those on tree seeds (Vanine and Kotchkina, 1932) and on rice (Tullis, 1936). Alcock (1931) provided a list of 53 organisms which she tentatively associated with seeds of roots, pulses, forage crops, vegetables, flax and flowers in the UK. Orton's (1931) bibliography of seedborne parasites, however, was a much more comprehensive record of fungi, bacteria, nematodes and viruses of cereals, grasses, beans, vegetables and many other seed species in North America. This compilation was updated by Porter in 1949. A similar comprehensive approach was taken in Europe with the publication of *An Annotated List of Seed-borne Diseases* by Noble *et al.* (1958). This lists comments on a very wide range of organisms and provides information on their disease status and transmission. It includes records from other lists including those of Orton (1931) and from the *Annual Reports from the Phytopathological Laboratory of J.E. Ohlsens Enke* in Copenhagen. It has been reprinted twice. The fourth edition by M.J. Richardson was published in 1990 by the International Seed Testing Association (ISTA). Such lists provide an excellent entry point for information on the identity and pathology of a wide range of organisms occurring on seeds.

Research on Seedborne Diseases

Research on the biology and epidemiology of seedborne diseases has been undertaken by scientists in universities, research stations, advisory and extension services and within private seed firms. The increase in research is associated with a developing appreciation that seedborne organisms transmit diseases which often cause partial or total crop loss. Neergaard's (1977) review of plant losses due to seedborne pathogens of the major cereal, legume, oil, vegetable and fibre crops illustrated the magnitude of such problems and identified the need for research. Losses of US$ 5.6 million in the Pacific North-West resulting from the downgrading of wheat due to smuttiness of the grain caused by the bunt fungus (*Tilletia tritici*) and famine in parts of Japan due to rice blast in the 1930s and 1940s are examples of extreme effects of diseases arising from the use of infected seeds.

2 Infection of Seeds

Introduction

Seeds and other forms of planting material are the starting units in any crop cycle; seeds also may be the end point. Infections which develop during the growth of a crop ultimately may affect the flowering parts of plants and thereby the seeds. The types of organism involved and the infection routes which are followed determine the location of organisms on and in the seed tissues, affect their capacity for survival in the seedborne phase and, ultimately, affect their transmission from seeds to seedlings. The process of infection is dynamic, but it is often viewed retrospectively by histopathological, electron microscopic and other techniques, which largely are interpretive and which individually may present slightly different perspectives of the stages of the process.

Terms and Definitions

Seeds

Generative organs (seeds, fruits, etc.) and vegetative units (tubers, rhizomes, corms, bulbs, etc.) ensure the survival and continuation of many crop plants.

This book is concerned with the problems of seeds. Seeds are fertilized ovules containing embryos surrounded by an integument. The embryo results from the fertilization of the egg cell; the seed coat originates from the mother plant, normally from the integuments of the ovule (Fig. 1A,B). Endosperm or nucellar tissue provide food reserves or, where these are rudimentary, cotyledons replace this function (Mayer and Poljakoff-Mayber, 1989). Embryos

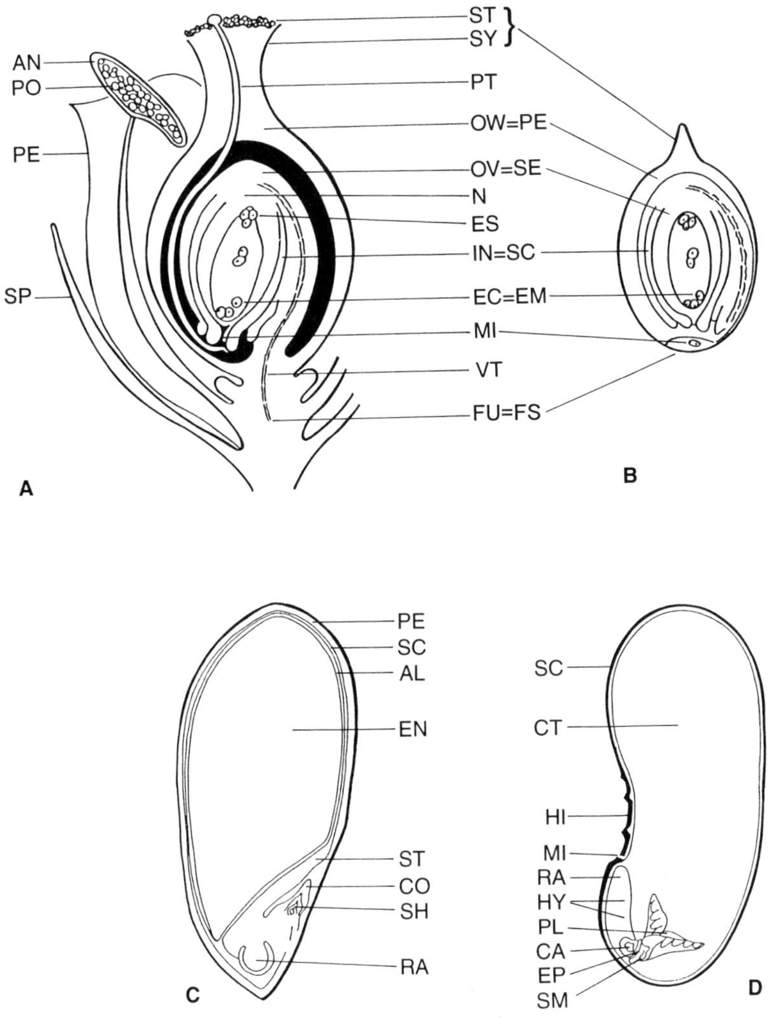

Fig. 1. Comparative anatomy (diagrammatic) of flower, fruit and seeds. (A) Longitudinal section (LS) of a dicot flower. AN, anther; PO, pollen; PE, petal; SP, sepal; ST, stigma; SY, style; PT, pollen tube; OW(PE), ovary wall (pericarp); OV(SE), ovule (seed); N, nucellus; ES, egg sac; IN(SC), integument (seed coat); EC(EM), egg cell (embryonic plant); MI, micropyle; VT, vascular trace; FU(FS), funicle (funicle scar = hilum). (B) LS of a newly formed fruit. Abbreviations as (A). (C) LS of monocot wheat fruit (caryopsis), a functional seed. PE, pericarp; SC, seed coat; AL, aleurone layer; EN, endosperm; ST, scutellum; CO, coleoptile; SH, shoot; RA, radicle. (D) LS through dicot phaseolus bean seed. SC, seed coat; CT, cotyledon; HI, hilum; MI, micropyle; RA, radicle; HY, hypocotyl; PL, plumule or shoot; CA, cotyledon attachment; EP, epicotyl; SM, shoot meristem. Note that the ovary wall forms a pod from which the bean seeds are shed.

comprise the storage organs and the rudimentary plant, which has a plumule (shoot) and radicle (root) connected by a hypocotyl (Fig. 1C,D). In addition to true seeds, for example, peas, brassicas, beans, etc., 'functional seeds', i.e. one-seeded (rarely two- or three-seeded) fruits are considered here. The fruit coat develops from the ovary of the flowering plant (Fig. 1A,B). Such seeds include carrot, celery (Umbelliferae), cereals (Gramineae; Fig. 1C) and others (see Kozlowski, 1972).

This book does not describe diseases of bulbs, corms, tubers and potato seed pieces.

Seedborne pathogens

Seedborne fungi are defined by Ingold (1953) as those 'which are dispersed in association with some kind of dispersal units of the host'. This description embraces all seed types and all associated microfungi. Baker (1972) limits the description to seedborne pathogens and seedborne diseases. He describes seedborne pathogens as those organisms which on or in seeds may or may not show symptoms (on the seeds) and seedborne diseases as those which actively attack seeds and may or may not show symptoms on the seeds. Many of the necrotrophic fungi which are seedborne pathogens causing diseases of emerging seedlings also attack and kill a proportion of the seeds, often producing symptoms on them, for example; *Mycosphaerella pinodes* leaf and pod spot with foot rot of peas. The suffix 'borne' relates to 'carrying' plant pathogens rather than 'expressing' disease or disease symptoms. Diseases which are wholly or mainly expressed on seeds may be better termed 'seed diseases'. The definition of seedborne pathogens as used here applies to those organisms (i.e. fungi, bacteria and viruses) carried with, on or in seeds, some of which may damage the seeds and all of which are transmitted by seeds to infect the crops which grow from them.

This book is concerned mainly with those pathogens which are seed transmitted and cause disease. Those organisms which are seedborne but not seed transmitted (McGee, 1990; 1992) are not considered. Similarly, disorders of seeds due to deficiency of certain elements, for example boron (hollow heart of groundnut) and manganese (marsh spot of peas), and other physiological disorders and disorders caused by extremes of temperature and humidity are described elsewhere (Perry and Howell, 1965; Neergaard, 1977).

Necrotrophs and biotrophs

All seeds bear microorganisms which function in different ways; some have saprophytic roles (Pugh, 1973) and others parasitic. Organisms are defined by Dickinson and Lucas (1977) and Byrde and Archer (1979) as (i) *biotrophs*, those which are parasitic in nature, cause minimal damage to the host, have a narrow host range and usually cannot be grown in axenic (i.e. pure) culture;

and (ii) *necrotrophs*, those which rapidly kill the host cells and then live saprophytically on the dead tissues. These frequently have a wide host range and can be grown in axenic culture.

Fungi, bacteria and viruses, seedborne or otherwise, are subject to these definitions.

All viruses and some fungi (i.e. the Peronosporaceae, Albuginaceae, Erysiphales, Uredinales, and Ustilaginales) are biotrophs. The majority of seedborne fungi are necrotrophs. Many bacteria, because they grow readily in culture, are technically necrotrophs but those such as *Rhizobium* spp., some of which are seedborne and which do not destroy host tissue, are biotrophic in character.

Necrotrophy and biotrophy are broad definitions between which are continuous gradations (Dickinson and Lucas, 1977). They are useful categories into which to place seedborne organisms since the capacity for biotrophy or necrotrophy in organisms greatly affects their life cycles and disease cycles. Necrotrophic fungi and bacteria which are seedborne, depending upon opportunity and timing, may inhabit the superficial or deeper tissues of seeds.

Contamination is used here to denote the occurrence of a pathogen on the surface of seeds either as cells (bacteria) or as mycelium with or without spores (fungi) or as virus particles. Contamination may be wholly superficial or it may be indicative of the presence of a deeper infection by the contaminating organism. *Infect* is used according to the definition 'to enter and establish a pathogenic relationship within an organism; to enter and persist in a carrier' (Anon., 1950). Entering here relates to the penetration of the seed or fruit coat by an organism thereby establishing an internal inoculum; the seed or fruit then becomes the carrier of that organism. By definition, a *carrier* 'is an organism harbouring a parasite without itself showing disease' (Anon., 1950). This definition only partially accommodates seedborne pathogens as symptoms may or may not occur on the seed. Many necrotrophic pathogens cause these types of infection; biotrophs generally establish more intimate relationships within the embryos of seeds (see Chapter 4, pp. 44–47).

Names of organisms

The names of the seedborne fungi, bacteria and viruses described in this book are given in Tables 1–4. Fungi are divided into necrotrophs and biotrophs; necrotrophs are further subdivided into broad categories reflecting the types of diseases they cause, e.g. leaf spots and stem rots, foot and root rots, etc. The common names of the diseases caused and the host crops are given. The current binomials for teleomorph (sexual stage) and anamorph (asexual stage) of necrotrophs and their synonyms are given where appropriate. As is the usual practice, the name of the teleomorph has been used to identify the organism. However, where a process is described involving the asexual stage, the name of the anamorph is given. In places the author has chosen to use the

Table 1. Names of fungal necrotrophs given in the text.

Teleomorph	Anamorph	Common name and host crop
LEAF SPOTS, STEM ROTS		
	Acroconidiella tropaeoli (syn. *Heterosporium tropaeoli*)	Leaf spot, nasturtium
	Alternaria alternata (syn. *Alternaria tenuis*)	Leaf spot, lobelia
	Alternaria brassicae	Dark leaf spot, brassicas
	Alternaria brassicicola	Dark leaf spot, brassicas
	Alternaria dauci	Leaf blight, carrot
	Alternaria padwickii	Stackburn, rice
	Alternaria radicina (syn. *Stemphylium radicinum*)	Black rot, carrot
	Alternaria sesamicola	Leaf blight, sesame
	Alternaria zinniae	Leaf spot, zinnia
	Ascochyta boltshauseri (syn. *Phoma boltshauseri*)	Leaf and pod spot, phaseolus beans
	Ascochyta fabae f.sp. *lentis* (syn. *Ascochyta lentis*)	Leaf spot, lentil
	Ascochyta pisi	Leaf and pod spot, peas
	Cercospora beticola	Leaf spot, beet
	Cercospora kikuchii	Purple stain, blight, soyabean
Cochliobus heterostrophus	*Bipolaris maydis* (syn. *Drechslera maydis*)	Leaf blight, maize
Cochliobolus miyabeanus	*Bipolaris oryzae* (syn. *Drechslera oryzae*)	Brown spot, leaf spot, rice
Cochliobolus sativus	*Bipolaris sorokiniana* (syn. *Drechslera sorokiniana*)	Spot blotch, cereals
Discosphaerina fulvida (syn. *Guignardia fulvida*)	*Aureobasidium lini* (syn. *Polyspora lini*)	Barley, rye browning, stem break, flax
Didymella fabae	*Ascochyta fabae*	Leaf spot, faba beans
Didymella lycopersici	*Phoma lycopersici*	Stem rot, tomato
	Gloeocercospora sorghi	Zonate leaf spot, sorghum
Itersonilia pastinacae		Canker, parsnip
Leptosphaeria maculans	*Phoma lingam* (syn. *Plenodomus lingam*)	Black leg, canker, brassicas
Magnaporthe grisea	*Pyricularia oryzae*	Blast, rice
Mycosphaerella fijiensis	*Paracercospora fijiensis*	Black Sigatoka, banana
Mycosphaerella linicola (syn. *Mycosphaerella linorum*)	*Septoria linicola*	Pasmo, flax
Mycosphaerella musicola	*Paracercospora musae*	Yellow Sigatoka, banana

Table 1. *continued*

Teleomorph	Anamorph	Common name and host crop
Mycosphaerella pinodes (syn. *Didymella pinodes*)	*Ascochyta pinodes*	Leaf and pod spot, peas
Mycosphaerella rabiei	*Ascochyta rabiei*	Leaf spot, chickpea
	Rhynchosporium secalis	Scald, leaf blotch
	Septoria apiicola	Leaf spot, celery
	Septoria petroselini	Leaf spot, parsley
	Septoria lactucae	Leaf spot, lettuce
Phaeosphaeria nodorum (syn. *Leptosphaeria nodorum*)	*Septoria nodorum*	Seedling blight, wheat
Phaeosphaeria avenaria f. sp. *triticea*	*Septoria avenae* f.sp. *triticea*	Leaf blotch, rye
Pleospora betae (syn. *P. bjoerlingii*)	*Phoma betae*	Black leg, sugar beet
Pyrenopeziza brassicae	*Cylindrosporium concentricum*	Light leaf spot, brassicas
Pyrenophora chaetomioides (syn. *Pyrenophora avenae*)	*Drechslera avenacea*	Leaf spot, oats
Pyrenophora graminea	*Drechslera graminea*	Leaf stripe, barley
Pyrenophora teres	*Drechslera teres*	Net blotch, barley
Venturia pirina	*Fusicladium pyrorum*	Black spot, scab, pears
FOOT AND ROOT ROTS, DAMPING-OFF AND WIRESTEM		
	Aphanomyces cochlioides	Black root, sugar beet
	Aphanomyces euteiches	Root rot, peas
	Aphanomyces euteiches f.sp. *phaseoli*	Root rot, beans
	Fusarium culmorum	Foot rot, cereals
	Macrophomina phaseolina (syn. *Rhizoctonia bataticola*)	Root rot, sunflower charcoal rot, cotton
Corticium rolfsii (syn. *Athelia rolfsii*)	*Sclerotium rolfsii*	Root rot, collar rot, peas, soyabeans
Gaeumannomyces graminis	*Phialophora radicicola*	Take all, cereals
Gibberella avenacea	*Fusarium avenaceum*	Foot rot, cereals
Gibberella fujikuroi	*Fusarium moniliforme*	Bakanae disease, rice
Gibberella intricans	*Fusarium equiseti*	Root rot, wheat
Gibberella zeae	*Fusarium graminearum*	Root rot, wheat, maize
Monographella nivalis (syn. *Micronectriella nivalis*)	*Microdochium nivale* (syn. *Fusarium nivale*)	Brown foot rot, cereals
Nectria haematococca	*Fusarium solani* f.sp. *phaseoli*	Foot rot, phaseolus beans
Nectria haematococca	*Fusarium solani* f.sp. *pisi*	Foot rot, peas
	Phoma medicaginis var. *pinodella* (syn. *Fusarium scirpi*)	Foot rot, peas

Table 1. *continued*

Teleomorph	Anamorph	Common name and host crop
Thanatephorus cucumeris	*Rhizoctonia solani*	Foot rot, wirestem, peas, lettuce
	Sclerotium cepivorum	White rot, onions
	Phytophthora infestans	Blight, potato
	Pythium ultimum	Damping-off, peas, lettuce
ANTHRACNOSE, ENDOPHYTES, MOULDS AND ROTS		
	Acremonium lollii	Ryegrass (endophyte)
	Acremonium coenophialum	Tall fescue (endophyte)
	Botrytis allii	Neck rot, onions
	Botrytis anthophila	anther mould, red clover
Gloeotinia granigena (syn. *Gleotinia temulenta*)	*Endoconidium temulentum*	Blind seed, rye
Glomerella cingulata f.sp. *phaseoli* (syn. *Glomerella lindemuthiana*)	*Colletotrichum lindemuthianum*	Anthracnose, French bean
	Colletotrichum dematium	Anthracnose, chilli
Glomerella gossypii	*Colletotrichum gossypii*	Anthracnose, boll rot, cotton
Glomerella graminicola	*Colletotrichum graminicola*	Anthracnose, maize, sorghum
	Colletotrichum lini (syn. *Colletotrichumlinicola)*	Anthracnose, flax
	Colletotrichum trifolii	Anthracnose, lucerne
	Lasiodiplodia theobromae	Stalk & kernal rot, maize
Sclerotinia fuckeliana	*Botrytis cinerea*	Grey mould, many hosts
Sclerotinia sclerotiorum		White mould, beans
Sclerotinia spermophila		Rot, white clover sunflower
Sclerotinia trifoliorum		Rot, red clover
Typhula ishikariensis		Snow mould, cereals
	Phomopsis spp.	Decay, soyabean
WILTS		
	Fusarium oxysporum	Wilt, many crops
	Fusarium oxysporum f.sp. *asparagi*	Wilt, asparagus
	Fusarium oxysporum f.sp. *betae*	Wilt, beet
	Fusarium oxysporum f.sp. *callistephi*	Wilt, asters
	Fusarium oxysporum f.sp. *ciceris*	Wilt, chickpea
	Fusarium oxysporum f.sp. *lycopersici*	Wilt, tomato

Table 1. *continued*

Teleomorph	Anamorph	Common name and host crop
	Fusarium oxysporum f.sp. *matthiolae*	Wilt, stocks
	Fusarium oxysporum f.sp. *pisi*	Wilt, pea
	Fusarium oxysporum f.sp. *vasinfectum*	Wilt, soyabean, cotton
	Phoma tracheiphila	Wilt (mal secco), citrus (lemon)
	Verticillium albo-atrum	Wilt, spinich
	Verticillium dahliae	Wilt, many hosts

Table 2. Names of fungal biotrophs given in the text.

Pathogen	Common name and host crop
POWDERY MILDEWS	
Erysiphe graminis f.sp. *hordei*	Powdery mildew, barley
Erysiphe graminis f.sp. *tritici*	Powdery mildew, wheat
Erysiphe polygoni	Powdery mildew, pea
BUNTS AND SMUTS	
Sphacelotheca cruenta	loose smut, sorghum
Sphacelotheca reiliana	Head smut, sorghum
Sphacelotheca sorghi	Covered smut, sorghum
Tilletia controversa	Dwarf bunt, wheat
Tilletia indica	Karnal bunt, wheat
Tilletia laevis (syn. *Tilletia foetida*)	Bunt, stinking smut, wheat
Tilletia tritici (syn. *Tilletia caries*)	Bunt, stinking smut, wheat
Urocystis agropyri (syn. *Urocystis tritici*)	Flag smut, wheat
Ustilago bullata	Head smut, prairie grass
Ustilago segetum var. *avenae* (syn. *Ustilago avenae*)	Loose smut, oats
Ustilago segetum var. *segetum* (syn. *Ustilago hordei*)	Covered smut, barley, oats
Ustilago segetum var. *segetum* (syn. *Ustilago kolleri*)	Covered smut, oats
Ustilago segetum var. *tritici* (syn. *Ustilago nuda*)	Loose smut, barley
Ustilago segetum var. *tritici* (syn. *Ustilago tritici*)	Loose smut, wheat

Table 2. *continued*

Pathogen	Common name and host crop
Ustilago zeae (syn. *Ustilago maydis*)	Loose smut, maize
RUSTS	
Phakopsora pachyrhizi (syn. *Uromyces sojae*)	Rust, soyabean
Puccinia antirrhini	Rust, snapdragon
Puccinia calitrapae (*syn. Puccinia carthami*)	Rust, safflower
Puccinia malvacearum	Rust, mallow
DOWNY MILDEWS	
Bremia lactucae	Downy mildew, lettuce
Peronospora ducometi	Downy mildew, buckwheat
Peronospora manschurica	Downy mildew, soyabean
Peronosclerospora maydis	Java downy mildew, maize
Peronosclerospora philippinensis	Philippine downy mildew, maize
Peronosclerospora sacchari	Downy mildew, sugarcane
Peronosclerospora sorghi	Downy mildew, sorghum
Peronospora viciae	Downy mildew, peas
Plasmopara halstedii (syn. *Peronospora halstedii*)	Downy mildew, sunflower
Sclerospora graminicola	Downy mildew, pearl millet
ERGOT	
Claviceps fusiformis	Ergot, millet
Claviceps purpurea	Ergot, barley, grasses
CLUB ROOT	
Plasmodiophora brassicae	Club root, brassicas

Table 3. Names of bacterial necrotrophs given in text.

Pathogen	Common name and host crop
Clavibacter michiganensis ssp. *insidiosus* (syn. *Corynebacterium michiganense* pv. *insidiosum)*	Wilt, lucerne
Clavibacter michiganensis ssp. *michiganensis (Corynebacterium michiganense* pv. *michiganense)*	Canker, tomato
Curtobacterium flaccumfaciens pv. *betae* (syn. *Corynebacterium betae)*	Silvering, beet
Curtobacterium flaccumfaciens pv. *flaccumfaciens* (syn. *Corynebacterium flaccumfaciens* pv. *flaccumfaciens)*	Wilt, bean
Erwinia carotovora ssp. *carotovora*	Soft rot, tobacco
Erwinia stewartii	Bacterial wilt, maize
Pseudomonas fuscovaginae	Brown sheath rot, rice
Pseudomonas glumae	Grain rot, rice
Pseudomonas solanacearum	Wilt, many crops
Pseudomonas syringae pv. *atrofaciens*	Leaf necrosis, wheat
Pseudomonas syringae pv. *glycinea*	Bacterial blight, soyabean
Pseudomonas syringae pv. *lachrymans*	Angular leaf spot, cucurbits
Pseudomonas syringae pv. *phaseolicola*	Halo blight, phaseolus bean
Pseudomonas syringae pv. *pisi*	Bacterial blight, peas
Pseudomonas syringae pv. *sesami*	Bacterial blight, sesame
Rhodococcus fascians (*Corynebacterium fascians*)	Gall
Xanthomonas manihotis	Bacterial blight, cassava
Xanthomonas campestris pv. *campestris*	Black rot, brassicas
Xanthomonas campestris pv. *carotae*	Bacterial blight, carrot
Xanthomonas campestris pv. *graminis*	Wilt, grasses
Xanthomonas campestris pv. *incanae*	Bacterial blight, stocks
Xanthomonas campestris pv. *malvacearum*	Bacterial blight, cotton
Xanthomonas campestris pv. *oryzae*	Bacterial blight, rice
Xanthomonas campestris pv. *phaseoli*	Common blight, phaseolus beans
Xanthomonas campestris pv. *tomato*	Bacterial blight, tomato
Xanthomonas campestris pv. *translucens*	Black chaff, wheat
Xanthomonas campestris pv. *vignicola*	Bacterial blight, cowpea
Xanthomonas campestris pv. *zinniae*	Leaf spot, zinnia

Table 4. Names of virus biotrophs given in the text.

Name	Host crop
Alfalfa mosaic virus (AMV)	Lucerne (alfalfa)
Arabis mosaic virus (ArMV)	Many hosts
Barley stripe mosaic virus (BSMV)	Barley
Barley yellow dwarf virus (BYDV)	Cereals
Bean common mosaic virus (BCMV)	Phaseolus bean
Beet curly top virus (BCTV)	Beet
Black raspberry latent virus (BRLV)	*Rubus* spp.
Capsicum mosaic virus	Pepper
Cauliflower mosaic virus (CaMV)	Cauliflower
Cherry leaf roll virus (CLRV)	Tobacco
Cherry yellows virus	Cherry
Cowpea banding mosaic virus	Cowpea
Cowpea mosaic virus (CPMV)	Cowpea
Cucumber green mottle mosaic virus (CGMMV)	Cucumber
Cucumber mosaic virus (CMV)	Cucumber
Lettuce mosaic virus (LMV)	Lettuce
Maize chlorotic dwarf virus (MCDV)	Maize
Maize dwarf mosaic virus (MDMV) (syn. sugarcane mosaic virus)	*Saccharum* spp.
Pea early browning virus (PEBV)	Pea
Pea enation mosaic virus (PEMV)	Pea, vicia bean
Pea seedborne mosaic virus (PSbMV)	Pea
Peanut mottle virus (PeMoV)	Peanut
Plum pox virus (PPV)	*Prunus* spp.
Prunus dwarf virus (PDV)	*Prunus* spp.
Prunus necrotic ringspot virus (PNRSV)	*Prunus* spp.
Southern bean mosaic virus (SBMV)	Phaseolus bean
Soyabean mosaic virus (SoyMV)	soyabean
Squash mosaic virus (SMV)	*Cucurbita* spp.
Strawberry latent ringspot virus (SLRSV)	Many hosts
Tobacco ringspot virus (TRSV)	Many hosts
Tomato mosaic virus (TOMV)	Tomato
Tomato ringspot virus (TRSV)	*Prunus* spp.
Tomato spotted wilt virus (TSWV)	Tomato
Turnip mosaic virus (TuMV)	Most brassicas

form of binomial more familiar to plant pathologists rather than that which may be taxonomically correct. Fungal biotrophs are grouped according to the types of diseases they cause, for example, powdery mildews, bunts and smuts, etc.; the viruses are listed in alphabetical order.

Seed Infection Routes

Direct: systemic invasion via mother plant tissues to the seed embryo

Many viruses, some fungi and a few bacteria are transmitted directly through the tissues of the mother plant to the developing seeds. This is systemic infection and with viruses it is thought that the route is direct to the female gametes (Fig. 2A) affecting the megaspore mother cell at a very early stage of its development (Walkey, 1991). As the ovule develops, the further passage of viruses may be prevented by differences in the growth rate of embryonic and endospermic tissues causing the rupture of the plasmodesmata linking these tissues with the nucellus (Caldwell, 1934; Walkey, 1991) (see Fig. 1A). Viruses, such as alfalfa mosaic virus (AMV), are present in the ovary walls and pistils of lucerne before fertilization, but decline in incidence as pistils develop into pods and as the pods mature (Bailiss and Offei, 1990). Further reduction occurs as pods dry, for example loss of southern bean mosaic virus (SBMV) from phaseolus bean pods (Cheo, 1955). AMV and SBMV located, respectively, in the testae of lucerne (Bailiss and Offei, 1990) and phaseolus bean seed (McDonald and Hamilton, 1972) are progressively inactivated as the seed coats mature. AMV is transmitted only by affected embryos (Bailiss and Offei, 1990).

Fifteen embryo-transmitted viruses are listed by Walkey (1991), including, for example, lettuce mosaic virus (LMV) (on *Lactuca sativa*), AMV (on *Medicago sativa*) and pea early browning virus (PEBV) (on *Pisum sativum*).

There are fewer fungi which infect by the direct route, but there is evidence that downy mildew (*Plasmopara halstedii*) infection of the capitulum and thereby of seeds of sunflowers may have arisen in this way (Cohen and Sackston, 1974; Sackston, 1981). The covered smuts of cereals, for example *Tilletia tritici* (stinking smut or bunt of wheat), invade the young coleoptiles producing a systemic mycelium which keeps pace with the developing tissues of the shoots of plants. The fungus grows into the flowers, replacing the tissues of the ovaries with teliospores (bunt balls) which are enclosed by the pericarp which remains intact. The pathogen is so contained until the grain is thrashed when the pericarps are ruptured and the teliospores released to contaminate healthy grain. When this grain is sown the teliospores germinate producing secondary sporidia and germ tubes from these invade young coleoptiles (Butler and Jones, 1949; Webster, 1986). These fungi and the viruses already described are biotrophs dependent to a greater or lesser extent upon the survival of the host for the completion of their life cycle. The biotrophic endophytic fungi of grasses represent a more advanced relationship where the fungal life cycle can be completed symptomlessly in symbiosis within the host plant and, ultimately, the seed tissues (Siegel *et al.*, 1987).

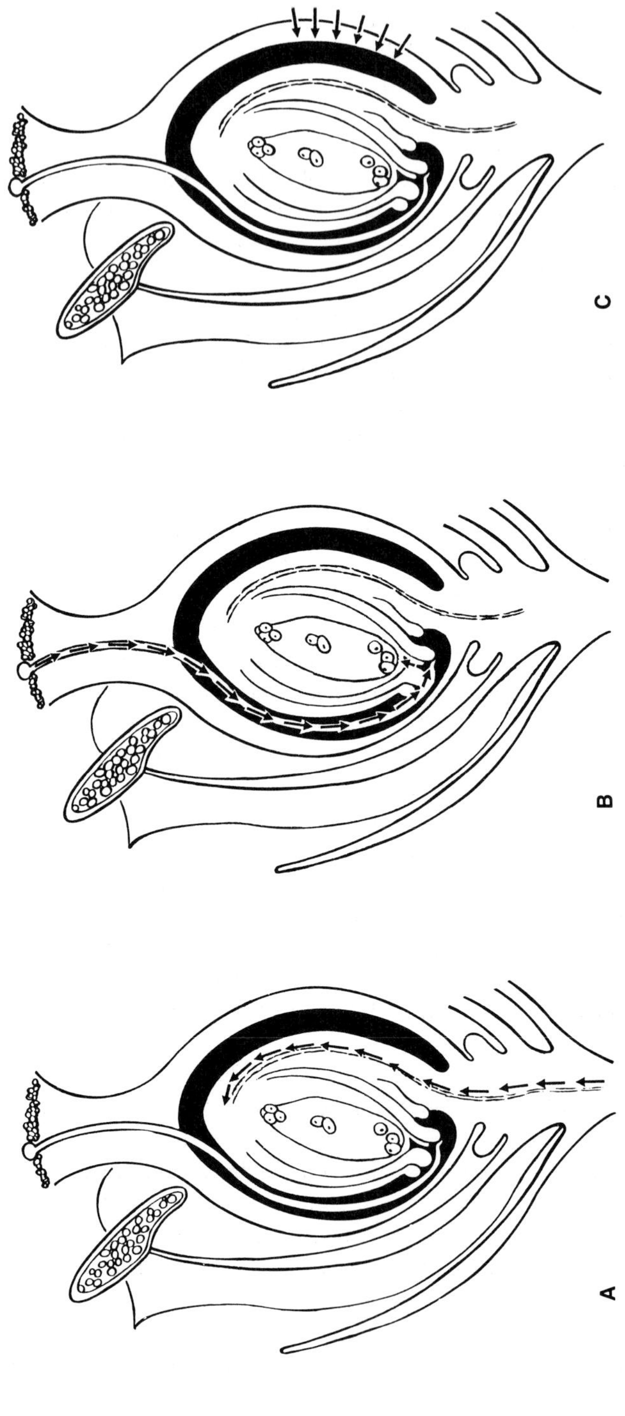

Fig. 2. Direct and indirect infection routes (arrowed). (A) Direct invasion via the xylem of the mother plant tissues—viruses, some bacteria and fungi. (B) Indirect invasion via the stigma to the embryo—viruses in pollen tubes; fungi as hyphae growing down through stylar tissues to invade the embryo. (C) Indirect, penetration of the ovary wall tissues—many organisms.

Although some necrotrophic fungal pathogens have been reported to infect seeds directly from the tissues of mother plants (Agarwal and Sinclair, 1987), it seems less likely that many of these organisms, which degrade tissues as they advance, would be transmitted to the embryonic seeds through the system of the parent plants. The wilt fungi (*Verticillium, Fusarium* spp.), which invade the vascular tissues of plants, are a possible exception to this. In pot experiments, van der Spek (1972) obtained limited evidence that *Verticillium albo-atrum* (spinach wilt) infected the developing fruits (functional seeds) by a vascular link with the mother plants, but this conflicts with Snyder and Wilhelm's earlier report (1962) that seedborne infection arose by the external contamination of the flowering heads. However, *Fusarium* spp. (*F. moniliforme, F. oxysporum* and *F. equiseti*: Rudolph and Harrison, 1945; *F. oxysporum* f.sp. *pisi*: Snyder, 1932; Macneill and Howard, 1959) also have been reported to invade the seeds via the xylem of the mother plant. Baker (1948) confirmed that this was an established route for the infection of garden stock (*Matthiola incana*) seeds by *Fusarium oxysporum* f.sp. *matthiolae* (wilt).

Bacteria with a systemic phase of development resulting from the invasion of the vascular system of plants may continue along this route into the tissues of the seeds without the production of external disease symptoms. *Xanthomonas campestris* pv. *phaseoli* is reported to infect bean seeds in this way (Zaumeyer, 1929; Lelliot, 1988). General systemic invasion of the inflorescence stems of cabbage by *X. campestris* pv. *campestris* (black rot) has occurred by the time the plants are in flower and the bacterium then advances in the xylem of the pedicels into the suture vein of the pods thereby reaching the funicle and occasionally the xylem of the seed coat (Walker, 1950; Cook *et al.*, 1952). Although the systemic route of seed infection has been invoked for *Pseudomonas syringae* pv. *phaseolicola* (Sherf and MacNab, 1986), and the toxin produced by this bacterium is systemic causing necrosis, Taylor *et al.* (1979a) found no evidence for translocation of bacterial cells from other parts of the plant to the seed. The vascular strand link with the mother plant is more substantial in fruits (functional seeds, e.g. beet, carrot, cereals) than in true seeds (e.g. brassicas, peas, etc.) and in this respect is a more accessible route for invasion.

Indirect: systemic infection via stigma to the seed embryo

Indirect infection occurs when organisms from the diseased inflorescences of mother plants are transported (by insects, wind-driven rain, etc.) to affect the flowers and ultimately the seeds of vegetatively healthy plants. For example, virus-infected pollen transported to the stigmas of flowering plants germinates to grow down the style and through the micropyle, probably delivering virus along with the male gametes into the embryo sac (Walkey, 1991) (Fig. 2B). Pollen-transmitted viruses are restricted to the developing embryo because of its surrounding callose layer and cannot escape to cause systemic infection of the parent plant (Mandahar, 1981). However, virus-infected pollen alighting on host plants may incidentally incite systemic infection in them, for example, prunus necrotic ringspot virus in cherry and squash, cherry yellows virus in cherry, etc. (Mandahar, 1981).

In some pathogen–plant relationships, infected pollen may increase the incidence of infected seeds (Walkey, 1991) but in others virus-affected pollen may not be viable and in some cases the pollen tubes are shorter and develop more slowly than healthy ones (Medina and Grogan, 1961; Cooper, 1976) resulting in poor fertilization. These factors may account for the lower rate of pollen transmission compared with ovule transmission (infected via the mother plant) reported for lettuce mosaic (Ryder, 1964) and several other viruses (Mandahar, 1981). Pollen transmission has been shown for those seedborne viruses, which are subsequently transmitted by soil-borne nematodes (nepoviruses) (Murant, 1981).

Certain biotrophic fungi are reputed to infect via the stigmatic route. This was accepted as the infection route for the loose smut fungi (*Ustilago segetum* var. *tritici*) of wheat and barley (Gaumann, 1950). The teliospores (brand spores) blown from the smutted ears of infected plants alight on the stigmatic surfaces of ears of healthy plants and there produce basidia from which hyphae invade the stigmatic tissue when it wilts. The hyphae grow between the cells down the stylar canal and enter the young seed through the micropyle or, if this is closed, the hyphae penetrate the inner integument to the nucellus and then penetrate between this and the young endosperm into the scutellum and thence to the embryo and its growing point (Lang, 1917; Gaumann, 1950). However, Lang (1917) also indicated that infection could occur through the ovary wall and this view has largely been substantiated by the work of Batts (1955) on wheat and of Pedersen (1956) and Malik and Batts (1960a) on barley. These studies show that infection hyphae of *U. segetum* var. *tritici* penetrate the ovary wall (pericarp) directly (Fig. 2C), cross the parenchyma and move along the integument (testa) on the ventral side of the grain, crossing the endosperm at the base of the grain to enter the scutellum from which they will permeate the hypocotyl and growing point regions (Fig. 3A). This, then, is probably the main route for infection and there is also some evidence to suggest the stylar tissue may have an inhibitory effect on fungal development (Jung, 1956). As the smutted grains mature the hyphae thicken and enter a resting phase; infection of the future plant is now assured (also p. 47). Subsequently, when the infected grain is sown and germinates, the pathogen will grow symptomlessly and intercellularly behind the vegetative growing points, ultimately replacing the flowers with a dark brown mass of spores contained by a membrane which ruptures to release them to infect the following season's crop.

The highly specialized necrotroph *Botrytis anthophila*, the cause of anther mould of red clover, also infects by a stigmatic route (Silow, 1934).

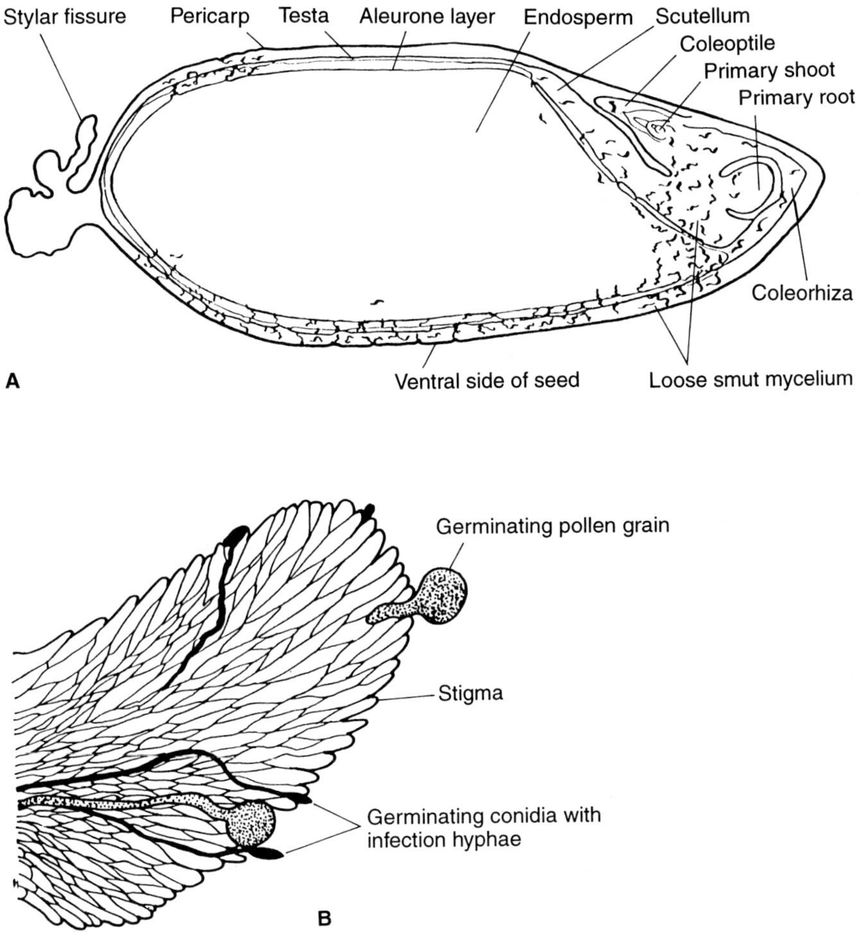

Fig. 3. Infection routes: (A) Longitudinal section (LS) showing the position of a loose smut (*Ustilago segetum* var. *tritici*) mycelium in an infected barley grain (redrawn from Malik and Batts, 1960a). (B) LS of a clover stigma showing germinating pollen grains and conidia of *Botrytis anthophila* (anther mould of red clover) (redrawn from Silow, 1934).

Conidia contaminate the pollen of diseased anthers and are probably transported by bees pollinating clover flowers. Spores deposited on the stigmatic surfaces of healthy flowers germinate into very fine hyphae which rapidly grow down the stylar canal (Silow, 1934) (Fig. 3B) to infect the young ovules (Bondarzew, 1914). Infected clover seeds rarely fail to germinate and the fungus lives almost as a symbiont in affected seedlings (Noble, 1948). *B. anthophila* appears to be confined to the intercellular spaces of the pith of plant stems (Silow, 1934).

The seed–plant–seed infection cycle is not characteristic of all indirectly transferred pathogens which invade via the stigmas of plants. In some, for example *Claviceps purpurea* (ergot), the female sex organs are the only part of the cereals and grasses which is attacked. The infection route, however, is not certain for this fungus (Webster, 1986). Early work by Engelke (1902) suggested that, at flowering, ascospores or conidia germinated on the stigmas to grow downwards in the manner of a pollen tube into the base of the ovule, completely changing the ovary into a dirty white pseudomorph which, after conidiation, is transformed into a dark horny sclerotium, the ergot. Campbell (1958) demonstrated that *C. purpurea* infection of barley (*Hordeum vulgare*) occurs through the ovary wall to penetrate the ovule and stylar invasion was rarely, if ever, a primary route. More recently, Willingale and Mantle (1987) have demonstrated that infection of pearl millet (*Pennisetum americanum*) by *C. fusiforme* in the field crop occurs through fresh receptive stigmas and the passage of infection hyphae down the stylodia closely follows the path normally taken by millet pollen. Colonization of the ovary by the fungus proceeds predominantly through the abaxial wall towards the vascular trace supplying the ovary (Willingale and Mantle, 1987). In both cases as the cereal host matures the ergot falls to the ground where it overwinters, discharging ascospores the following year when the host is in flower. *Claviceps* spp. cause a disease of seeds, which is not seed transmitted, but the pathogen may be transported as ergots mixed with samples of harvested seeds.

Indirect: infection via flower/fruit parts to the ovary and ovule tissues

Necrotrophic pathogens invade, and kill, the local host tissues as they advance. Progress of infection, therefore, is not strictly systemic but is a more limited invasive process (Bonman and Gabrielson, 1981). Weakly pathogenic necrotrophs, such as *Botrytis cinerea* (teleomorph *Sclerotinia fuckeliana*, grey mould), commonly infect the petals of many plants (Jarvis, 1977). Stamens and styles may be attacked and the fungus may become quiescent in these tissues as they rot. However, *B. cinerea* has not been found to penetrate the style as far as the ovary (Jarvis, 1977). More usually, affected senescing flower parts remain attached to the developing fruit and provide a food base for the saprophytic growth of the fungus which may later attack that fruit, for example, pea and bean pods, causing them to rot and in some cases affecting the seeds (Jarvis, 1980). More aggressive necrotrophs, for example *Ascochyta pisi* (leaf and pod spot) of peas and *Alternaria brassicicola* (dark leaf spot) of brassicas, attack the floral parts directly, where pathogenicity may be enhanced by the presence of pollen which provides nutrition for the rapid germination and growth of spores (Channon, 1970). From this location invasion of the pistillate end of developing pods may occur sometimes affecting the pod suture (Fig. 4A, 1 and 2). The suture contains the main vascular strand and when this is

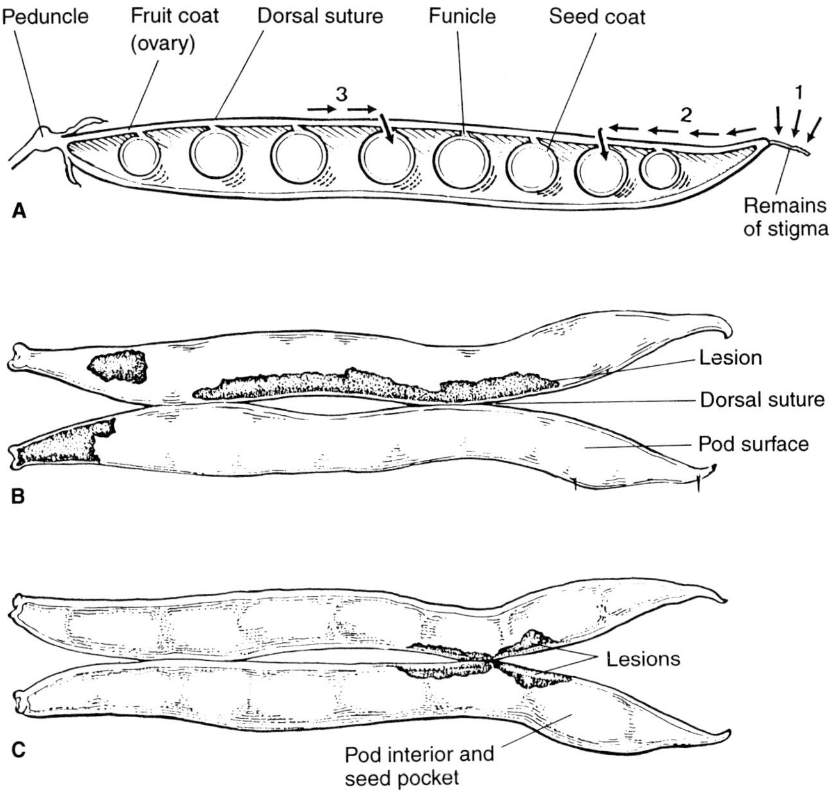

Fig. 4. Pod and seed infection routes. (A) 1, Stigmatic (pistillate) end of pea pod subject to attack by many necrotrophs (fungi and bacteria). 2 & 3, Invasion of the dorsal suture may result in pentration of developing seeds via the funicle, by fungi and bacteria. (B,C) Infection of phaseolus bean pods and seeds by the bacterium *Pseudomonas syringae* pv. *phaseolicola*, showing (b) external lesions from which penetration of the pod tissues occurs, and (c) internal pod lesions from which seeds are infected. (Beans redrawn from Taylor *et al.*, 1979a.)

invaded the infection route may be via the funicles (the stalks which attach the sutures to the ovules) into the putative seeds (Fig. 4A, 2 and 3). In dry dehiscent fruits (Hayward, 1951), for example the pods of peas and beans and the silicas of brassicas, the suture is a pronounced main vein running along their adaxial surfaces into the pedicels on which these fruits are borne on the main inflorescences. When seeds are detached from the funicles a scar, the hilum, remains which has a valvular action balancing the moisture levels of the seeds. Beside the hilum there is a pore, the micropyle, which may be occluded or even obliterated. The funicle and the micropyle (when open) are natural entry points for fungi (see Fig. 1 A,B and D), and the hilum is for bacteria (Agarwal and Sinclair, 1987). For example, necrotrophs including

Alternaria spp., *Cercospora kikuchii* and *Phomopsis* spp. have been observed in the hilar area and micropyles of soyabean seeds, which may represent further infection routes (Singh and Sinclair, 1986; Vaughan *et al.*, 1988). It is not certain, however, whether the hilum can be considered as a functional access point in most seed types. The hilum of peas, for example, is protected by two layers of heavy-walled palisade cells underlain by sclerenchymatous cells which may be regarded as defensive against fungal invasion (Hayward, 1951).

The aerial dispersal of fungal spores or bacterial cells results in a random distribution of inoculum on the surface of fruits, such as pods, where fungi may invade through cracks in the pod walls; for example, *Phomopsis sojae* of soyabean (Kulik, 1983) or infection of the surface of the pods produces lesions (Fig. 4,B,C). The pod wall is penetrated directly, thereby contaminating and infecting the underlying seeds.

Depending on the initial dispersal of inoculum a necrotroph may invade seeds by one or more routes. Direct penetration of the pod wall has been demonstrated for the bacterium *Pseudomonas syringae* pv. *phaseolicola* (halo blight) of bean (Taylor *et al.*, 1979a) (Fig. 4B,C) and for *Leptosphaeria maculans* (cabbage black leg) where Bonman and Gabrielson (1981) found that the asexual *Phoma lingam* state of the fungus was isolated more frequently from the stigmatic end of silicas than from the pedicel end. Bacteria or fungi which affect seeds sometimes produce 'symptoms', for example, a buttery deposit, discoloration or dried bacterial film produced by *P. syringae* pv. *phaseolicola* on the surface of bean seeds or staining by a range of fungi (*B. cinerea*; Ogilvie, 1969; *Ascochyta pisi*, *Mycosphaerella pinodes* (leaf and pod spot with foot rot of pea): Maude, 1966a; *Colletotrichum lindemuthianum* (anthracnose of bean): Tu, 1983) which results from early penetration of the seed coat tissues. Not all seed types which are contaminated bear visible symptoms, even of those which do a proportion of seeds will be symptomless; for example, 48% in one sample of halo blight-infected beans (Taylor *et al.*, 1979a) and 25% and 49%, respectively, of unstained pea seeds from pea pods slightly and heavily infected with *M. pinodes* (Maude, 1966a). Similar observations have been made for fungal pathogens of soyabean seeds (Sinclair, 1991).

Necrotrophic fungi can penetrate intact seed coats or pass through cracked parts of the seed coats. For example, *Alternaria* spp. penetrated the cuticle of soyabean seeds directly (Vaughan *et al.*, 1988) and *Cercospora kikuchii* and *Phomopsis* spp. invaded through natural pores or through cracks in the surface of soyabean seeds (Ilyas *et al.*, 1975; Singh and Sinclair, 1986). Skoric (1927) reported that the bacterium *Pseudomonas syringae* pv. *pisi* (bacterial blight of pea) invaded via the dorsal suture of pea pods to the funicle (hilum) and sometimes the micropyles of pea seeds (Fig. 4A, 2 and 3). A similar infection route was reported for *Xanthomonas campestris* pv. *phaseoli* in beans (Zaumeyer, 1929).

Pods (peas, beans) and silicas (brassicas) are representative of dry dehiscent fruits (Hayward, 1951) where the seeds are arranged in a single row, each seed connected by a funicle to the dorsal suture of the pod (Fig. 4A). In fleshy fruits, for example berries (Solanaceae), seeds are attached by their funicles to a central placenta or placentas. Here the invasion route can be via the calyx into the placenta and from the placenta via the funicles to the seeds. The fungus *Phoma lycopersici*, the asexual state of *Didymella lycopersici* (tomato stem and fruit rot) (Fisher, 1954) and the bacterium *Clavibacter michiganensis* subsp. *michiganensis* (bacterial canker) (Orth, 1939) infect by this route (Fig. 5a). Fruiting structures such as pods and berries, having

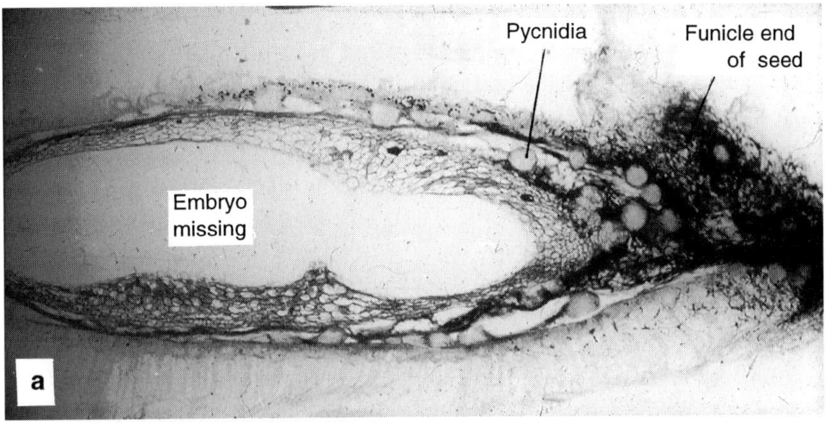

Fig. 5. Seed infection routes. (a) *Didymella lycopersici*: hyphae and pycnidia of *Phoma lycopersici* in the funicle region of a tomato seed (from Fisher, 1954). (b) *Itersonilia pastinacae*: infected flowers and aborted seeds on the partial umbels of parsnip (from Channon, 1969). Reproduced with the permission of Horticulture Research International.

enclosed environments, provide ideal incubators for seed-infecting pathogens.

The capsular fruit of onion and the schizocarps of celery, carrot and parsnip are borne in flowering heads (umbels), in which the developing flowers are exposed *en masse* to the surrounding environment. Fungal necrotrophs such as *Botrytis allii* (onion neck rot) may attack the floral parts of onion, including the anthers, stigmas and pedicels of individual flowers (Munn, 1917). Severe attack results in flower abortion (Presly, 1977) causing 'blasting' of seed umbels and reduction of, or complete loss of, the seed crop (Yarwood, 1938; Blodgett, 1946; Netzer and Dishon, 1966; Ellerbrock and Lorbeer, 1977). *B. allii* was readily recovered from the ovary walls (capsules) of onions where less than 5% of the inflorescence was damaged; fungal attack was restricted to the surface of the seeds (Presly, 1977). However, the fungus also has been recovered from surface-sterilized seeds, indicating a degree of penetration of the seed coat tissues (Maude and Presly, 1977). Similar modes of infection occur with umbelliferous crops. The fungus *Itersonilia pastinacae* (canker) causes greatest damage to young inflorescences of parsnip often with complete destruction of their flowers and pedicels (Fig. 5b). Young fruits are damaged or destroyed, though older fruits are not visibly affected (Channon, 1969). Inoculation studies confirmed that unexpanded umbels, young flowers, the perianth and stamens were attacked but isolations revealed that the fungus was fairly shallowly situated in naturally infected seeds. Soteros (1979a) recovered *Alternaria radicina* (black rot) from the bract leaves, bractlets, pedicels, floral segments and developing ovules, but not from the endosperm or embryos, of developing carrot seeds in flowering umbels. Strandberg (1983) described a similar pattern for *Alternaria dauci* (leaf blight) infection of carrot inflorescences. He found that the fungus was normally confined to the pericarp tissues of viable seeds, but that it could pentrate through to the endosperms and embryos producing severely affected, shrivelled and non-viable seeds.

Location of Inoculum

Superficial inoculum: seed/fruit coats

The coats of certain seeds, for example cereals, peas, brassicas and onions, may appear smooth and intact to the eye but when examined by scanning electron microscopy are seen to be fissured and broken to some degree. Such crevices give protection to fungal spores and bacterial cells (Splittstosser and Mohamed-Yasseen, 1991). Seeds such as celery, carrot and lettuce provide rougher, sculptured and often spiny fruit coats (pericarps) which also facilitate the adherence of bacterial and fungal propagules.

The cutting and windrowing (field drying) of seed crops places them in contact with infected leaf debris and soil at the base of the crop in a moist microclimate conducive to fungal growth. Seeds may become contaminated by various pathogens, for example leaf spotting fungi such as *Alternaria brassicicola* (Maude and Humpherson-Jones, 1980a) and vascular wilts such as *Fusarium oxysporum* spp. (Gambogi, 1983). Field drying of seeding white cabbages on tripods appears to have had a similar effect, resulting in an increased incidence of seedborne *Phoma lingam* (teleomorph *Leptosphaeria maculans*) (Cruger, 1979). Threshing of seed crops in the field physically dislodges considerable numbers of spores causing even greater contamination. The conidia of *Fusarium* wilt spp. including *F. oxysporum* f.sp. *callistephi* (aster wilt), *F. oxysporum* f.sp. *asparagi* (asparagus wilt) and *F. oxysporum* f.sp. *betae* are primarily external contaminants smeared onto seeds during threshing (Gambogi, 1983).

It is also suggested (Cruger, 1979) that the threshing and cleaning of cabbage seeds causes the dissemination of pycnidiospores of *P. lingam* which then contaminate the seeds. Contamination of seed dressing equipment used during the cleaning of infected bean seeds is postulated as the way by which the delta race of *Colletotrichum lindemuthianum* was transferred to many commercial phaseolus bean seed samples, the possible sources of the anthracnose epidemic in Canada in 1976 (Tu, 1992). Pathogenic bacteria can also contaminate seeds from crop debris, for example *Xanthomonas campestris* pv. *manihotis*, bacterial blight of cassava (Persley, 1979), and from dust produced by thrashing or cleaning procedures, for example *Pseudomonas syringae* pv. *phaseolicola* (Grogan and Kimble, 1967). Bacteria surviving as a dried slime on the surface of seeds following an invasion of the pods, for example *P. syringae* pv. *phaseolicola*, halo blight of beans (Taylor *et al.*, 1979a) and *P. syringae* pv. *pisi*, pea blight (Skoric, 1927), are well placed to contaminate other seeds when the pods are threshed.

With certain biotrophic fungi, contamination of seeds is necessary for the transmission of disease (see Chapter 4, pp. 45–47). For example, the thick-walled adhesive teliospores of ustilaginaceous bunt and smut fungi are capable of surviving exposure on the surface of seeds for a considerable time (see Chapter 3, p. 35) and then of germinating to infect the developing seedlings (see Chapter 4, pp. 45–47). With others, such as many downy mildew species, the oospores which encrust seeds in various states of maturity may be infectious but lose this capacity quickly and there is controversy concerning the seed transmission of these fungi (Spencer, 1981) (see Chapter 4, p. 46).

With the exception of tomato mosaic virus (ToMV) (Broadbent, 1965), which contaminates the external layers of tomato testae and is infectious in that location, most viruses, although they occur superficially on immature seeds, cannot be recovered once the seeds are mature and dried (Walkey, 1991). Necrotrophic fungi which have been recovered from the seed coat surface occur there as mycelia and also as fruiting structures (conidia, pycnidia, etc.).

Mycelium and conidia of *Alternaria brassicicola* were readily observed microscopically (× 200, by reflected light) on the surface of commercial cabbage seed samples (Maude and Humpherson-Jones, 1980a). Conidia removed by washing were always ungerminated; washing did not remove mycelium. The number of *A. brassicicola* spores per cabbage seed were significantly correlated with internal infection of the seeds. High populations of spores occurred on the seed coats of severely infected seed samples and populations were low where seedborne infection was slight or absent (Maude and Humpherson-Jones, 1980a). Conidia, conidiophores and mycelium of *A. dauci* were common on the ridges and spines of carrot mericarps (Strandberg, 1983). Similarly, the *Phoma lycopersici* state of *Didymella lycopersici* (stem rot of tomato) is known to be on and within tomato seeds extracted from infected fruit (Derbyshire, 1961). Agarwal and Sinclair (1987) list 37 pathogenic fungi found in the seed coats of a range of crop seeds. Many of these also will have occurred as superficial inoculum contaminating the surface of seeds.

Internal inoculum: seed/fruit coat tissues and embryos

Necrotrophic fungi and bacteria generally are located in the seed/fruit coat tissues. Deeper penetration of the storage tissues (cotyledons, endosperm, see Fig. 1A–D) is infrequent and these organisms are rarely present in the generative tissues (radicle, hypocotyl, plumule, see Fig. 1C,D) of the embryos of seeds. Biotrophic pathogens, including almost all viruses and some specialized fungi, are usually located within the embryos of seeds (Table 5). Seed coat infections are important in the transmission of necrotrophic fungi and bacteria and ToMV (see pp. 50–52). Generally, the seed coat (peas, beans, brassicas) or pericarp (carrots, parsnip, cereals) are relatively thin structures

Table 5. The location of pathogens on or in seeds.

Organism	Status	Location on/in seeds
Viruses	Biotrophs	Embryonic axes (most species); testa and endosperm occasionally
Fungi	Biotrophs	Embryonic axes (most smut species); superficial (most bunts, a few rusts and downy mildews)
Fungi	Necrotrophs	Superficial; in testa/pericarp tissues (most species); in endosperm/cotyledon tissues (some species); in embryonic axes (very few species)
Bacteria	Necrotrophs	As for necrotrophic fungi but almost none in embryonic axes

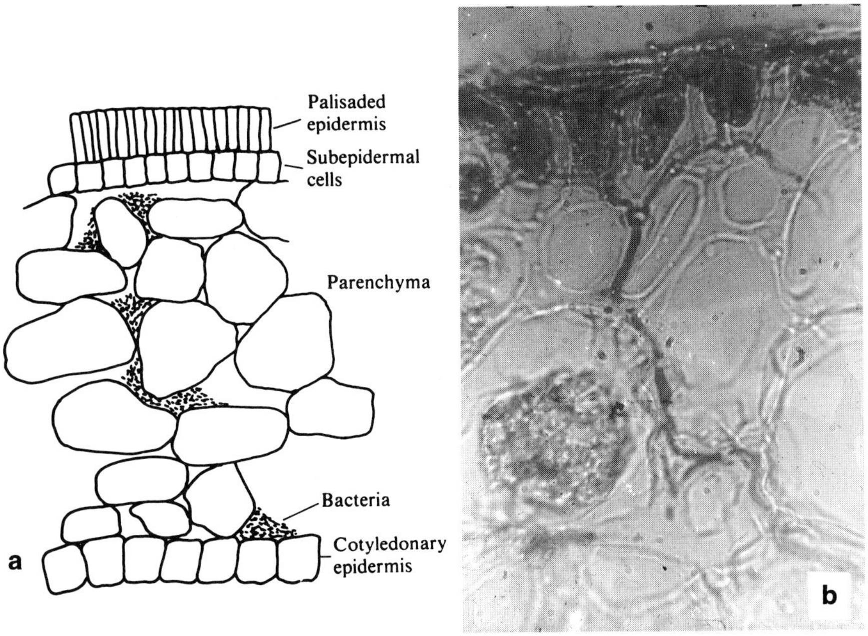

Fig. 6. Seed infection routes. (a) Diagram of a section of a phaseolus bean seed coat illustrating the typical location of *Pseudomonas syringae* pv. *phaseolicola* bacteria (cotyledon absent). (b) Longitudinal section through the cotyledon of a pea seed (seed coat missing) showing the deep-seated intercellular hyphae of *Ascochyta pisi*. Reproduced with the permission of Horticulture Research International.

which, in addition to protecting the internal tissues of seeds, also protect the pathogenic organisms which may have invaded and are contained in those tissues.

Those organisms which have invaded cruciferous or leguminous seeds directly through the pod wall can be found as intercellular mycelium (fungi) or cells (bacteria, e.g. of *Pseudomonas syringae* pv. *phaseolicola*; Fig. 6a and Table 5) beneath the outer palisaded epidermal cell layers, between the spongy parenchymatous cells inside the seed coat. Deeper penetration results in hyphae and bacterial cells being located in the tissues of the embryo, in these cases the cotyledon. These largely are intercellular (e.g. hyphae of *Ascochyta pisi*; Fig. 6b and Table 5) in location, but where infection has been severe there is invasion of cotyledon cells. Intracellular mycelium of *A. pisi* and *A. pinodes* in peas and *P. syringae* pv. *phaseolicola* bacteria in beans have been found in the cotyledonary tissues.

Infection is located mainly in the pericarp tissues of umbelliferous seeds, for example carrots invaded by *Alternaria dauci* (Netzer and Kenneth, 1969; Strandberg, 1983) and by *A. radicina* (Soteros, 1979a). Mycelium of

Colletotrichum graminicola (anthracnose of sorghum) is located mainly in the pericarp and endosperm of seeds and more rarely in the embryo (Basu Chaudhary and Mathur, 1979; Prasad *et al.*, 1985). Where infection has occurred via the funicle, for example *Phoma lycopersici* (stem rot of tomato), hyphae and pycnidia of the fungus are located in the gaps left by the resorption of the large thin-walled parenchymatous cells of the middle zone of the seed coat (Fig. 5a) (Fisher, 1954). The fungus was not found in the embryos.

The location of biotrophic and necrotrophic organisms (fungi, bacteria) in the area of the hilum of seeds is common. This may be the result of local invasion of the funicle or by systemic infection within the vascular system of the plant continuing through the xylem tissues of the funicle to the developing seeds, for example the vascular wilt fungi and bacteria. The vascular tissues of the funicle end at, or pass through, the seed hilum. Location of the mycelium of the biotroph *Plasmopara halstedii* was heaviest in the hilum region of sunflower seeds (Cohen and Sackston, 1974). Local penetration of the micropyles of peas by the bacterium *Pseudomonas syringae* pv. *pisi* results in water-soaking symptoms in the hilum region (Skoric, 1927). Conidia of the necrotroph *A. brassicicola* (dark leaf spot) were recovered more frequently from the hilum of cabbage seeds than from the remainder of the surface of the testa (Knox-Davies, 1979); mycelial growth of *Phoma lingam* (teleomorph *Leptosphaeria maculans*) was often found near the funicle attachment (Jacobsen and Williams, 1971); heavy aggregation of the mycelium of *A. sesamicola* in the hilum of sesame seeds is considered indicative of the penetration of the fungus at this point (Singh *et al.*, 1980).

The location of biotrophic viruses and fungi in the embryo tissues of seeds depends on the type of organism, the infection route (described above) and to some extent on the duration of the infection period. Viruses infect the female egg cells (gametes) before flowering via the mother plant tissues and during flowering by pollen transmission (Walkey, 1991).

Mycelium and sporangia of the downy mildew fungus *Plasmopara halstedii* were located in most parts of sunflower seeds including the embryos (Novotelnova, 1963; Cohen and Sackston, 1974) but not in developing apical cells (Novotelnova, 1963). The downy mildews of maize (*Sclerophthora* and *Peronosclerospora* spp.) are reported from similar locations in seeds (see McGee, 1990). Fungal endophytes of perennial rye grass (*Lolium perenne*) and tall fescue (*Festuca arundinacea*) also are located within embryos as mycelium in the region of the aleurone layer or between it and the nucellus (Welty, 1986; Siegel *et al.*, 1987). Specialized fungi, for example *Tilletia tritici* (covered smut, bunt) and *Claviceps purpurea* (ergot), where disease transmission is dependent on the production and release of spores, may completely replace the embryo tissues within the ovary with the pathogen. In others, for example the loose smut pathogens of barley (*Ustilago segetum* var. *tritici* syn. *U. nuda*) and wheat (*U. segetum* var. *tritici* syn. *U. tritici*) the fungus is located in the scutellum (see Fig. 3A) close to the growing point behind which it will travel as the young plant develops and grows.

Necrotrophic pathogens, for example the vascular bacterium *Curtobacterium flaccumfaciens* pv. *flaccumfaciens* (bean wilt) and *Xanthomonas campestris* pv. *phaseoli* (bean black rot) are not located in the embryo of seeds but occur on the surrounding cotyledons, from which, at germination, the developing seedlings are infected (Zaumeyer, 1932). Similar observations have been made for the location of the halo blight bacterium *Pseudomonas syringae* pv. *phaseolicola* (Zaumeyer, 1932; Taylor *et al.*, 1979a). Examples of pathogens which infect the outer layers of embryo tissues are the *Ascochyta* spp. (*A. pisi* and *A. pinodes* of peas: Maude, 1966a), *Phoma lingam*, black leg of crucifers (Jacobsen and Williams, 1971; Gabrielson, 1983), and *Lasiodiplodia theobromae*, stalk and kernel rot of maize (Kumar and Shetty, 1983).

Necrotrophic fungi are infrequently situated in the embryonic axis (radicle, hypocotyl, plumule) of seeds. However, hyphae of *Phoma lingam* (black leg) have been recorded from the radicles of sectioned cabbage seeds (Jacobsen and Williams, 1971). Severe infection of seeds by necrotrophic pathogens, though not usual, can result in invasion of the whole embryo system. Such infections probably have happened early in seed formation and the resultant seeds are most likely to be small and often are not viable.

Outcome of Infection

Infection of the flowers and fruits of plants completes the disease cycle of most seedborne pathogens and re-establishes them on or in the host seed tissues from which the new life cycles and disease cycles will begin. The infection routes of the more specialized biotrophic pathogens of seeds generally result in the invasion of the internal tissues of seeds where the organisms are protected by the host tissues and less susceptible to environmental effects when the seeds are sown. Those which do not succeed in infecting the developing seeds may remain on the surface of the seed coats where they fail to survive (see Chapter 3). The more opportunistic and directly invasive processes of necrotrophic organisms result in progressive penetration of seeds so that severity may follow a distribution curve with the extremes represented by superficial contamination and internal infection. Superficially placed necrotrophs may survive (see Chapter 3) to transmit disease (see Chapter 4), but in general they are susceptible to adverse antibiotic effects of organisms in the soil environment.

When seeds are returned to the soil soon after harvest the inoculum they bear is likely to be highly infective. However, much commercial seed is stored for a period before re-use and conditions of storage may affect the infectivity of seedborne organisms (see Chapter 3).

3 Longevity of Seedborne Organisms

Introduction

The survival of seedborne organisms is intimately linked to the survival of the host seeds. In that respect the conditions under which seeds are stored may affect the survival of the pathogens they bear, depending on the intrinsic structure of those organisms and their location on and in the seed tissues (see Chapter 2).

Factors Affecting Longevity

Lifespan of seeds

The lifespan of 'orthodox' seeds, i.e. those of most arable and horticultural species, is a function of time, temperature and moisture content and their longevity can be increased by the reduction of temperature and moisture content (at least down to 5%) in store (Cromarty *et al.*, 1982).

Cool dry storage

Modern commercial seed stores are designed with temperature and humidity control facilities (George, 1980) largely based on the maxim (James, 1967) that 'where percentage relative humidity and degrees Fahrenheit total 100 or less, conditions are suitable for seed longevity'. Thus a seed store operating at 50°F (10°C) and 50% relative humidity (RH) will provide suitable storage conditions for prolonging the 'useful' life of temperate orthodox seeds. The term useful life relates to the retention of high viability (germination capacity) and vigour (speed of germination) by seeds over a commercial storage period which usually falls short of their biological lifespan. Seedborne pathogens, particularly those which are not wholly superficial, persist at least as long as the useful life of the seed. Seeds can survive in a viable state from less than 3 years to over 100 years (Bewley and Black, 1994), but in practice most vegetable and cereals seeds are not stored for longer than 3 years in temperate climates. Rice, wheat and maize are stored for less time in the tropics (Neergaard, 1977).

Low temperature storage

Where seeds are retained as a genetic resource, storage facilities are designed to preserve seed viability in the medium term of up to 10 years (stored at 0–10°C) or in the long term for at least several decades (stored at sub-zero temperatures) (Cromarty *et al.*, 1982). Under such conditions the continuity of certain seedborne organisms may be secure for a considerable period of time.

Storage of recalcitrant seeds

There is another group of seeds called recalcitrant seeds which are damaged by drying and low temperatures (Chin and Roberts, 1980; Cromarty *et al.*, 1982; Farrant *et al.*, 1988). These include tropical fruits and timber species of temperate and tropical origin. Wet storage of these seeds is possible only for a few weeks or months (Cromarty *et al.*, 1982; Corbineau and Come, 1988). In such seeds the longevity of pathogens is necessarily brief and they are affected mainly by organisms which cause seed deterioration rather than by those which are transmitted by seeds.

Longevity of pathogens

Gross survival patterns

There is considerable information on the longevity of seedborne organisms (Neergaard, 1977; Agarwal and Sinclair, 1987), on the survival of phytopathogenic bacteria including those which are seedborne (Schuster and Coyne, 1974) and on seed-transmitted plant viruses (Bennett, 1969; Mandahar, 1981). Based on combined information from the first two sources, Fig. 7 gives the incidence of seedborne fungi, bacteria and viruses with increasing storage time of different agricultural and horticultural seeds. It is apparent that the capacity for survival over the first 5 years of storage is high with many seedborne fungi, bacteria and viruses. After 5 years of storage there is an increasing decline in survival of the three types of organism. However, the optimum incidence of seedborne organisms generally occurs within the time limits for commercial storage of seeds.

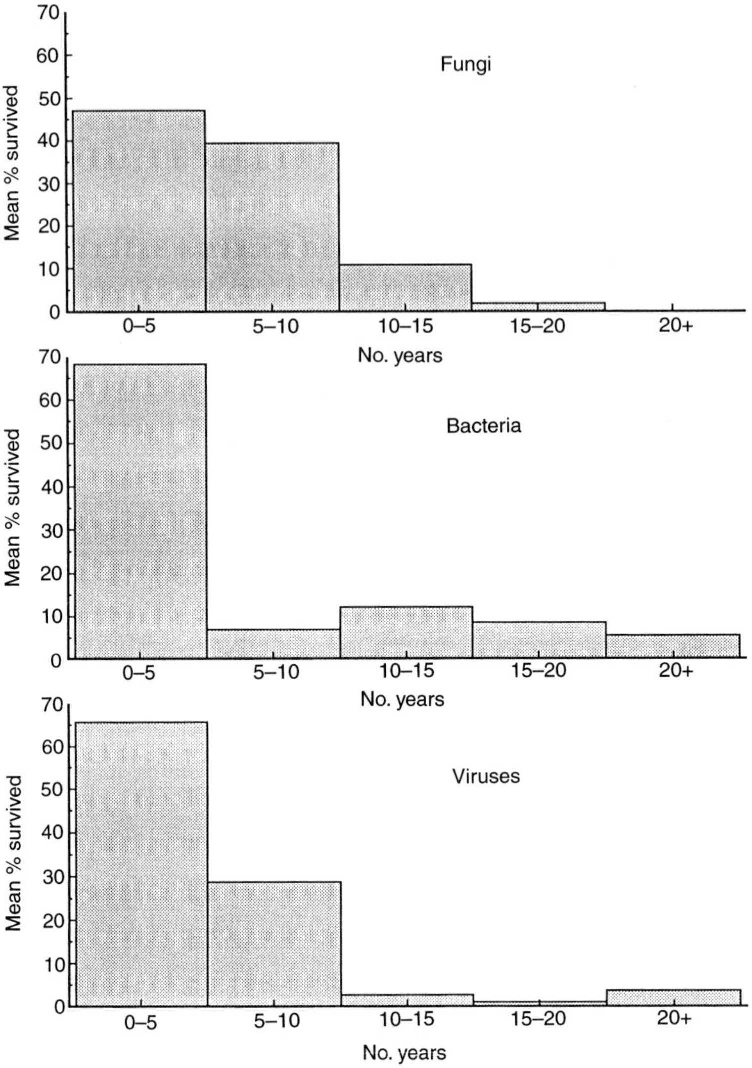

Fig. 7. Longevity of seedborne fungi, bacteria, and viruses (data from Neergaard, 1977 and from Agarwal and Sinclair, 1987).

Intrinsic factors affecting survival

Types of organism and survival

Neergaard (1977) divides fungi into five groups that, based on their on their intrinsic morphology, have different survival capacities in the seedborne phase.

1. Hyaline fungi with thin-walled conidia, i.e. moniliaceous Hyphomycetes including *Botrytis*, *Cercospora*, and *Gloeotinia* and most *Fusaria*.
2. Strongly pigmented fungi with thick-walled conidia, i.e. dematiaceous Hyphomycetes including *Drechslera* and *Alternaria*.
3. Pycnidial species within the Coelomycetes (i.e. Sphaeropsidales, Melanconiales) where the spores are fully protected by the fruiting bodies.
4. The deep-seated resting mycelium and the heavily pigmented spores of some smut fungi (Ustilaginales).
5. Resting mycelia of many species, particularly if thick-walled.

Groups 1 and 2 represent opposite extremes with the more fragile hyaline fungi surviving for shorter periods of time than dematiaceous fungi. Using this approach, Neergaard found it possible to identify taxonomic groups of fungi as short living, fairly long living and mostly long living on and in seeds. Table 6 is a condensation of his information, which is limited by the available data. Presented in this form it suggests that the Uredinales are short lived (mean, 1.8 years), the moniliaceaous Hyphomycetes, Melanconiales and Sphaeropsidales are of medium-term persistence (means 4.3, 5.0 and 5.0 years, respectively) and Ascomycetes, dematiaceous Hyphomycetes and Ustilaginales are longer lived (means 7.8, 7.2 and 14.9 years, respectively). These should be regarded as empirical guidelines only on the longevity of taxonomic groups of fungi on or in seeds. Although these intrinsic morphological differences may influence the persistence of seedborne fungi they are subordinate to the effects of the location of the organism on or in seeds in terms of duration of survival (Neergaard, 1977) and to the conditions under which the seeds are stored.

Neergaardian groupings have not been extended to bacteria and viruses, presumably because of their generally more uniform morphology. Most plant pathogenic bacteria are not spore formers so there are virtually no alternative structures which might extend the survival period of those species. Seedborne

Table 6. Order of survival of taxonomic groups of fungi on or in seeds (data from Neergaard, 1977).

		Longevity (years)	
Persistence	Taxonomic group	Range	Mean
Short term	Uredinales (rusts)	1.5–2+	1.8
Medium term	Moniliaceous Hyphomycetes	2–13.5	4.3
	Melanconiales	1–13.5	5.0
	Sphaeropsidales	1.5–9.0	5.0
Long term	Ascomycetes	3–13.0+	7.8
	Dematiaceous Hyphomycetes	1–10.0	7.2
	Ustilaginales	1–64.0	14.9

viruses also have common characteristics resulting in uniformity of survival (Bennett, 1969).

Location of organisms

The location of organisms on and within seed tissues (see Chapter 2) influences longevity. The better an organism is protected by the tissues of the seed host, the greater the potential for its survival. The location of the pathogen, therefore, is an important component of longevity.

Superficial inoculum. Certain seedborne biotrophs survive least well when superficially situated. Thus, though many downy mildews and viruses are present on developing seeds few, if any, remain when the seeds have matured (Sackston, 1981; Bailiss and Offei, 1990; McGee, 1990, 1992; Walkey, 1991). An exception is the tomato mosaic virus (ToMV) which can survive the drying out of the seed coat as maturation occurs (Broadbent, 1965).

Surface-borne bacteria may survive for 2–3 years (Neergaard, 1977). Bacterial exudates in general extend the longevity of many bacteria in various environments (Schuster and Coyne, 1974). Xanthan gums, for example, produced by certain xanthomonads (e.g. *Xanthomonas campestris* pv. *campestris*) can provide a protecting matrix for bacteria on the surface of seeds, thereby extending their period of survival. Surface-borne fungal pathogens persist for varying periods of time from months to years depending on the taxonomic type, the severity of contamination, the morphology of the contaminating propagules and the external environment (i.e. the conditions of storage and the microflora of the seeds).

Internal inoculum and differential survival. Similar precepts can be applied to the internal inoculum of organisms which, because it is more deeply situated in the tissues of seeds and more protected from adverse microbiological and weathering effects in the environment, should survive for longer periods of time than the superficial inoculum of the same species. Usually this happens. Within the tissues of seeds the actual location (pericarp or seed coat, or deeper in the endosperm or cotyledon, ultimately in the embryonic axes) of organisms (fungi, bacteria, viruses) bears some relationship to their persistence but identity of the organism and the storage environment have a considerable effect in the determination of longevity.

Differential survival of fungal pathogens may occur between surface-borne and internal inoculum. However, there may be morphological differences between the two forms of inoculum, for example superficial conidia and internal mycelium, which may affect the rate of survival of either form. Examples are given by Agarwal and Sinclair (1987) for *Pyricularia oryzae*, the imperfect state of *Magnaporthe grisea*, on rice (Suzuki, 1930) and *Cercospora beticola* on sugarbeet (Wenzl, 1959). Maude and Humpherson-Jones (1980a) demonstrated a linear relationship between the concentration of surface-borne spores of *Alternaria brassicicola* and the internal inoculum of that fungus in cabbage seeds. They found that total seed infection remained high for the first 7 years of storage (at 10°C, 50% RH) but only 1.1% of surface-borne spores were viable after that time, demonstrating that even under relatively uniform environmental conditions surface-borne conidia survive less well than internal inoculum. The reverse, suggested by Krout (1921), that internal mycelium of *Septoria apiicola* survived for a shorter period than the pycnidiospores of the fungus, was not confirmed by Maude (1964) who found that the thallus and pycnidia of the fungus in the seed (fruit) and the pycnidiospores persisted for a similar length of time.

Conditions of storage

Storage conditions are a major factor in determining the length of survival of seedborne pathogens. It is recognized that by reducing temperature and moisture content seeds may be stored, without loss of viability, for increased periods of time. Generally, the same is true for the parasites they bear. It is also a reason, taking into account the location of and type of pathogen involved, why there is such divergence in the literature over the longevity of specific seedborne pathogens. Seeds stored over a range of temperatures and humidities will perform differently in terms of their own germination and also in terms of the longevity of the pathogens they bear. For example, the fungus *S. apiicola* infects the pericarp of celery seeds and in these tissues produces pycnidia containing pycnidiospores. Because storage conditions for the seeds have not been reported, or have varied, the stated survival period for this pathogen on celery seeds also varies. Krout (1921) found that spores from the pericarps and peduncles of seeds were mostly dead after 2 years of storage; Maude (1964) found that spores from pericarps survived for 11 months and Sheridan (1966) for 10–12 months; Gabrielson (1962) reported that the fungus was not viable in 1 year or older commercial seed samples; Hewett (1962) obtained infected seedlings from 1-year-old seed; and Stirrup and Ewan (1931) obtained infected plants from 5-year-old celery seeds.

Krout (1921) suggested that one method of disinfecting celery seeds was to store them for 3–4 years, by which time the fungus would be dead. Although this strategy is supported by the majority of the evidence given here it is apparent that little is known of the effect on the pathogen of storage conditions of the seeds. From Fig. 8 it can seen that survival of *S. apiicola* pycnidiospores is dependent upon the humidities and temperatures at which the seeds are stored. Pycnidiospore viability was retained at a high level for 3 years in infected seeds maintained under good storage conditions (at moisture contents (mc) of 7.3% and temperatures of 5°C and 15°C); however, spore survival was reduced considerably by seed storage at higher humidities and temperatures.

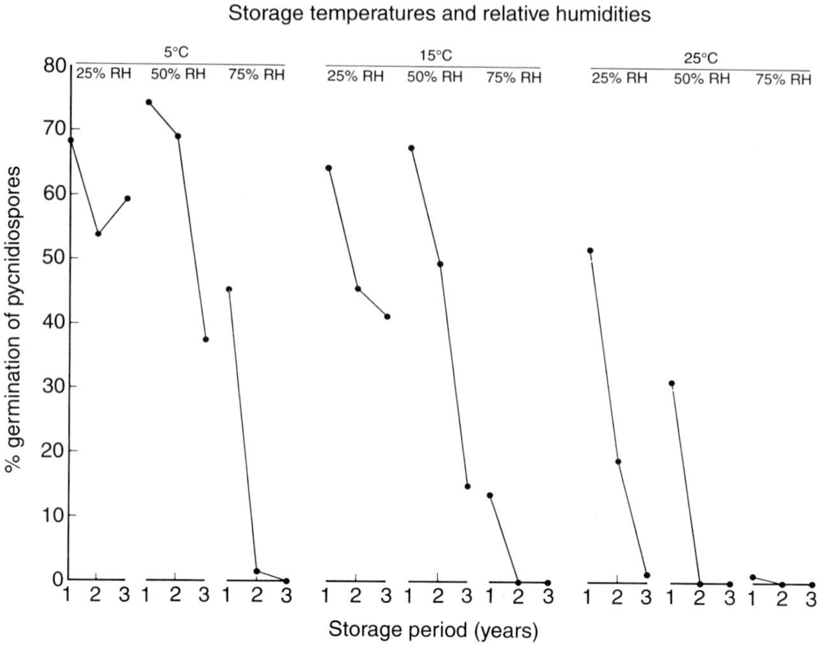

Fig. 8. Survival of seedborne *Septoria apiicola* pycnidiospores from celery seeds stored at different temperatures and relative humidities. Note 25% RH = 7.3% seed moisture content (mc), 50% RH = 9.7% mc, 75% RH = 12.4% mc; equivalencies are based on data from Roberts and Roberts (1972). (Adapted from Maude, 1973.)

The variability in length of survival of this seedborne pathogen, as reported in the literature, thus probably resides in the variability of storage environment of the seeds; the strategy of storing celery seeds to rid them of the fungus would only be possible under conditions which would not be suitable for maintenance of the germination capacity of the seeds. If James' good storage conditions for seeds (°F + % RH ≤ 100) is applied to Fig. 8 it can be seen that conditions which are favourable for the storage of celery seed are favourable for the survival of the pathogen and vice versa. Similar relationships were established by Arndt (1946) for cotton seeds infected by *Colletotrichum gossypii* where increases in storage temperatures and moisture content of seeds reduced seed germination and infection (Table 7). Likewise, endophytes in perennial ryegrass seeds (*Acremonium lolii* and *A. coenophialum*) remained viable for at least 15 years when stored at 0–5°C and 25–50% RH (Latch, 1983) but for only 2.25 years under non-controlled conditions (Neill, 1940).

There are many examples of the maintenance of pathogen longevity by the storage of seeds under cool dry conditions, but unless the storage para-

Table 7. Effect of seed moisture content and storage temperature on the germination and infection of cotton seeds (Deltapine 11a) by *Colletotrichum gossypii* after storage for 17 months (from Arndt, 1946).

Seed moisture content (%)	Storage temperature (°C)	Germination (%)	Seedlings (%) Infected	Seedlings (%) Dead
8	1	97	100	93
	21	88	20	7
	Air*	90	11	4
	33	80	9	0
10	1	89	90	71
	21	90	2	0
	Air	85	6	5
	33	42	0	0
12	1	90	91	70
	21	75	12	0
	Air	83	7	0
	33	0	0	0
14	1	85	88	71
	21	50	4	4
	Air	12	33	0
	33	0	0	0
16	1	90	81	48
	21	0	0	0
	Air	0	0	0
	33	0	0	0

Air = −20°C to + 36°C.

meters of temperature and relative humidity have been thoroughly tested against seedborne inoculum, the policy of ageing seed to free it of seedborne fungi may not be fully effective and therefore is of doubtful value. Probably for this reason, ageing of seeds for the elimination of pathogenic bacteria (Burkholder, 1948) also is considered too variable and unreliable to use as a control method (Schuster and Coyne, 1974). Nonetheless, the holding over of infected seeds to free them from pathogenic fungi has been followed in at least one commercial situation. The reports (Neill and Hyde, 1939; Calvert and Muskett, 1945; Noble and Gray, 1945) that storage of infected perennial ryegrass seeds for 2 years or more would free them from the fungus *Gloeotinia granigena* (syn. *G. temulenta*) (blind seed disease) were confirmed by Hardison (1948) who introduced the use of 2-year-old seed as part of a programme of cultural practices to control the disease in commercial crops of ryegrass in Oregon, USA. This method of disease control is proposed by a number of

Table 8. Storage of seeds to free them from seedborne pathogens.

Seed type	Pathogen	Duration of storage (years)	Reference
Celery, *Apium graveolens*	Septoria apiicola	3–4	Krout, 1921
Perennial ryegrass, *Lolium perenne*	Gloeotinia granigena	2	Neill and Hyde, 1939; Hardison, 1948
Field bean, *Vicia faba*	*Botrytis fabae*	< 1	Harrison, 1978
Sugarbeet, *Beta vulgaris*	Cercospora beticola	2	Aebi and Rapin, 1954
Flower seeds and flax, (*Linum usitatissimum*)	*Botrytis cinerea*	< 3	Neergaard, 1977

workers who have stored infected seeds and measured the decline in inoculum (Table 8). However, too often advice on pathogen survival in infected seeds has been based on laboratory storage, sometimes at room temperature, which is not an accurate measure of the longevity of the same organisms when seeds are stored under commercial conditions.

Cool temperature storage

Other examples where reduction in temperature and humidity in storage increased longevity include (i) seedborne *Cercospora kikuchii* where the fungus lost viability more rapidly at > 10.8% mc than it did at 6.3% mc and also at storage temperatures of 16–28°C than it did at 8°C or 12°C (Lehman, 1952); (ii) *Colletotrichum gossypii* (cotton anthracnose) where after 13.5 years storage at 1°C and 8% mc cotton seed had a 93% emergence of which 70% were infected by the fungus compared with 28% emergence and no infection of seed when seeds were stored at 21°C and 8% mc (Arndt, 1953); (iii) *Pseudomonas syringae* pv. *phaseolicola* where bacteria were dead in most naturally infected seed samples stored for 3 years in a partially heated building (at 10–27°C) but survived for 6 years in some of the samples stored under controlled conditions (7–10°C and 45–50% RH) (Taylor *et al.*, 1979a).

Low temperature and refrigerated storage

Reduction to low storage temperatures may also increase the length of survival of seedborne organisms. Fifty nine percent of soyabean samples remained infected with *Sclerotinia sclerotiorum* after storage for 1.5 years at 3 ± 1°C compared with only 6% when the seeds were stored at 22 ± 3°C for 2 years (Nicholson *et al.*, 1972). Similarly, the infectivity of the seedborne teliospores of safflower rust (*Puccinia carthami*) was preserved by storage at 3°C for 18 months and caused infection in 45–60% of seedlings in three cultivators of safflower compared with 0–10% infection from the same seed lots stored at 20–25°C for the same period of time (Halfon-Meiri, 1983). *Colletotrichum dematium* persisted for only 9 months in chilli seeds stored at room temperature (Smith and Crossan, 1958) but at 5°C survival was extended to over 8 years in chilli, and to over 10 years in mungbean and soyabean (Siddiqui *et al.*, 1983). *C. lindemuthianum* persisted for at least 5 years in seeds which had been air dried and stored at 4°C (Tu, 1983).

Although low temperature storage may reduce the rate of decline of many seedborne fungi and bacteria, others, for example certain viruses, may not respond differentially, at least in the short term, to different storage temperatures. Thus alfalfa seeds (20% infected with alfalfa mosaic virus) showed no decrease in virus activity over 5 years at any of the three storage temperatures (–18°C, 4°C and 21–27°C) (Frosheiser, 1974). However, critical end points can be reached where, with further storage, there is a considerable drop in infectivity. Over a period of 4 years seeds of *Prunus pennsylvanica* bearing cherry necrotic ringspot virus stored at 2°C produced 60–70% infected seedlings, but by the sixth year there were less than 5% diseased seedlings although plant emergence had not decreased over that period (Fulton, 1964).

There are other examples of pathogens recovered from seeds which have been stored at low temperatures for different lengths of time but, because, in many cases, no comparisons have been made with storage under ambient conditions, the significance of any increase in duration of survival is diminished. There is clear evidence, however, that where seeds are placed in deep freeze storage (–20°C) the fungal pathogens they bear are preserved with little if any reduction in viability for considerable periods of time (Hewett, 1987). Pathogens on over 80 samples, mostly cereals and pulses, retained at least 75% viability for periods up to 14 years; this represented the end of the test or tests (Table 9). Thus, infected seeds placed in gene banks at sub-zero temperatures (–18°C is recommended by the International Board for Plant Genetic Resources (Cromarty *et al.*, 1982)) have the potential to perpetuate disease over a considerable timespan.

Storage at reduced moisture content

Hermetically stored seeds in foil packs, tins or jars have a prolonged viability. Under airtight conditions zinnia seeds infected with *Xanthomonas campestris* pv. *zinniae* remained infectious for more than 4 years compared with 2 years for those stored in paper bags (Strider, 1979). Similarly, hermetic sealing of red clover seeds at 5% mc preserved internal *Botrytis anthophila* infection for 6 years whereas the fungus had virtually disappeared from seeds stored at 7% mc and from those in unsealed linen bags for the same period of time (Narkiewicz-Jodko, 1975).

Table 9. Viability of seedborne fungal pathogens after storage at –20°C (from Hewett, 1987).

Fungus	Host	No. of samples	Storage period (years)	Seeds infected (%) Start	End
Ascochyta pisi	Pea	8	8–11	18	14
Ascochyta fabae	Vicia bean*	5	9–13	14	14
Pleospora betae	Sugarbeet	2	14	30	23
Leptosphaeria nodorum	Wheat	9	9–14	50	39
Micronectriella nivalis	Wheat, rye, barley	5	9–12	19	19
Cochliobolus sativus	Wheat, barley	9	8–12	43	33
Pyrenophora teres	Barley	5	8–12	24	16
Pyrenophora graminea	Barley	4	11–12	47	44
Colletotrichum lindemuthianum	Phaseolus bean†	1	12	99	93
Ascochyta boltshauseri	Phaseolus bean	1	12	52	41
Leptosphaeria maculans	Cabbage	1	11–13	13	12
Alternaria dauci	Carrot	4	9–14	22	21
Alternaria radicina	Carrot	3	14	37	28

* *Vicia faba.*
† *Phaseolus vulgaris.*

Effect of seed storage on pathogen viability and infectivity

The successful recovery of fungi, bacteria and viruses in a viable condition (as mycelia, spores, cells or particles) from stored seeds by various test methods over a period of time may be considered indicative of a continuing pathogenicity. Although in many instances, and particularly with fungi and bacteria, this relationship has not been tested, where comparisons have been made a decline in pathogen viability is correlated with a reduced infectivity (Maude, 1964; Hewett, 1987). A progressive reduction in the number of plants which become infected over time can be associated with a concomitant loss of viability of seedborne fungal spores (Maude, 1964; Halfon-Meiri, 1983) and bacterial cells (Taylor *et al.*, 1979a). The close correspondence between the embryo mycelium test for the incidence of the loose smut fungus (*Ustilago segetum* var. *tritici*) and the percentage of infected barley plants in the field ceases after 6–9 years of seed storage (Russell, 1961). Because it is a biotroph, the viability of the fungus cannot be tested in culture and the numbers of infected embryos (as assessed by the embryo detection method) do not decline with time. Russell (1961) considered that the more likely explanation for the decline of disease was that progressively more smut-infected seeds failed to germinate after longer periods of storage, thus reducing the number of smutted plants in barley crops. His data did not support the alternative conclusion that the viability of *U. segetum* var. *tritici* mycelium in the barley embryos declined with prolonged storage of the seeds (Russell, 1961). Calvert and Muskett (1945), however, observed that as the incidence of *Gloeotinia granigena* (blind seed disease) in stored infected ryegrass seeds decreased with time so did the vitality of the seedborne mycelium which was still alive.

General effects of storage on survival of seedborne inoculum

The longevity of a seedborne pathogen is modified by a number of factors including the conditions of storage of the seeds, the intrinsic nature of the organism and its location on or in the tissues of seeds. Because of the influence of these variables there is little room for generalization from the wealth of information which is available in the literature. What can be stated is that pathogens usually survive for as long as the commercially useful life of seeds and probably as long as their biological lifespan. Advances in the use of controlled temperature and humidity conditions for the storage of agricultural and horticultural seeds and of genetic resource seeds are conducive to extension of the longevity of many seedborne pathogens. Control of seedborne organisms by the storage of seeds would appear to have very limited application, and a decline in viability of organisms with time generally does not cause a concomitant reduction in their infectivity.

4 Seed Transmission of Disease

Introduction

Seed transmission of disease occurs by the transfer of inoculum from infected seeds to the germinating seeds and seedlings. A large number of fungal species belonging to about 90 genera in several hundred plant hosts are seed transmitted; as are about 60 species of bacteria belonging to five genera in over 100 hosts (Phatak, 1980); and 18% (Phatak, 1980) to 20% (Tomlinson, 1987) of viruses in over 500 hosts (Phatak, 1980). In a recent review, Mink (1993) suggested that 140 viruses were erroneously listed as pollen or seed transmitted, leaving 108 viruses for which there was clear evidence for transmission through seed or pollen.

Where seedborne infections are deep seated and the embryonic axes of seeds are affected, as by some biotrophic organisms (e.g. certain viruses and fungi), each infected seed may produce a diseased seedling and environmental factors may have little effect on transmission. The large majority of necrotrophic organisms, and some biotrophs, are superficial or located in the testa/pericarp or outer embryo tissues and do not affect the female gametes that form the new plant. As a result, transmission is discontinuous and is influenced to a greater extent by a number of intrinsic and environmental parameters which affect inoculum transfer from seed to plant.

Treatment of Transmission by Various Authors

The seedborne organisms which transmit disease have been divided into groups based on different criteria by several authors. Baker's (1972) classification was based on location of organism producing seven, but in reality only four, main types of transmission. Neergaard (1977) related transmission to location of organism and outcome of infection, i.e. local or systemic, and in so doing created nine separate categories. Agarwal and Sinclair (1987) have used systemic and non-systemic infection as a primary subdivision of transmission and then the location of infection to create further sub-sections and have reduced transmission to seven categories. In this book the author has used the type of organism, i.e. biotroph or necrotroph, as a primary division and then location of organism to produce further groupings.

Transmission by biotrophs causing systemic infection of plants

By contaminated seeds

Teliospores, otherwise known as chlamydospores, brand spores, melanospores or ustospores (Moore, 1972), are the seed-contaminating agents of certain of the Ustilaginales fungi which parasitize grasses and cereals. The bunt fungi, *Tilletia tritici* and *T. laevis* (common bunts of wheat), *T. indica* (Karnal bunt of wheat), certain smuts (*Ustilago segetum* var. *segetum* (covered smut of barley and oats), *Sphacelotheca sorghi* (covered smut of sorghum) and *S. cruenta* (loose smut of sorghum) all infect through the seedborne phase. In others, for example *T. controversa* (dwarf bunt of winter wheat), *S. reiliana* (head smut of sorghum) and *U. zeae* (loose smut of maize) teliospores in the soil are the main source of infection (Smith *et al.*, 1988).

Teliospores of *T. tritici* are surrounded by the pericarp of the wheat grain and are not released until this is broken when the grain is thrashed at harvest and they are disseminated over the surface of surrounding healthy seeds. Teliospores are short lived in the soil (4 weeks as single spores, 8–10 weeks in spore balls) (Weltzien, 1957) but they overwinter on the grain and germinate at the same time as the grain to infect the emerging coleoptiles. In *T. tritici* transmission is by dikaryotic infection hyphae from the pericarp of seeds which penetrate between and through the cells of the emerging coleoptiles into the base of the first and second leaves and intercellularly into the deeper tissues, thereby reaching the area immediately below the growing point (Swinburne, 1963). The transmission phase is completed after about 12 days (Swinburne, 1963) and the infection cycle proceeds (Fig. 9). The transmission phase is very similar in most species of the Ustilaginales; teliospores germinate to produce a pro-mycelium (basidium, meta-basidium) in which the diploid nucleus undergoes reduction division after which primary sporidia (basidiospores) are released. Primary sporidia may produce secondary sporidia, but in either case sporidia of different mating types fuse to produce dikaryotic infection hyphae which penetrate the germinating coleoptiles of cereals (Fig. 9).

Puccinia carthami is the only seedborne rust (Uredinales) known, so far, to be transmitted directly from seed to seedling (Halfon-Meiri, 1983). Telios-

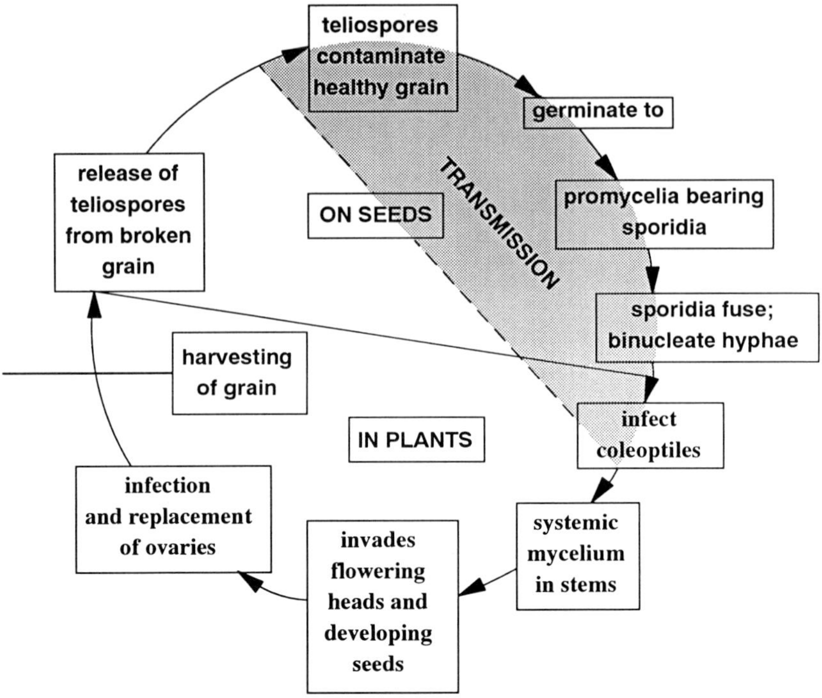

Fig. 9. Transmission of *Ustilago hordei* (covered smut of barley).

pores contaminating the surface of safflower seeds give rise to sporidia whose germ tubes penetrate the seedling tissues during germination, producing intercellular hyphae and intracellular haustoria.

In other biotrophic fungi, for example the downy mildews, oospores contaminating the seeds, or in the pericarps of seeds, germinate to systemically invade emerging seedlings producing stunted or infected but symptomless plants (e.g. *Plasmopara halstedii* of sunflower: Cohen and Sackston, 1974; *Peronospora ducometi* of buckwheat: Zimmer *et al.*, 1990) or to cause the death of seedlings (*P. manshurica* of soyabean: Dunleavy, 1981). Oospores present in the glumes surrounding sorghum seeds represent an external but primary source of inoculum for the transmission of *Peronosclerospora sorghi* (Yao *et al.*, 1990). The seed transmission phase of many downy mildews, if it occurs in field soil, is erratic and at a very low incidence (Novotelnova, 1963; Delanoe, 1972; Cohen and Sackston, 1974; Zimmer *et al.*, 1990). These fungi are extremely transitory and although seeds of maize and soyabean may bear an active downy mildew pathogen (e.g. *Peronosclerospora maydis*, *P. philippinensis*, *P. sorghi* and *P. sacchari*) when freshly harvested, and can transmit disease at that time, the viability of this inoculum is quickly lost when seeds are dried in various ways (McGee, 1990, 1992).

Transmission of tomato mosaic virus (ToMV) from contaminated seed coats occurs only if these are in direct contact with damaged seedling tissues, such as those of hypocotyls or roots injured during transplanting (Broadbent, 1965). Transmission results in systemic invasion of the tomato but plants grown from affected seeds and not transplanted, i.e. not injured, rarely contract the disease (Broadbent, 1965).

By infected seed coats or pericarps

Surface-borne biotrophic pathogens can be located deeper in the seed tissues from which transmission may be possible. Shetty *et al.* (1980) considered that embryo-borne mycelium of *Sclerospora graminicola* (downy mildew of pearl millet) probably was the main source of transmission of that disease. However, location of a downy mildew fungus in the embryo of seeds is indicative of severe infection which also may kill the seed host, resulting in non-transmission of the disease. Although the ToMV virus can occur in the internal layers of the testa, endospermic transmission is probably uncommon (Broadbent, 1965).

By infected embryonic axes

The mycelium of the loose smut fungi (*Ustilago segetum* var. *tritici*) is located in the scutellum of cereal seeds (Batts, 1955) (Fig. 10A). When seeds germinate the growing point moves to form the crown node near the soil level. The mycelium is activated by seed germination and moves into the crown node with the growing point (Fig. 10B) and permeates the node to enter the growing points of the young tillers (Malik and Batts, 1960b) (Fig. 10C). Not all tillers become infected, for example in wheat it is not unusual for a plant with up to five or six tillers to have only one or two tillers infected (W.J. Rennie, pers. comm., 1993). The proportion of diseased tillers probably relates to the extent of the mycelium in the embryo (W.J. Rennie, pers. comm., 1993). With the infection of tillers transmission of mycelium within the embryonic axis is then complete and mycelium is carried up with the growing points by the elongation of the tiller internodes.

Virus particles are present in embryo sacs and eggs of inflorescent plants at an early stage of seed development (see Chapter 2). Transmission is effected when the infected fertilized egg divides and begin to develop into the rudimentary embryonic axes. Symptoms may be apparent at seedling emergence (on cotyledons or the first pair of leaves) but disease expression varies considerably with different viruses and different host plants (Bennett, 1969). Bennett (1969) suggests that most seed-transmitted viruses have certain characteristics in common. These include juice transmissibility affecting the parenchymatous tissues of plants causing mottling, or necrotic or chlorotic

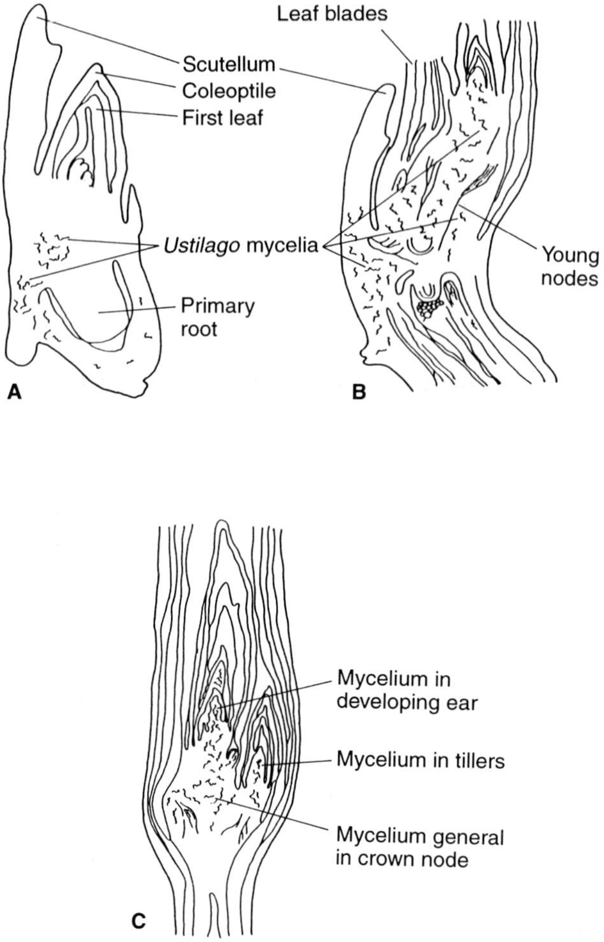

Fig. 10. Transmission of infection by *Ustilago segetum* var. *tritici* (loose smut of barley and wheat). (A) Longitudinal section (LS) of a wheat embryo 33 days after inoculation. (B) LS of a 3-week-old barley seedling. (C) Crown node of a 5-week-old barley seedling. (Redrawn from Batts 1955 and Malik and Batts, 1960b.)

lesioning. Several non-persistent aphid-transmitted viruses and most nematode-transmitted viruses are seedborne.

Nematode-borne viruses belong to two taxonomic groups, the tobraviruses and the nepoviruses which are mostly transmitted through the seeds of many host species (Shepherd, 1972; Murant, 1983). Mink (1993) lists 25 taxonomic groups which contain seed-transmitted viruses; in only five of the 25 groups are all the members known to be transmitted through seeds.

Four of these are single member groups (alfalfa mosaic, maize chlorotic dwarf, pea enation mosaic and tomato spotted wilt) (Mink, 1993).

Transmission of necrotrophs causing local and vascular infections

Many seed-transmitted necrotrophs infect shoots and/or roots of germinating seedlings causing seedling blights and leaf spots and/or foot and root rots. There are also those which invade the vascular tissues of seedlings and become systemic causing wilts (see pp. 52–54). Fungi and bacteria mainly are implicated.

Seed transmission causing local infections

Contaminated seeds

Contamination of seeds by necrotrophic pathogens may occur during the threshing and cleaning of crops or it may be part of an infection process which finished at that point, which if continued would have resulted in the penetration of the seed coat and possibly the embryo tissues (see Chapter 2). There is probably little difference in the mechanics of transmission of a pathogen present on the surface of the seed and that of the same pathogen slightly more deeply seated within the seed coat tissues.

Contaminated seeds do transmit disease. A reduction in viable, superficial fungal inoculum (conidia and mycelium of *Alternaria brassicicola*) on cabbage seeds by surface sterilization results in a reduction in disease transmission (Maude and Humpherson-Jones, 1980a). Cruger (1979) reports that pycnidiospores adhering to the surface of cabbage seeds may be one source responsible for the transmission of *Phoma lingam* (asexual state of *Leptosphaeria maculans*). Bacterial cells of *Pseudomonas syringae* pv. *phaseolicola* contaminating the surface of bean (*Phaseolus vulgaris*) seeds (Grogan and Kimble, 1967; Guthrie, 1970) transmit infection to the cotyledons of emerging seedlings as do symptomless but affected bean seeds, some of which may have been superficially contaminated by the bacterium (Taylor *et al.*, 1979a).

Transmission is by contact. Where that contact is made depends to some extent on how seedlings germinate and on the nature of specific diseases. There are two broad types of germination, epigeal and hypogeal in seeds of monocotyledonous and dicotyledonous plants. In epigeal germination the cotyledons emerge above the soil surface and in hypogeal germination they remain below. Thus epigeal germination favours the transmission of cotyledonary pathogens which are thereby suitably located for aerial dissemination in crops, and hypogeal germination aids the transmission of those which cause foot and root rots beneath the soil. This is a general rule and there are many exceptions, for example transmission of pea blight (*Pseudomonas syringae* pv. *pisi*: Skoric, 1927) happens by contact of the plumule during its emergence from the soil with the bacterially contaminated cotyledons which remain beneath the soil (hypogeal germination). Transmission in epigeal seeds, for example brassica seeds contaminated with *A. brassicicola* is by contact of the cotyledons with affected parts of the seed coat which is moved aside by the emergence of the cotyledons. The seed coat may remain attached to the tip of a cotyledon or be deposited in the soil where a chance contact with the seedling hypocotyl may also result in disease transfer (causing damping-off). In this example conidia and/or mycelium may transfer infection.

Infected seeds

Transmission from seed coat and pericarp tissues of epigeal vegetable seeds may cause progressive invasion of cotyledons (*Phoma* of cabbage: Walker, 1969; *Didymella* of tomato: Maude, 1962 (Fig. 11); *Botrytis* of onion: Maude and Presly, 1977 (Fig. 12); *Aspergillus* of onion: Hayden and Maude, 1992), or may cause leaf lesioning and hypocotyl infection *Alternaria* of brassicas: Maude and Humpherson-Jones, 1980a; *Septoria* of celery: Maude, 1964). Bacteria also are transmitted by contact in this way, for example *Pseudomonas syringae* pv. *phaseolicola* (Taylor *et al*., 1979a).

In hypogeal vegetable seeds, plumule infection may result from contact with diseased seeds which remain below the soil during germination (e.g. *Ascochyta pisi* of peas (Fig. 13a, 2) and *Colletotrichum lindemuthianum* of phaseolus beans (Fig. 13b)), but with other pathogens invasion may occur via the seed attachment into the hypocotyl of plants to cause foot and stem rotting (e.g. *Mycosphaerella* of pea: (Fig. 13a, 1): Maude, 1966a). Similar transmission causes foot rotting in cereals (*Fusaria* of barley and wheat: Colhoun and Park, 1964) and root, hypocotyl and cotyledon infection in oil seeds (*Macrophomina phaseolina* of sunflower: Raut, 1983).

Transmission of certain seedborne necrotrophs of cereals may result from the direct penetration of the base of the emerging coleoptile by mycelium from the pericarps of seeds causing, for example, seedling blight of wheat (*Septoria nodorum* state of *Leptosphaeria nodorum*: Holmes and Colhoun, 1971) and leaf spot of oats (*Pyrenophora avenae*: Old, 1968) and barley (*P. graminea*: Teviotdale and Hall, 1976). Baker (1980) gives examples of seedborne pathogens of flowers, and Schuster and Coyne (1974) cite seedborne bacteria which are restricted or favoured by the method of germination of the seeds. In many cases the mycelium of fungi borne within the seed coat or pericarp tissues transmits infection but in some the asexual fruiting stages can be present and may contribute to transmission. For example, pycnidiospores of *Septoria apiicola* (leaf spot, late blight) released in a water film from pycnidia located in the pericarps of celery fruits attached to emerging cotyledon tips infect the cotyledon blades causing leaf spots (Maude, 1964) (Fig. 13c, d).

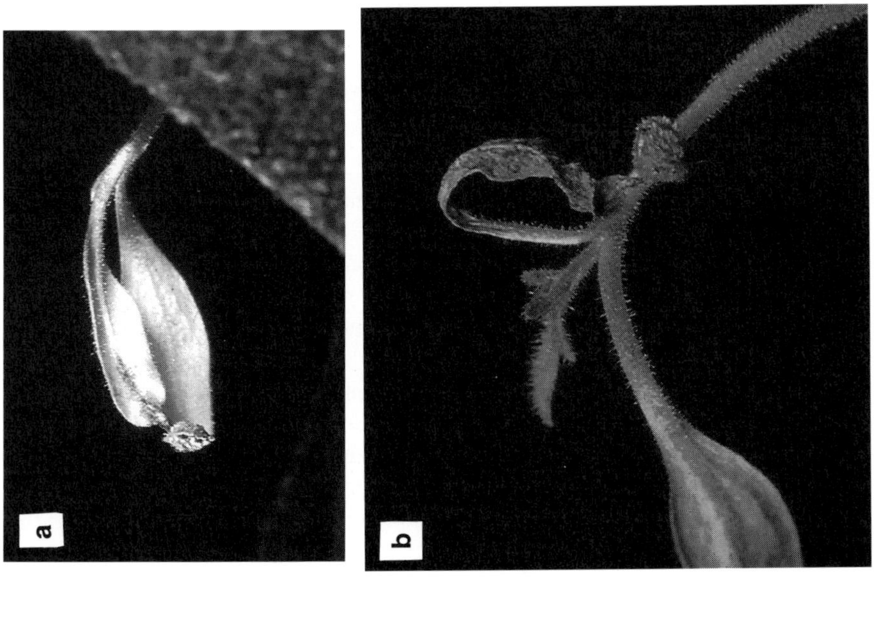

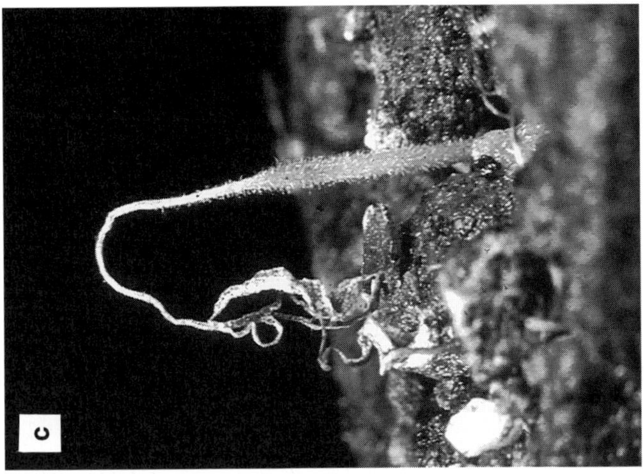

Fig. 11. Seed transmission of infection (*Didymella lycopersici*) to a tomato seedling. (a) Fungal invasion of the cotyledon midrib from the affected seed coat (week 2). (b) Progress of infection involving most of the cotyledon (week 3). (c) The crown and upper stem are invaded and the seedling is dead (week 5). Reproduced with the permission of Horticulture Research International.

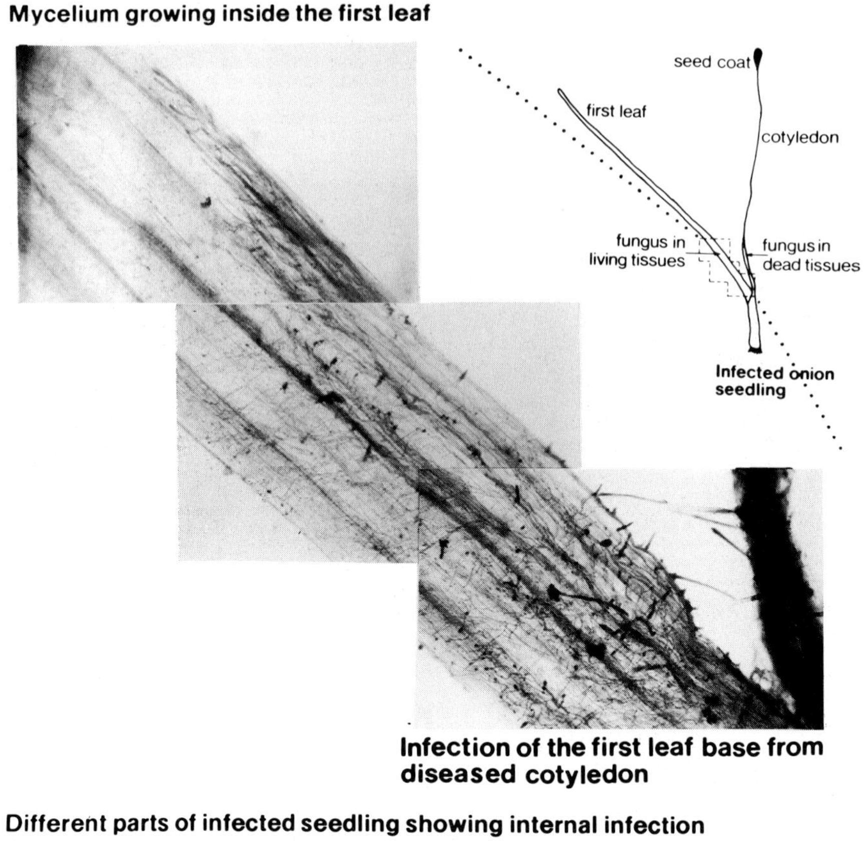

Fig. 12. Transmission of seedborne *Botrytis allii*. Reproduced with the permission of Horticulture Research International.

Seed transmission causing vascular infections

The seed transmission of wilt-causing forms of *Fusarium oxysporum* is reviewed by Gambogi (1983). Most of these fungi are persistent soil-borne pathogens favoured by high soil temperatures (Walker, 1969), and there are 26 formae speciales (f.sp.) of this fungus listed by Richardson (1990) affecting 33 seed hosts. *F. oxysporum* f.sp. are transmitted to the vascular systems of developing plants from contaminated seed coats (e.g. *F. oxysporum* f.sp. *callistephi* of China aster) or from deeper seated infections of seeds (e.g. *F. oxysporum* f.sp. *matthiolae* of garden stock). Where there is epigeal emergence the fungus may grow from the seed coat into the cotyledons then the stem and finally the roots of plants. Where, however, emergence is hypogeal

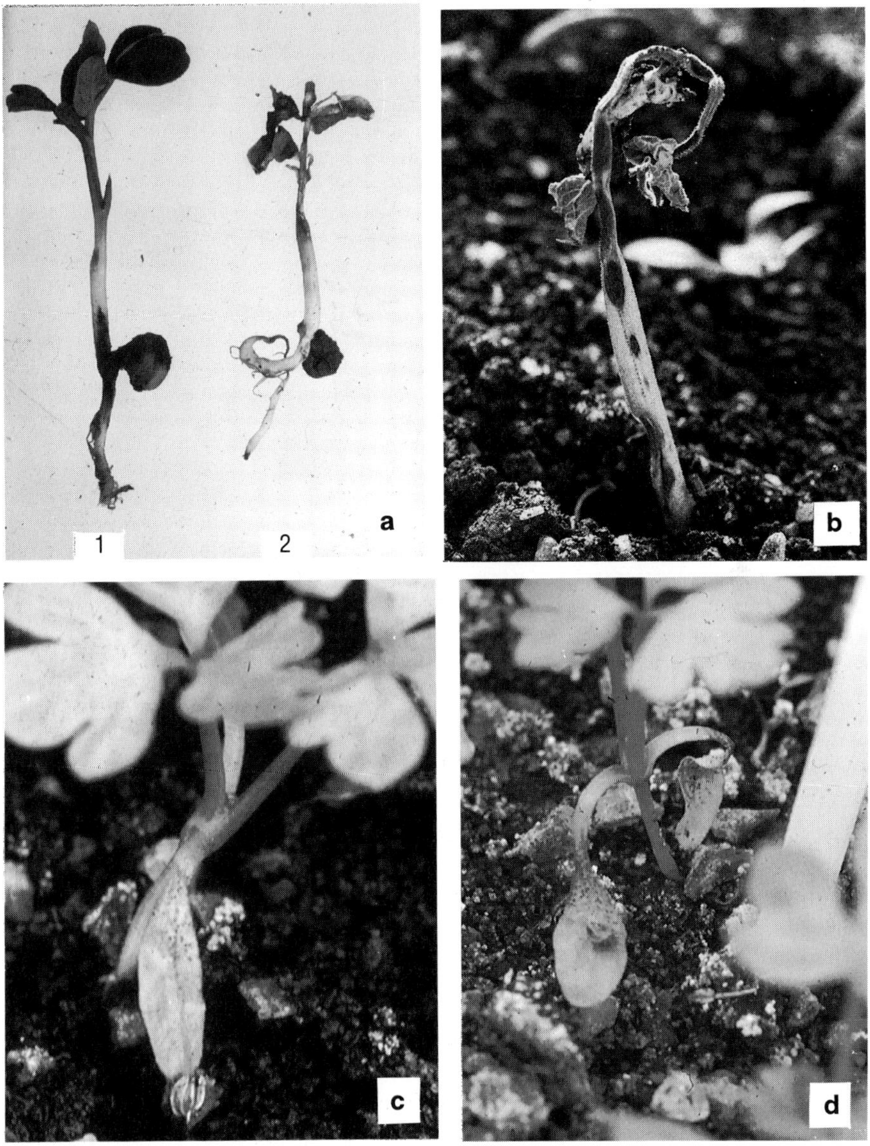

Fig. 13. Seed transmission of infection. (a) Hypogeal germination of a pea. 1, *Mycosphaerella pinodes*—direct transmission from the seed to the hypocotyl causing foot rot. 2, *Ascochyta pisi*—contamination from the seed coat causing leaf spot infection of the emerging plumule. (b) Hypogeal germination of a phaseolus bean. *Colletotrichum lindemuthianum*—contamination from the hypogeal cotyledons causing anthracnose infection of the emerging plumule. (c,d) Epigeal germination of celery. *Septoria apiicola*—pycnidiospores from seed coats liberated into water drops to cause leaf spot infection of the cotyledon blades. Reproduced with the permission of Horticulture Research International.

(e.g. *F. oxysporum* f.sp. *ciceris* of chickpea) invasion is direct from the cotyledons via the seed attachment into the hypocotyl extending downwards into the roots (Gambogi, 1983). The fungus permeates the vascular system and wilt is caused by the blockage of those tissues.

Seed transport of the mainly soil-borne *Verticillium* spp. (*V. albo-atrum* and *V. dahliae*) which cause vascular wilting in some 300 plant species around the world, is likely and seeds may, therefore, become infected but seed transmission of these fungi is not well established and they may be unimportant in the crop situation (Sackston, 1983)

Bacterial wilts, particularly *Curtobacterium flaccumfaciens* pv. *flaccumfaciens* of beans (*Phaseolus vulgaris*), are unlikely to be transmitted by bacteria contaminating the surface of seeds (Schuster and Smith 1983). Transmission may occur when the germinating radicle penetrates the seed coat causing damage to itself and releasing bacterial cells from within the seed coat or later from bacteria infecting the emerged cotyledons reaching the xylem of the stem causing wilting and other symptoms (Schuster and Smith, 1983).

Transmission by organisms accompanying seeds

When seeds are threshed, plant and inflorescence dust and debris and some soil is mixed in with the seeds. Much of this material is removed by subsequent cleaning processes (dressing) but some remains mixed in loosely with the seeds or more firmly attached where cleaning has incompletely removed the former flower parts. Potentially, many organisms could be transported mixed with seeds. Viruses are most unlikely to survive in these situations (see Chapters 2 and 3) but most fungi and bacteria, which have a limited or a more extended survival capability, can accompany seeds mixed with them as resting organs, reproductive organs, dried bacterial cells, on or in infected debris, etc. When infested seeds are sown, the accompanying inoculum is returned to the soil. Although transmission by organisms accompanying seeds is recorded in the literature (Baker, 1972; Neergaard, 1977; Agarwal and Sinclair, 1987), the significance of this for fungi and bacteria is difficult to estimate. The concentration of inoculum present is reduced by the seed-cleaning processes (Ball and Reeves, 1992) and hence the rate of transmission of many of the organisms accompanying seeds is probably low, in addition to which it may be further reduced by adverse biological effects of the soil microflora.

Menzies and Jarvis (1994) report contamination of tomato seeds at rates of 0.01–0.1% in fruits of plants with stem infections caused by *Fusarium oxysporum* f.sp. *radicis-lycopersici*, but much higher levels of seed contamination occurred when clean seeds came into contact with hands which had previously handled sawdust contaminated by the fungus. The pathogen survived on seed sent across Canada and stored for up to 12 weeks (Menzies and Jarvis, 1994).

Transmission by fungal resting bodies and spores

Sclerotia of fungi, particularly those of species in *Sclerotinia* and *Claviceps*, are transported mixed with seeds. Transmission is usually indirect. Sclerotia deposited in the soil germinate after a cold stimulus to produce ascocarps from which ascospores are discharged to infect crop plants. Where sclerotia are in contact with the stem bases of plants, mycelial transmission may occur (Scott and Evans, 1984). The large irregular sclerotia of *S. sclerotiorum* (white mould, cottony soft rot, watery soft rot) can occur mixed with many crop seeds (Neergaard, 1977) and have an extremely wide host range (Smith *et al.*,1988). Some 60% of the small clover seed-sized sclerotia of *S. trifoliorum* (clover rot) germinated to produce apothecia (Scott and Evans, 1984) so there is considerable potential for the introduction of the disease into new areas by sclerotia mixed in with the seeds. *S. spermophila* is widespread on white clover seed but rarely causes any damage (Smith *et al.*, 1988). The sclerotia (ergots) of *C. purpurea* (ergot of cereals) can overwinter mixed with grain for sowing, producing the ascigerous (sexual) stage after receiving a cold stimulus in the soil (Smith *et al.*, 1988).

Chlamydospores of certain foot rot fungi, for example *Fusarium solani* f.sp. *phaseoli* (Nash and Snyder, 1964), and *Phoma medicaginis* var. *pinodella* (Baker, 1972), occurring in soil particles may be transported with seeds. Direct infection occurs by mycelial invasion of plant tissues. Spores of certain Uredinales (rust) fungi (*Puccinia antirrhini, P. malvacearum*) may also be transported mixed with seeds (Neergaard, 1977).

Transmission by organisms on and in plant debris

Diseased pieces of plant debris are most frequently cited as the means of introduction of diseases transported with seeds. The exact methods of transmission are not always apparent but it is presumed these are related to the location of the debris after the seeds are sown and to the biology of the organisms involved. Some of the fungi transferred in this way include *Gaeumannomyces graminis* (take-all of cereals) and *Itersonilia pastinacae* (parsnip canker) and imperfect (pycnidial) states including *Ascochyta rabiei* (leaf spot of chickpea), *Septoria linicola* (pasmo of flax) and *S. apiicola* (celery leaf blight) (Neergaard, 1977). Some of the bacteria include *Clavibacter michiganensis* subsp. *insidiosus* (bacterial wilt of lucerne), *Pseudomonas syringae* pv. *phaseolicola* (halo blight of bean) (see Chapter 2, p. 27) and *C. michiganensis* subsp. *michiganensis* (canker) (Baker, 1972; Neergaard, 1977).

In addition to seedborne organisms which are transported mixed with seeds it is also possible that soil-borne diseases could be distributed in this way. Two reports (Warne, 1943; Maklakova, 1964) suggest that *Plasmodiophora brassicae* (club root of brassicas) may be disseminated with infested seeds. Similarly, it is suggested that *Pseudomonas solanacearum* which causes bacterial wilt in a wide range of crops in hot climates (latitudes 40° north and south) may be introduced into new growing areas with infested seeds (Frison *et al.*, 1990). Substantiating research is required in both cases.

Environmental Factors Affecting Transmission

There are many host and environmental factors which directly or indirectly affect the transmission rates of seedborne pathogens. Temperature, moisture, light and pH are the more important environmental factors which influence the infection of plants (Tarr, 1972). These parameters, as they apply in air or soil or both media, also modify the transmission of seedborne diseases. In addition they may affect the action of the soil microflora in the reduction and even suppression of transfer of inoculum from seed to seedling. Temperature and moisture are critical integral factors determining transmission rates of seedborne diseases.

Temperatures which are optimal for the growth of pathogens *in vitro* may also be optimal for seed transmission of disease *in vivo*. Seed transmission of temperate and tropical pathogens will relate to the temperature profiles of their respective environments. For example, the incidence of *Aspergillus niger* was greatest on onion seeds produced in hot (desert) climates and its transmission rate highest (Hayden and Maude, 1992) at temperatures close to the optimum for growth of the fungus (32.5°C) (Maude, 1990a). Equally, the pathogenic microflora of sown seeds may remain quiescent under non-optimal conditions, becoming active when temperatures and moisture (namely relative humidity and rainfall) are appropriate for their expression. For example, zinnia seeds infected with *Alternaria zinniae* planted in the dry interior valleys of California produce seed crops virtually free of the pathogen (Baker and Davis, 1950a) (see Chapter 6). In such situations moisture is the critical factor in determining the level of disease transfer; with certain diseases free water is needed for transmission to be completed.

Temperature and moisture effects on seedborne biotrophs

Because seedling susceptibility of cereals is short lived, effective seed transmission of certain Ustilaginales fungi, for example the common bunts of wheat (*Tilletia tritici* and *T. laevis*), requires that the teliospores should germinate more or less coincidentally with the grain (Neergaard, 1977). Soil temperatures and, possibly less critically, moisture, control spore germination and as a result the seed transmission rates of both pathogens can vary considerably. Teliospores of *T. tritici* germinate optimally in soils at 10–15°C and at moisture levels above 11% w/w (field capacity is equivalent to 24% w/w at 34 kPa) but spore germination is severely reduced by high temperatures (20–25°C) and dry soil conditions (Purdy and Kendrick, 1957). Conditions which are optimal for spore germination in soil are also optimal for seed transmission of bunt; 10°C and 13% soil moisture are the most favourable conditions (Kendrick and Purdy, 1959). The first part of this statement also can be applied to other members of the Ustilaginales in temperate and tropical climates where teliospores contaminating the external surfaces of cereal seeds are the agents of infection (Neergaard, 1977). Very wet soils (Leukel, 1936), very dry soils, high soil temperatures and acid soils (Neergaard, 1977) generally reduce transmission of seedborne members of the Ustilaginales.

The effect of the environment on the transmission of the cereal smuts has been reviewed comprehensively (Tapke, 1948). Environmental parameters also control the transmission of other superficially borne biotrophs. For example, soyabean seeds encrusted with oospores of *Peronospora manshurica* sown in soil at 13°C produced 40% infected seedlings, but no infection occurred in soil at 18°C (Dunleavy, 1981). Where, however, the pathogens are internally borne, as with loose smut of wheat (*Ustilago segetum* var. *tritici*), provided there is sufficient mycelium within the embryo, each infected seed will produce an infected plant in susceptible cultivars and the effect of the soil environment on transmission may be slight (Batts and Jeater, 1958a). In addition, Batts and Jeater (1958b) showed that the date of sowing or the locality of the crop did not change the incidence of loose smut of wheat (*U. segetum* var. *tritici*), but Doling (1965a) demonstrated that one stock of infected wheat produced different proportions of smutted ears when grown at different centres in the UK. These differences he related to the effects of different crop environments on tillering.

There is little information on the effect of temperature on the rate of transmission of seedborne viruses to crop plants but the temperatures at which the seeds are produced may determine whether or not a virus is seed transmitted (Shepherd, 1972). For example, 95% of seeds harvested from plants grown at 16–20°C transmitted southern bean mosaic virus but only 55% of seeds from plants grown at 28–30°C did so (Crowley, 1959). Similarly, transmission of cherry leaf roll virus was 100% and 0% in seeds produced by *Nicotiana rustica* (scented tobacco) plants grown at 20 and 30°C, respectively (Cooper, 1976).

Temperature and moisture effects on seedborne necrotrophs

Affecting pathogen transmission to hypogeally germinating seeds at or below soil level

Many of the cereal diseases common to temperate regions of the world, and of varying importance in these areas, are initiated by infection of the coleoptile (Colhoun, 1971). Whether the source of inoculum is soil-borne or seedborne, soil temperature and moisture limit the incidence of this primary infection. As for the ustilaginaceous fungi (see pp. 56–57), transmission is facilitated by relatively dry cool soils and reduced by wet warm soils. Much of this information has been obtained in experimental systems but it agrees in the main with field observations of disease transmission and applies to a range of diseases. Pre- and post-emergence deaths caused by seedborne *Fusarium culmorum* and *F. graminearum* (teleomorph *Gibberella zeae*) (*Fusarium* foot rots and seedling blights) were greatest in dry soils (30–37% maximum water-holding capacity (mwhc)) at temperatures of 16–23°C, with little disease in wet soils (50% mwhc) at any temperature (Colhoun and Park, 1964; Duben and Fehrmann, 1979; Halfon-Meiri *et al.*, 1979). There was 56% seed transmission of *Pyrenophora avenae* (leaf spot of oats) in cold dry soils and 15% in warm wet soils over a temperature range (6.6–13°C) likely to occur in a normal spring (Muskett, 1937). Likewise, *Leptosphaeria nodorum* (seedling blight) infected coleoptiles in dry soils at 12°C but wetter soils and temperatures of 8°C or 17°C permitted less transmission (Holmes and Colhoun, 1971). Temperatures of 10–12°C (Johnson, 1925; Teviotdale and Hall, 1976) favoured infection by *P. graminea* (leaf stripe of barley) but little occurred above 15°C (Teviotdale and Hall, 1976) or 20°C (Johnson, 1925). Similarly, temperatures of 20°C and above severely reduced seed transmission of *Rhynchosporium secalis* (leaf blotch, scald); 16°C was optimal at various sowing depths (Skoropad, 1959). Primary infection by *P. teres* (net blotch) occurs at temperatures of 10–15°C (Singh, 1958). Net blotch increased in importance in 1981 in the UK as a result of an increase in area of winter barley previously sown with susceptible cultivars; although sources other than seed may be more important in most countries, seed may have been implicated in that year when infection levels in harvested UK stocks ranged from 2% to 40% (Rennie, 1990).

In hot climates, lower rather than higher temperatures may restrict transmission. In India, there were no post-emergence losses in sunflower seedlings grown at 25°C from *Macrophomina phaseolina* (root rot)-infected seeds compared with 74% losses in seedlings grown at 35°C from the same seed sample (Raut, 1983).

In addition to the effects of these edaphic factors on transmission, high atmospheric humidities may be necessary to establish and maintain water films for the transmission of seedborne bacteria. For example, transfer of *Pseudomonas syringae* pv. *atrofaciens* (leaf necrosis of wheat) from artificially contaminated seeds to the coleoptiles occurred at relative humidities (RH) of 80–98% and air temperatures of 10°C (Fryda and Otta, 1978). This supports the earlier work of Leben (1965) who demonstrated that, in a controlled environment, non-pathogenic marker bacteria used to contaminate cucumber seeds migrated to the seedling leaves under conditions of high (> 90% RH) but not low (30–40% RH) atmospheric humidities. Movement of bacteria by swimming to the upper parts of seedlings in a water film or by the simultaneous division of bacterial and plant cells at the growing point, were postulated as possible mechanisms of transmission (Leben and Daft, 1966). For unexplained reasons, contaminated seeds gave a more uniform distribution of bacteria on cucumber (epigeal) seedlings than on soyabean (epigeal) or maize (hypogeal) seedlings.

Affecting pathogen transmission to the aerial parts of hypogeally or epigeally germinating seeds

Seed transmission of many pathogens affects the above-ground parts of plants (e.g. the cotyledons and first leaves) and the disease then spreads across the foliage. Environmental conditions in the soil influence disease transmission from the time seeds begin to germinate to the time their cotyledons or first leaves emerge from the soil. Where there is hypogeal germination, and the seed cotyledons remain in the soil, the incidence of the pathogen on the plumule or first leaf to appear above the soil depends on the frequency of contact between the emerging plumule and the infected seed, which is affected by the rate of germination which is dependent on soil moisture and temperature conditions (see p. 58). As moisture levels in compost increased from 15% (−0.29 MPa) to 29% (−0.028 MPa) the transmission of bacterial pea blight (*P. syringae* pv. *pisi*) increased from 0.75% to 23.3% infected seedlings (Roberts, 1992); temperatures of 12–18°C had no effect on the transmission rate.

Where there is epigeal emergence, the cotyledon sometimes emerges with the seed coat or the remains of it still attached. Once emerged, further transmission from the seed to the cotyledon is modified by climatic conditions. Free moisture is important in this respect. In dry seed beds there was less than 1% transmission of *Septoria apiicola* from infected celery seed cases which had remained attached to seedling cotyledon tips after emergence. However, over 20% of plants became infected where free water enveloped the cotyledon blades and pycnidiospores were released from the pycnidia embedded in the seed cases (Maude, 1964). Water transfer of conidia of *Colletotrichum lindemuthianum* (anthracnose) and of pycnidiospores of *Phoma lingam* (teleomorph *Leptosphaeria maculans*, canker) was necessary to establish primary cotyledon infections of phaseolus beans and brassicas, respectively (Walker, 1969).

Temperature and moisture effects on vascular pathogens

There is little information on the effects of these parameters on seed transmission of *Fusarium* and *Verticillium* wilts but there is a considerable literature relating to their effect on disease transmission from soil (Nelson, 1981; Schnathorst, 1981; Snyder and Nash Smith, 1981; Sherf and Macnab, 1986). *Verticillium* wilts occur mainly in temperate to subtropical regions, infecting optimally in moist soils, at temperatures between 21°C and 25°C and with neutral pH (Sherf and Macnab, 1986). *Fusarium* wilts (e.g. *F. oxysporum* f.sp.

lycopersici) attack in warmer, drier soils (28°C) under conditions of low light intensity and acid pH (Sherf and Macnab, 1986).

Soil Microfloral Effects on Seed Transmission of Disease

Interactions between the environmental parameters can alter the percentage transmission of disease by their direct effects on pathogens or indirectly by affecting the soil microflora, which in turn modify the transfer of disease from seed to seedling. For example, 35% of pea seeds which were superficially contaminated with *Ascochyta pisi* (leaf and pod spot) failed to produce infected seedlings in field soil (communication from A.L. Wharton in Maude, 1973). There is considerable information showing that the seed transmission rate of pathogens in unsterile soil is reduced by comparison with their rate of transfer in more sterile media (e.g. compost, vermiculite, etc.) and that seedling infection is considerably less than the initial incidence of the individual seedborne pathogens as determined by laboratory tests (Table 10). The pre-emergent death of severely infected seeds and seedlings only partially accounts for this reduction. Otherwise, soil types and their concomitant microflora are implicated but the exact modes of action are not certain.

Cunfer (1983) suggests that the greater colonization of wheat coleoptiles by *Leptosphaeria nodorum* in dry soils is because, according to Griffin (1972), most bacteria are inhibited at water potentials only slightly below field capacity which would restrict their antagonistic and competitive activities and possibly that of other similarly acting microorganisms in the rhizosphere. Cunfer (1983) postulates that because several fungi, bacteria and yeasts have been identified by authors as masking or having been antagonistic to *L. nodorum* in different test situations, the chance for similar relationships exists in the soil.

In addition to these commonly occurring partial reductions in seed transmission there are suppressive soils which have considerably greater effects by reducing vascular diseases caused by *Fusarium* wilt fungi (Mace *et al.*, 1981; Gambogi, 1983). These are the clay and organic soils of parts of South, Central and North America which are disinfective of mainly soil-borne *F. oxysporum* inoculum (Nelson, 1981); in certain instances their action against seedborne inoculum has also been reported (Baker, 1980, 1981; Gambogi, 1983). Baker (1980, 1981) cites the example of suppressive Californian soils into which hundreds of tons of pea seeds of susceptible varieties contaminated with *F. oxysporum* f.sp. *pisi* have been planted over the years with outbreaks of pea wilt limited to two small areas in that time. If soil from suppressive areas is steamed then pea wilt can be established in them (Baker, 1981); rhizosphere bacteria are probably involved and can cause both suppressive and conducive effects, but the mechanisms are complex (Smith and Snyder, 1972). Conversely, some US soils are so conducive to *F. oxysporum* f.sp. *pisi* that only resistant cultivars of peas can be grown there (Baker and Cook, 1974).

Table 10. Seed transmission relationships.

Pathogen	% infection in Lab/glhse	% infection in Field soil	Transmission ratio	Crop	Reference
Polyspora lini	15.0	1.7	9:1	Flax	Henry and Campbell, 1938
Colletotrichum lini	66.3	17.0	4:1	Flax	Henry and Campbell, 1938
Pyrenophora graminea and *P. teres*	50–75	5–10	10:1 to 7.5:1	Barley	Jorgensen, 1977
	50–65	0–11	0:0 to 6.0:1	Barley	Jorgensen, 1977
Ascochyta pisi	11.2	3.3	4:1	Peas	Maude and Kyle, 1970
	6.3	0.4	16:1	Peas	Maude and Kyle, 1970
	34.0	6.5	5:1	Peas	Maude and Kyle, 1970
Alternaria brassicicola	62.0	11.0	6:1	Cabbage	Maude and Humpherson-Jones, 1980b
	11.5	1.2	10:1	Kale	Maude and Humpherson-Jones, 1980b
	1.5	0.0	0:0	Kale	Maude and Humpherson-Jones, 1980b
Alternaria brassicae	28.0	9.3	3:1	Cabbage	R.B. Maude, pers. comm., 1991
	10.5	1.2	9:1	Oilseed rape	R.B. Maude, pers. comm., 1991
	14.0	1.2	12:1	Oilseed rape	R.B. Maude, pers. comm., 1991
Pseudomonas syringae pv.	9.2	0.84	11.0:1	Phaseolus bean	Taylor, 1970b
phaseolicola	3.5	0.37	9.5:1	Phaseolus bean	Taylor, 1970b
	1.4	0.15	9.3:1	Phaseolus bean	Taylor, 1970b
	0.1	0.0	0:0	Phaseolus bean	Taylor, 1970b
Pseudomonas syringae pv.	16.1	1.80	8.9:1	Phaseolus bean	Taylor *etal.*, 1979b
phaseolicola	1.1	0.13	8.5:1	Phaseolus bean	Taylor et al., 1979b
	2.4	0.22	10.9:1	Phaseolus bean	Taylor *et al.*, 1979b
	2.4	0.42	5.7:1	Phaseolus bean	Taylor *et al.*, 1979b
	5.4	0.57	9.5:1	Phaseolus bean	Taylor *et al.*, 1979b

Lab/glhse, laboratory/glasshouse tests

Microbiological activity probably affects the viability of the *Fusarium* chlamydospores (Smith and Snyder, 1972) which may be carried with the seeds (Snyder, 1932).

Effects of Host Genotype on Disease Transmission and Resistance

Both horizontal and vertical resistance may operate through the mother plant of the seed and through the seeds to modify and sometimes prevent transmission of disease. Horizontal (mechanical) resistance by a closed flowering habit (Pedersen, 1960; Hewett, 1972) is a method by which the developing seeds of some barley cultivars prevent infection by the loose smut fungus, thereby stopping subsequent seed transmission. Mechanical resistance such as this is not susceptible to the development of new races of the pathogen but it is affected by infection pressure, and increased levels of airborne inoculum at flowering may result in increased proportions of embryo-infected seeds which freely transmit the disease (Hewett, 1968). The period required for millet (*Pennisetum americanum*) stigmas to become infected by *Claviceps fusiformis* (ergot) is usually 36–48 h in the tropics (Willingale and Mantle, 1987). Disease escape is mediated by the development of a localized stigmatic constriction which may occur 6 h after self-pollination providing a unique mechanical barrier to invasion of the fertilized ovary by fungal pathogens (Willingale *et al.*, 1986). Consequently, breeding at the International Crops Research Institute for the Semi-Arid Tropics (ICRISAT) for ergot resistance in millet under conditions of high inoculum pressure is based on the selection of individuals in which the stigmas emerge for only a few hours before self-pollen is shed and stigmatic constriction ensues (Willingale *et al.*, 1986).

Vertical (physiological) resistance to wheat and to barley loose smut (*U. segetum* var. *tritici*) also operates through the seeds. Popp (1951, 1958) describes vertical resistance for some cultivars that is characterized by the reaction of the embryos which are (i) completely resistant to infection; (ii) susceptible except for the plumular bud (the growing point and the surrounding convolute second and subsequent leaves); and (iii) completely susceptible in all tissues. Hewett (1979a) also reports a number of spring barley cultivars with high embryo infection counts which produced adult plants with few smutted ears. A product of vertical resistance has been the development of a complex race structure in loose smut of barley (Mohajir *et al.*, 1952; Konzak, 1953; Thomas, 1974; Thomas and Metcalfe, 1984) and in loose smut of wheat (Nielsen, 1983; Huang and Nielsen, 1985). As a result, the predictive use of embryo testing to determine field infection levels of smut which was extremely effective with susceptible cultivars (Batts and Jeater, 1958a; Marshall, 1959;

Hewett, 1980; Neergaard, 1986) has little value where the host growing point is resistant to transmission (Hewett, 1979a).

Heald (1921) obtained a correlation between inoculum concentration and cultivar susceptibility, demonstrating that seeds of the stinking smut (*Tilletia tritici*) resistant cv. Marquis required between 542–5043 teliospores per grain to transmit the disease compared with 104 spores per grain for the susceptible cv. Jenkins Club.

Both the host and the virus genotype control the capacity for and seed transmissibility of, a virus in a host. The capability for seed transmission of a particular virus varies greatly in different species or even varieties and cultivars of a particular host plant (Mandahar, 1981). This has been correlated with the susceptibility or resistance of the mother plant of the seed, which if severely infected and therefore susceptible, is more likely to transmit disease through the seed than less severely infected plants which are tolerant of the disease. Pea seedborne mosaic virus (PSbMV) is a recent and serious problem in some pea cultivars in the UK (Matthews *et al.*, 1981). However, resistance to the transmission of PSbMV from seeds to seedlings occurs widely in domestic pea cultivars in the UK and potentially provides a strategy for disease control (Wang *et al.*, 1993). The mechanisms by which seed transmission of the virus is blocked remain to be discovered. Resistance to seed transmission of barley stripe mosaic virus (BSMV) has been shown to be controlled by a single recessive gene (Carroll *et al.*, 1979).

Host resistance to virus infection may be due either to the suppression of virus multiplication and therefore lower concentrations of virus in leaves or to a failure of the virus to move out of infected leaves into the developing seeds.

The utilization of resistance in commercial practice as a means of excluding seedborne diseases is considered in Chapter 6.

Relationship Between Inoculum Potential and Disease Transfer

Inoculum can be considered to be any material capable of producing infection (Dimond and Horsfall, 1960); in this account it comprises the fungal mycelium, propagules, spores, etc., the bacterial cells and the virus particles which contaminate and infect seeds. Horsfall (1932) originally defined inoculum potential as the number of infective particles present in the environment of the host. This view was concerned with the amount or virulence of inoculum rather than with the influence of the environment on the severity of infection. Modifications were made to this term by others; inoculum potential was a concept devised for the study of root diseases and soil-borne pathogens but equally it can be applied to seeds. When Heald (1921) first used the term spore load to describe the number of bunt spores per wheat kernel he was describing inoculum potential. Inoculum potential can be regarded as the driving force required to cause disease.

It is described as the product of the quantity of inoculum present and the capacity of the environment (in the broad sense) to produce disease in a host of given susceptibility with a pathogen of stated characteristics (Dimond and Horsfall, 1960).

If inoculum potential is the product of 'quantity' of inoculum × the environment, then in terms of successful transmission of seedborne pathogens it is also the product of 'quality' of inoculum × the environment. Too much or too little inoculum of many pathogens can result in the non-transmission of disease (de Tempe, 1968). For example, early infections of flowering plants may cause severe infections of the seeds, and the abundant inoculum reduces their viability so that by failing to germinate they fail to transmit disease (e.g. poor quality light seeds of cotton infected with *Xanthomonas campestris* pv. *malvacearum*: Brinkerhoff and Hunter, 1963; and lettuce seeds similarly affected with lettuce mosaic virus: Tomlinson and Faithfull, 1973). In contrast, late infections of flowering plant infloresences may result only in superficial contamination of seeds which is eliminated by the soil microflora when the seeds are sown.

It is the concentration of inoculum between these extremes which determines the 'transmission rate' in conjunction with the germination capacity of the seeds and the environmental conditions which apply at the time the seeds are sown. Some of the environmental conditions which modify the transmission rate have been described above; inoculum is much more difficult to quantify and Colhoun (1983) suggests that workers on seedborne diseases have only been successful in measuring inoculum when it is carried on the surface of seeds as spores. However, where inoculum is borne internally in seeds it is the transmission potential of the infected seed which is operative and this can be defined as the number of seeds in a sample capable of transmitting infection.

A meaningful assessment of inoculum potential should determine its relationship to transmission of infection, preferably in the field crop using naturally infected seeds sown under optimal environmental conditions for disease expression. This relationship also can be established in the glasshouse provided unsterile field soil is used and the appropriate environmental conditions obtain. Not all of these requirements will have been fulfilled in any one situation and as a result reported transmission relationships are not consistent.

Methods of assessing seed inoculum levels

Many of the standard international laboratory seed testing methods have been devised for the detection of seedborne pathogens and not for the determination of inoculum levels (Colhoun, 1983). However, some methods have been described (Colhoun, 1983) which have been used to measure seed inoculum levels including seed washing to determine spore loads (Lobik, 1926; Maude and Humpherson-Jones, 1980a), surface sterilization of seeds to evaluate deep-seated inoculum (Neergaard, 1977), use of special media to correlate with seed inoculum levels (Malalasekera and Colhoun, 1969), blotter tests of seeds correlating fungal growth types with seedling infections (Aulakh *et al.*, 1974), extraction of bacteria by grinding seeds to a flour (Taylor, 1970b), and embryo staining and assessment after extraction of cereal seeds (Popp, 1958; Hewett, 1970). Most of these have been tested in practice to relate inoculum load to transmission of infection to plants.

Disease transmission by superficial inoculum on seeds

Inoculum–disease relationships have been established by the addition of increasing numbers of spores to seeds which when sown produce increasing quantities of infected seedlings or plants. The relationship is usually linear and it illustrates the fact that very often large numbers of pathogen (biotrophs or necrotrophs) spores per seed, spanning a considerable numerical range, are needed to transmit infection at a maximum level to the field crop.

Teliospores of the biotrophic bunt and smut fungi can occur naturally at a high incidence on the exterior of cereal seeds. Thus Heald (1921) reports that the number of bunt (*Tilletia tritici*) spores per grain varied between 0 and 45,416 and Leblond (1948) recorded 16,000 spores per seed of a mixture of *Ustilago segetum* var. *avenae* (syn. *U. avenae*) and *Ustilago segetum* var. *segetum* (syn. *U. kolleri*) spores on oat seeds. When bunt spores were dusted onto seeds 36,000–150,000 spores per seed were required to produce a maximum number of infected ears in a wheat crop (Heald, 1921). Leblond (1948) obtained a linear response to increasing numbers of spores of *U. segetum* var. *avenae* and *U. segetum* var. *segetum* (loose and covered smut of oats) achieving maximum transmission at 50,000 or more spores per seed. Pre-emergence death of wheat in unsterile soil in boxes by transmission of the necrotroph *Fusarium culmorum* (foot rot) was greatest at concentrations of 5 million to 50 million spores per seed (Malalasekera and Colhoun, 1969).

Disease establishment, however, can be initiated by much lower concentrations of spores. Leblond (1948) obtained an average of 3.4% smut infected plants in the field from 0 to 100 spores per seed and Malalasekera and Colhoun (1969) demonstrated a 1.3% pre-emergence death of wheat seeds from 500 *F. culmorum* spores per seed. Although Heald's work (1921) with *T. tritici* discounted the previously held view that one bunt spore was sufficient to cause infection, Millar and Colhoun (1969) showed that, at low temperatures, every seedling could be infected by *F. nivale* (teleomorph *Monographella nivalis*, brown foot rot of cereals) when an average of only one spore was present on 50% of the seeds.

There are many factors, not the least the individual host–pathogen interactions, influencing and perhaps accounting for the variability in transmission of infection by seedborne spores. Thus higher spore loadings of *F. avenaceum* (a weaker pathogen of wheat) were required to achieve similar levels of seedling mortality to those of *F. culmorum* (a stronger pathogen) (Colhoun, 1970). Temperature and moisture are important (see pp. 57–58) in determining transmission rates but in certain circumstances high spore loadings of seeds may override and reduce the adverse effects of low temperature on transmission (e.g. *F. culmorum*) (Colhoun *et al*., 1968).

Disease transmission by internal inoculum in seeds

Severity of seed infection

With many necrotrophic pathogens there can be a continuity of inoculum extending from the surface of the seed into the tissues of the testa and possibly even deeper into the seed embryo. A relationship between superficial and internal inoculum can be established. For example, there is a significant correlation between the number of superficially borne conidia of *Alternaria brassicicola* per brassica seed and internal infection (Maude and Humpherson-Jones, 1980a) (Fig. 14A). The causal relationship, however, tested by sowing infected seeds in unsterile soil in the glasshouse, showed that disease transmission was more closely correlated with those seeds bearing the fungus internally than with total inoculum (superficial and internal) and that the sowing of seed samples containing increasing numbers of internally infected seeds resulted in a corresponding increase in numbers of infected seedlings (Fig. 14B). In this case differentiation between internal and total inoculum was achieved by surface sterilization of seed to eliminate external conidia. Bassey and Gabrielson (1983) confirmed the role of internal inoculum in transmission of this pathogen in controlled environments maintained at air temperatures of 20°C but found that total inoculum was more important at 30°C.

With increasing severity of infection of seeds, for example, as measured by the intensity of *Colletotrichum lindemuthianum* symptoms on bean seeds, the rate of transmission of anthracnose in phaseolus beans was increased (Tu, 1992) (Fig. 15A). Taylor *et al*. (1979a) established a similar relationship under glasshouse conditions for phaseolus bean seeds with increasingly severe symptoms caused by the bacterium *Pseudomonas syringae* pv. *phaseolicola*, but in field sowings the most severely affected seeds had a reduced emergence rate, and higher rates of transmission of halo blight were obtained from symptomless seeds and those with slight to moderate symptoms. However, field sowings of surface-sterilized wheat samples with increasing percentages of internally infected (*Fusarium graminearum*) seeds gave strong positive correlations ($r^2 = 0.96$ and $r^2 = 0.89$ at $P < 0.05$) for transmission of seedling blight to the stems of winter wheat (Duthie and Hall, 1987) (Fig. 15B). In that relationship transmission rates were consistent and high, possibly indicative of a more predictable infection transfer process and higher seed viability.

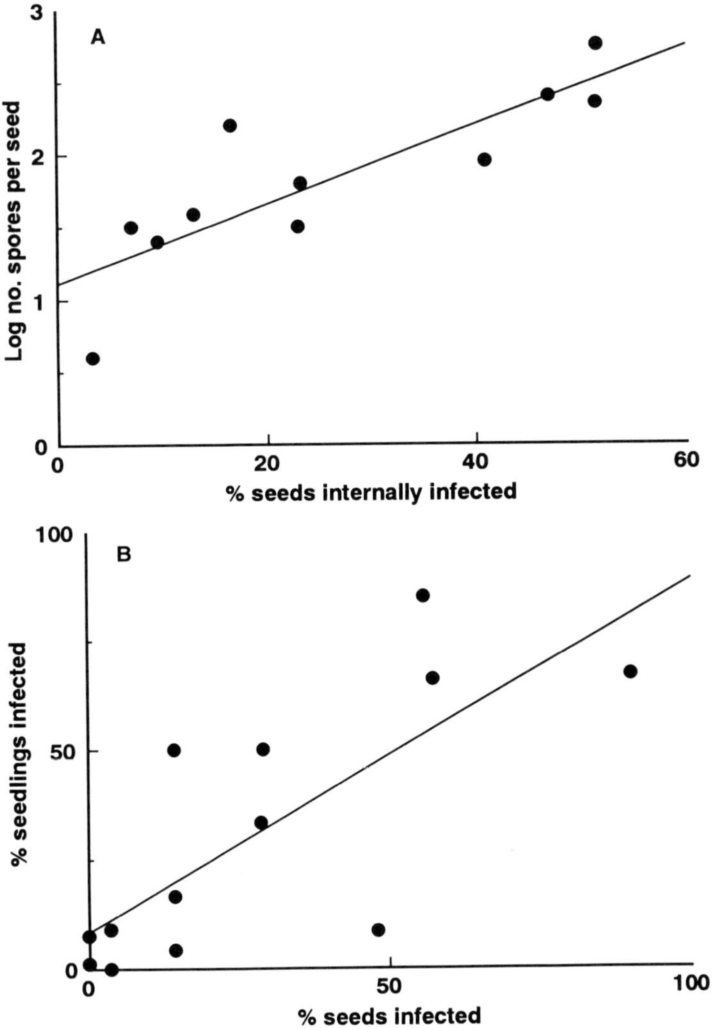

Fig. 14. Seed infection relationships of *Alternaria brassicicola.* (A) Relationship between superficial conidia and internal infection of seeds. (B) Relationship between internally infected seeds and diseases seedlings. (Redrawn from Maude and Humpherson-Jones, 1980a.)

Transmission rate

A reduction in transmission rate is the more usual effect where seeds infected by necrotrophic pathogens are sown in unsterile soil (see Table 10). A 10% reduction is sometimes quoted but this varies considerably for fungal pathogens as a result of the influencing effects of the various edaphic, climatic, seed

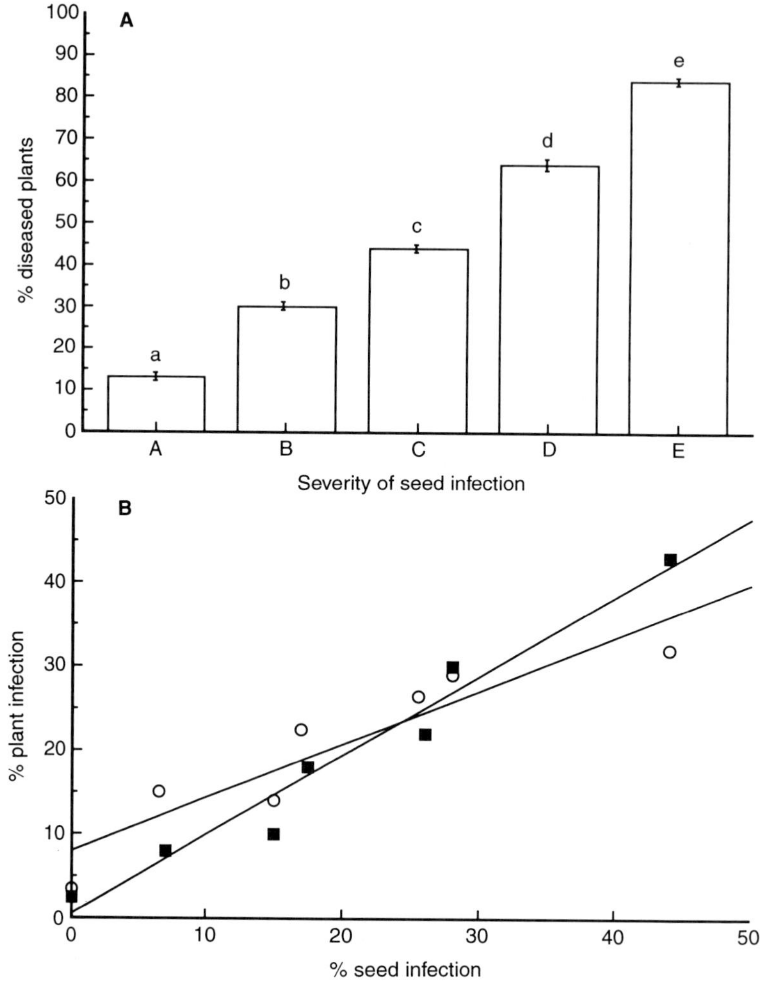

Fig. 15. Seed infection relationships. (A) *Colletotrichum lindemuthianum*—relationship between the severity of seed infection and seed transmission of bean anthracnose: A, bean seed without any apparent symptoms; B, seed with brown spot but no acervulus; C, seed with pitted brown streaks extending from hilum; D, seed with brown spots and acervuli; E, seed with pitted lesions extending into the cotyledon. Each figure is derived from between 109 and 127 plants. Means bearing the same letter are not significantly different ($P > 0.05$) according to Duncan's multiple range test (Redrawn from Tu, 1983.) (B) *Fusarium graminearum*—relationship between the incidence of internally infected seeds and disease transmission: ■, infection of shoot bases ($r^2 = 0.96$); ○, infection of tiller bases ($r^2 = 0.89$). (Redrawn from Duthie and Hall, 1987.)

infection factors, etc. already discussed and also because in many cases insufficient data have been collected to determine an average transmission rate. However, where laboratory and glasshouse (Taylor, 1970a) and field experiments (Taylor *et al.*, 1979b) were repeated over time it was possible to fix the laboratory : field transmission ratio of seedborne *P. syringae* pv. *phaseolicola* (halo blight of bean) at approximately 10:1, i.e. one infected seed in ten transmitted the disease in field soil (see Table 10). In contrast, certain biotrophs (e.g. *Ustilago segetum* var. *tritici*, loose smut of wheat and barley) may achieve a 100% transmission rate because the inoculum is present in the embryonic axes of seeds of susceptible cultivars and is ideally positioned to infect the putative seedling, yet is fully protected against adverse environmental effects (Batts and Jeater, 1958a; Marshall, 1959; Neergaard, 1977; Hewett, 1980).

Seed transmission of viruses depends on host genotype and virus strain. Transmission rates varying from less than 1% to 100% result and are tabulated for seedborne viruses in general (Mink, 1993) and for those of legumes worldwide (Frison *et al.*, 1990). Virus location and survival in seeds, the factors affecting seed transmission and transmission frequency and other effects described in Chapters 2–4 are summarized by Mink (1993).

Many factors influence the host–pathogen relationship involved in the transmission of infection from seed to seedling. What is reasonably certain in the majority of cases is that one infected seed may not produce one diseased plant. The determination of the transmission rate is essential for the establishment of the principles which govern the setting of tolerance levels for infected seeds in foundation and commercial seed stocks (see Chapter 6).

Transmission escapes

Transmission escapes occur where contact is not made between the emerging seedling and the infected host seed tissues. Thus 20% of barley seedlings escaped infection by *Rhynchosporium secalis* because the coleoptile, instead of emerging at the top of the affected seeds and making contact with lesions which develop on the lemma at the base of the awn, instead broke through the pericarp and lemma at a point near the base of the kernels (Skoropad, 1959). Deeper infections of testae or pericarps may result in the external layers of embryos being penetrated. In such examples seed surfaces, seed coat and embryo tissues together constitute the inoculum mass which transmits infection directly, and probably more severely to one or more parts of the germinating seeds as described above.

5 Spread and Survival of Seedborne Pathogens

Introduction

Disease epidemiology is a highly specialized part of plant pathology. General principles which govern disease spread and the development of epidemics have been derived by the application of mathematical techniques, principally those of modelling (Van der Plank, 1963, 1975; Horsfall and Cowling, 1978; Zadocks and Schein, 1979; Kranz and Hau, 1980; Teng, 1985). These principles equally apply to the spread of disease initiated by the dissemination of seedborne inoculum. Only those aspects of epidemiology which are relevant to the substance of this book will be considered here.

Relationship Between Seed-transmitted Inoculum and the Disease Cycle

Disease increase in field crops from primary foci of infection of seedborne origin will depend upon macro- and microclimatic influences, particularly those of temperature, moisture and light, which also affect transmission (see Chapter 4, pp. 56–60), and on the type of disease cycle of individual pathogens. There are monocyclic and polycyclic diseases.

Monocyclic diseases

Monocyclic diseases, also called simple interest diseases (Van der Plank, 1963), are those caused by pathogens with relatively uncomplicated life cycles, such as the loose smut of barley fungus (*Ustilago segetum* var. *tritici*)

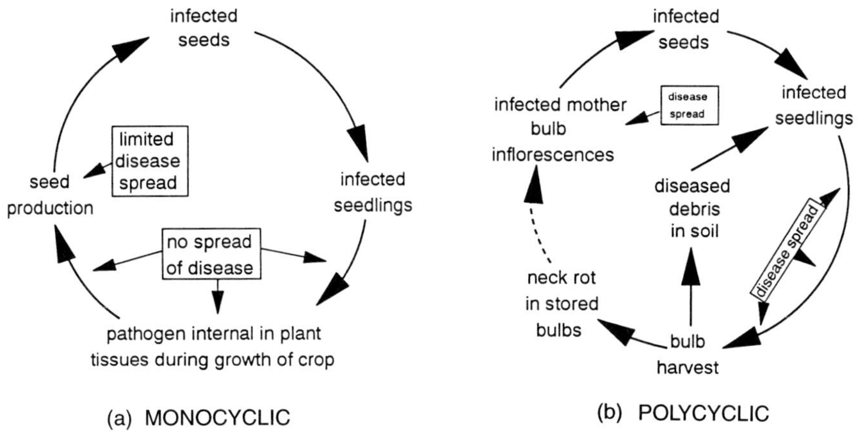

Fig. 16. Monocyclic and polycyclic diseases initiated by infected seeds. (a) monocyclic (*Ustilago segetum* var. *segetum*, loose smut of barley); (b) polycyclic (*Botrytis allii*, neck rot of onions).

(Fig. 16a), which remains inside its host for most of its existence (see Chapter 4, p. 47) and as a result there is no increase or spread of disease during this part of the life cycle. The biology of transmission of this biotroph (see Chapter 4) results in a relatively constant relationship between infected seeds and infected plants until anthesis in field crops (see Chapter 6) when the spores of the fungus are released over a very limited period of time (about 3 weeks) to establish the next generation of infected seeds. It is only at this time that weather conditions can affect subsequent disease levels by influencing the rate of infection of the cereal inflorescences (Hewett, 1978). Infection is greatest in wet seasons, partly because pollination is delayed by the conditions and the florets remain open (Tapke, 1931; Batts, 1956); light rain may also increase infection as raindrops collect the teliospores from the air and carry them into the open florets (Malik and Batts, 1960c).

Blind seed disease (*Gloeotinia granigena*) and ergot (*Claviceps purpurea*) can both be regarded as simple interest diseases (Hewett, 1978); infection in both occurs only at flowering when the incidence of infection may be modified by weather conditions (Hardison, 1960; Latch, 1966).

Polycyclic diseases

Polycyclic diseases, also called compound interest diseases (Van der Plank, 1963), are those which once established repeatedly multiply and spread whenever conditions are suitable during the growth of the host crop. With the majority of organisms, lesions bearing fruiting bodies are produced on host plants and provide inoculum for the increase and spread of infection. Other sources, including crop residues, volunteer plants and alternate hosts, increase the complexity of the disease cycle. The majority of seedborne necrotrophs are polycyclic in character. Examples include onion neck rot (*Botrytis allii*) (Fig. 16b), seedling blight of wheat (*Leptosphaeria nodorum*), canker or black leg of crucifers (*Leptosphaeria maculans*) and leaf spot of celery (*Septoria apiicola*), as well as many bacterial pathogens including *Pseudomonas syringae* pv. *phaseolicola* (bean halo blight), *Xanthomonas campestris* pv. *campestris* (brassica black rot) etc. Viruses can be allocated to both cycles (Thresh, 1974).

Dispersal and Spread of Polycyclic Diseases

Dispersal of viruses

The majority of the seed- and pollen-transmitted viruses listed by Walkey (1991) are spread by aphids. The rate of increase of disease may depend on the type of vector or mode of transmission of the virus and factors that govern the movement and abundance of the vector. For example, non-persistent seed-transmitted and aphid-spread viruses such as alfalfa mosaic virus (AMV), lettuce mosaic virus (LMV) and soyabean mosaic virus (SoyMV) will only be retained and spread for short periods of time over relatively short distances. Winged vectors (e.g. aphids) are wind-borne and tend to be carried downwind; the nature of the resultant infection gradients and virus spread are discussed by Thresh (1974, 1976, 1978). Although the virus causing foot-and-mouth disease of cattle may be spread in aerosols no rain-spread plant viruses have been found (Fitt *et al.*, 1989).

Dispersal of bacteria and fungi

Wind-driven rain is responsible for the dissemination of the majority of bacterial, and many of the fungal, pathogens of seeds which after transmission affect the aerial parts of seedlings. The action of rain in the dispersal of pathogens has been reviewed, as have models for the description of inoculum dispersal by rain and wind (Fitt *et al.*, 1989).

Bacterial cells are surrounded by mucilage which protects them from desiccation and loss of viability during dry weather and confines dispersal to periods of rainfall when conditions favour splash spread and free water is available for germination and infection of the host. Mucilage also protects and controls the release of many asexual fungal spores, including those of many seedborne Fungi Imperfecti of the class Coelomycetes. The spores in these fungi are contained in pycnidia (*Ascochyta, Phoma, Septoria* spp.), acervuli (*Colletotrichum* spp.) or sporodochia (*Fusarium* spp.) where moisture causes swelling and extrusion of mucilage containing the spores, which are then distributed by rain splash (Webster, 1986). Generally, such spores are not present in the air except during precipitation (rain or overhead irrigation, etc.) (Fitt *et al.*, 1989).

The forcible release of ascospores (sexual spores) usually follows a few hours after periods of rain (*Leptosphaeria maculans*: Maude and Humpherson-Jones, 1979; *Gibberella zeae*: Ye, 1980) which swells the mucilage of the asci, causing them to burst and discharging the ascospores which are then wind disseminated as dry spores. There is also evidence which suggests that ascospore discharge in *G. zeae* is initiated as the perithecia begin to dry out (Tschanz *et al.*, 1975). Overhead irrigation can also cause the release of ascospores, for example those of *Mycosphaerella pinodes* (leaf and pod spot with foot rot of peas) in pea fields in south Australia (Carter and Moller, 1961). Precipitation initiates the dissemination of the 'wet' fungal spores and bacterial cells of the organisms described above.

Many fungi produce conidia of the 'dry' spore type which are liberated passively into the air (Ingold, 1953). Often spore release is stimulated by falling humidities and restricted by high humidity creating a diurnal periodicity with maximum release into the air within and above crops when humidity is lowest (Hirst, 1953). Further dispersal of spores in the air within the canopy of, and above, source crops occurs by the action of wind. The effect of intermittent wind and models to describe its action upon fungal spores within and above the canopy of crops has been reviewed (Aylor, 1990). Fewer seedborne pathogens are dispersed by 'dry' spores. Conidia of *Drechslera graminea* (teleomorph *Pyrenophora graminea*) (leaf stripe of barley) are spread as dry spores by wind as are those of *Pyricularia oryzae* (teleomorph *Magnaporthe grisea*) (blast) which is most destructive of rice crops (Hewett, 1978; Mantle, 1988a). The conidia of the latter fungus float under the canopy of rice plants and then scatter out into the air above it and may be present in the air throughout the year in tropical countries, where several rice-growing cycles are possible each year (Suzuki, 1975; Castellani, 1988). Other passively released spores, for example those of *Alternaria brassicicola* (dark leaf spot) (Humpherson-Jones and Maude, 1982; Humpherson-Jones, 1992), remain in the air of the canopy for the duration of the growth of cabbage crops for seed production and are only physically displaced into the air in great numbers by harvesting operations (Fig. 17). These operations include the sequential practices of cutting, windrowing, threshing and the burning of old seed stalks.

Thus, rain and/or wind can be involved in the removal and transport of inoculum and its eventual deposition on a new host.

Transport of Inoculum

Depending on the wind velocity, splash-dispersed pathogens cause infection over relatively short distances from the source of inoculum at any one time.

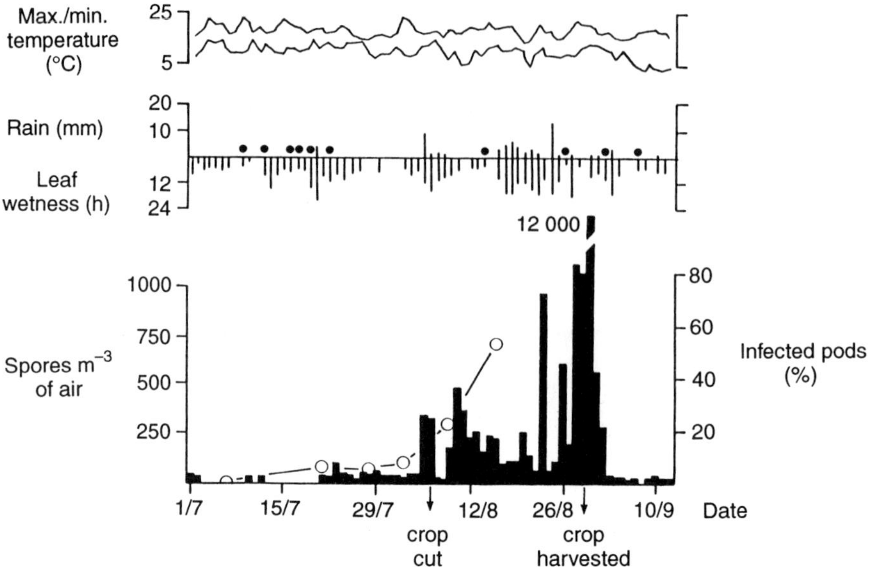

Fig. 17. Effects of climatic factors and harvesting practices on the mean daily concentration of *Alternaria brassicicola* conidia in the air of a cabbage seed production crop. ●, > 0.2 mm to < 1 mm of rain; ○—○, infected pods. (Redrawn from Humpherson-Jones and Maude, 1982.)

According to Gregory *et al.* (1959) the number of spore-carrying droplets increases as the size of the incident drop and the height of fall increase and the depth of liquid decreases, indicating that the effectiveness of a splash is partly determined by the impacting droplet, so that rain with large drops is more effective in transporting spores. In the experience of many workers, few inoculum-carrying droplets travel further than 1 m from the source during rainfall (Griffiths and Hann, 1976; Fitt *et al.*, 1989). Pathogens so dispersed tend to infect and travel along rows of plants more quickly than across them. Splash spread is in the direction of the prevailing wind, for example *Pseudomonas syringae* pv. *phaseolicola* (Walker and Patel, 1964; Taylor, 1972) and *P. syringae* pv. *pisi* (Grondeau *et al.*, 1991) produce characteristic fan shaped areas of disease in crops with dilution of inoculum from the source under conditions of wind-driven rain (Fig. 18). Although the rate of spread of bacteria by disease splash is slow, by the end of a growing season infection of halo bright will have travelled 20–24 m from the source (Walker and Patel, 1964), and distances of 5–10 m in a season are given for splash dispersal of some pycnidial fungi (*Ascochyta fabae*: Hewett, 1973; *Septoria nodorum*: Griffiths and Hann, 1976) and up to 30 m (radius) for the acervular pathogen *Colletotrichum lindemuthianum* (Tu, 1992).

Fig. 18. Aerial photograph of a pea crop showing diseased fan-shaped areas caused by wind-driven rain dispersal of *Pseudomonas syringae* pv. *pisi* bacteria.

Ballistic infection droplets (> 100 μm diameter) may be deposited very close to the source of inoculum but under conditions of heavy rainfall or sprinkler irrigation very fine secondary droplets (aerosols *c.* 5 μm in diameter) are produced. For example *P. syringae* pv. *glycinea* (bacterial blight of soyabean: Vennette and Kennedy, 1975) droplets have the potential to remain suspended in the air above crops for a considerable time and to deposit infective bacterial cells at considerable distances from the inoculum source (Graham and Harrison, 1975; Graham *et al.*, 1977). Conidia of *Pyrenopeziza brassicae* (light leaf spot) normally distributed in ballistic splash droplets (100–2000 μm) were trapped 20 m downwind of an oilseed rape crop during periods of dry weather, which is suggestive of dry spore dispersal or dispersal in very fine water droplets, i.e. aerosols (McCartney *et al.*, 1986). Proof of the infectivity of spores dispersed in this way is required to establish a risk pattern.

Wind-blown dry conidia of *Drechslera graminea* (teleomorph *Pyrenophora graminea*, leaf stripe) from diseased spring barley crops infected 5% of seeds in crops 200 m downwind of the source (Yarham and Jones, 1992). Probably because of its dry spore distributon, this was the only seedborne pathogen of cereals to cause significant reinfection in the very dry year of 1976 in the UK (W.J. Rennie, pers. comm., 1993). Ascospores, which are transported as dry spores in the air, can travel considerable distances. Those of *Mycosphaerella pinodes* have been trapped 304 m (Hare and Walker, 1944) to 388 m (Carter and Moller, 1961) downwind of infected pea material. Thus, the patterns of infection in crops as a result of the deposition of fungal spores or bacterial cells carried in the air over distance will be more diffuse than those created by localized rain splash, but they will exhibit disease gradients relating to the dilution of spore or bacterial concentrations with distance from the field source of inoculum (see pp. 82–83 and Fig. 19).

Relationship Between Seed-transmitted Inoculum and the Rate of Increase of Disease

The relationship between seedborne inoculum of monocyclic pathogens, such as *Ustilago segetum* var. *tritici* of barley, and disease increase is relatively constant (see pp. 101–102); that of polycyclic diseases from a similar inoculum source is not, for the reasons already given in this chapter. However, where infected seeds are the main source of a polycyclic disease, and secondary sources are unimportant or can be avoided, the establishment of a relationship between seed inoculum and the rate of disease increase is necessary to the development of strategies for its control.

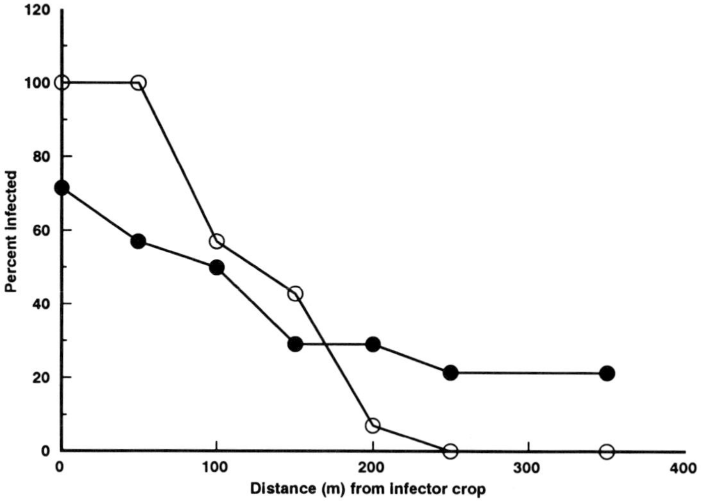

Fig. 19. Cross infection study of *Altemaria brassicicola*. ○, incidence of infected plants in a vegetative brassica crop, August 1978; ●, incidence of infected seeds from the same crop in July 1979. (Redrawn from Humpherson-Jones and Maude, 1982.)

Seedborne diseases develop relatively slowly (Hewett, 1978); disease increase is logarithmic with time, the rate of increase is unaffected by the initial levels of inoculum and the maximum infection rates of plants by pathogens are also low (Hewett, 1978). The infection rate (*r*), the rate at which the population of the pathogen increases (Van der Plank, 1963), is expressed as the unit increase of infected plants per day and is influenced by climatic factors and the susceptibility of the host plant material. Examples of infection rates for splash-spread organisms include 0.07 day^{-1} for the fungal pycnidiospores of *Ascochyta fabae* (Hewett, 1973); 0.12 day^{-1} for those of *Septoria avenae* f.sp. *triticea* on rye (Shearer and Wilcoxson, 1977) and 0.15 day^{-1} for the bacteria of *Pseudomonas syringae* pv. *phaseolicola* (Taylor *et al.*, 1979b). To put this into perspective, the infection rate quoted for a disease such as potato blight (*Phytophthora infestans*) which is initiated from field sources of inoculum is 0.42 day^{-1} (Van der Plank, 1963). Nonetheless, seedborne pathogens do cause disease epidemics and crop loss, in some cases from extremely small numbers of infected seeds (Table 11). Both empirical and modelling approaches have been applied to establish the relationships between seed inoculum and disease increase.

Complete loss of crop is the final outcome of disease increase but in most commercial situations practical constraints are imposed which determine the acceptability or otherwise of a diseased crop before this stage is reached. For example, with phaseolus beans grown for freezing or canning in the UK, those having more than 5% of *P. syringae* pv. *phaseolicola* blemished pods would be rejected by factory grading lines (Maude, 1988a) and this incidence sets the ceiling for infection in crops grown for such outlets. Similarly, stored bulb onions cannot be processed by grading lines if there are more than 10% of bulbs with neck rot (*Botrytis allii*). Since this disease is latent until the harvested bulbs have been in store for 8–10 weeks, the constraint applied to prevent high levels of neck rot is treatment of the seeds before sowing (Maude, 1983a).

Table 11. Inoculum thresholds and crop losses.

Crop	Pathogen	No. of affected seeds/ seedlings causing economic loss
Lettuce	Lettuce mosaic virus	1/30,000
Bean	*Pseudomonas syringae* pv. *phaseolicola*	1/10,000 to 1/16,000
Cabbage	*Leptosphaeria maculans*	1/10,000
Celery	*Septoria apiicola*	1/7000
Onion	*Botrytis allii*	1/100
Peas	*Ascochyta pisi*	> 5/100
Field bean	*Didymella fabae*	> 2/100

The establishment of relationships between infected seeds and the increase of disease in crops have been made for the practical purpose of applying disease control strategies, and for that reason this part of epidemiology has been considered in greater detail in Chapter 6.

Importance of Secondary Sources of Infection in Polycyclic Diseases of Seedborne Origin

Infected seeds in commercial stocks introduce diseases into new locations and into locations where the diseases already exist. The primacy of the seed as the main source of inoculum in either situation is modified by a number of factors which include the duration of the crop in the ground, the cropping systems used, alternative hosts, climatic factors affecting the survival of plant debris, survival structures of individual organisms, etc. Some of these are discussed here.

Duration of crops in the soil

Commercial production of food crops eaten as roots, vegetation or fruits (seeds) in temperate climates may be confined to a single 12-week summer growing season in which one or several crops of the same species may be produced. After harvest the land is ploughed and the crop residues are incorporated in the soil and the land may be left fallow until the following spring, a period of approximately 6 months. In some cases (Carter and Moller, 1961) the same crop species may be planted, but generally there will be a rotation of crops which ensures a break from that species of several years. Alternatively winter crops, for example cereals and crops for seed production (e.g. oilseed rape), are sown after autumn ploughing for harvest in the following summer. Again, when the crop is taken, the land may be left fallow until the following spring or a further similar or different winter crop may be planted. These practices can have a considerable effect on the survival of organisms initially transmitted from a seed source. In the first example the short duration of the crop followed by a relatively long break between like crops makes survival of secondary sources of infection on debris or on volunteer plants less likely, but in the second example recently harvested overwintering crops may provide infected plant residues which are sources of infection for newly emerged crops of the same or of a related species in the early autumn (see pp. 82–83).

In tropical countries, where a succession of crops of the same species is possible in 12 months, the problem is one of containing the aerial spread of increasing concentrations of inoculum from crop to crop although the original outbreak may have been initiated from a seed source.

Persistence of seedborne pathogens on plant residues and in the soil

Those seedborne necrotrophic fungal and bacterial pathogens which infect the aerial parts of plants survive in the saprophytic phase on crop residues in the soil, persisting for as long as the residues remain; as free organisms their capacity for survival is strictly limited (Table 12). The rate of decomposition of plant tissues in the soil will depend on the toughness of the tissues, the activity of microorganisms and animals affecting breakdown and the influence of other edaphic factors including temperature and humidity. In moist climates (temperate or tropical) plant remains can be broken down within months whereas in semiarid or arid climates residues may persist for longer periods. If debris is left on the surface of the soil its duration of survival may be prolonged further. In semiarid or arid climates, although the dissemination of splash-borne pathogens from surface debris is curtailed, the survival of certain fungal pathogens of seedborne origin, for example *Phoma lingam* (teleomorph *Leptosphaeria maculans*), may continue on the subterranean remains of crucifer plants and can cause problems where land is continuously cropped (Snyder and Baker, 1950). Gabrielson (1983) has reviewed the survival of this pathogen in other soil environments. Generally, the fungus survives for about 2 months in the absence of host tissues but it will transmit disease from affected debris for as long as this persists in the soil, a period of up to 5 years for oilseed rape residues (Alabouvette and Brunin, 1970; Gabrielson, 1983). The fungus may survive in crop debris as special thick-walled pycnidia of the asexual *Phoma lingam* state (Gabrielson *et al.*, 1978) or as perithecia of the sexual state, *Leptosphaeria maculans* (Alabouvette and Brunin, 1970). Similar comparisons in the survival of necrotrophs of seedborne origin in surface and buried residues can be made for many crop species (see Table 12).

Soyabean pathogens, for example *Phomopsis* spp. (seed rot) and *Colletotrichum dematium* (anthracnose) may overwinter on unploughed stubble (Kmetz *et al.*, 1979; Hartman *et al.*, 1986) as may cereal pathogens of seedborne origin such as *Rhynchosporium secalis* (scald) which may survive from year to year on barley debris along headlands (Reed, 1959); similarly *Monographella nivalis* (foot rot) can survive for 12 months on infected cereal stubble (Colhoun, 1971). Surface straw, stubble and volunteer plants are the main overwintering sources of net blotch of barley (*Pyrenophora teres*) in Canada (Piening, 1968), the USA (Singh, 1958), the UK (Jordan, 1981) and New Zealand (Sheridan *et al.*, 1983). By comparison, the saprophytic survival of *Drechslera teres* (anamorph of *P. teres*) in the soil is limited to a few weeks (Ammon, 1963). Other examples of fungi and bacteria of seedborne origin which have a limited survival on residues of maize and soyabean are given by McGee (1990; 1992)

Most plant pathogenic bacteria are not successful soil inhabitants (Burkholder, 1948; Wallace, 1978). They survive longer in surface residues than they

Table 12. Survival of necrotrophs of seedborne origin on debris and in soil.

Organism	Pathogen	Duration of survival (months)			Reference
		Inoculum in surface debris	Inoculum in buried debris	Free inoculum (buried)	
Fungus	*Septoria apiicola*	8+	7–9	–	Maude and Shuring, 1970
	Botrytis allii	–	24	19 (sr), 6 (vs)	Maude *et al.*, 1982
	Mycosphaerella pinodes	7	1+	–	Carter and Moller, 1961
	Mycosphaerella pinodes	–	18+	–	Sheridan, 1973
	Mycosphaerella pinodes	–	–	4 (s,c)	Dickinson and Sheridan, 1968
	Alternaria dauci	–	< 6	–	Maude, 1966c
	Colletotrichum lindemuthianum	–	9 (cd)	–	Tu, 1983
	Colletotrichum lindemuthianum	–	*c.* 5.5 (ps)	–	Tu, 1983
	Colletotrichum graminicola	> 8	8	–	Lipps, 1983
	Colletotrichum trifolii	3+	–	–	Lukezic, 1974
	Fusarium culmorum	–	*c.* 24	–	Garrett, 1963, 1966
	Gibberella zeae	3.5	–	–	Burgess and Griffen, 1968
Bacterium	*Xanthomonas campestris* pv. *malvacearum*	7.8	1.3–2.3	0.26	Brinkerhoff and Fink, 1964
	Xanthomonas campestris pv. *campestris*	–	8	0.5*, 1.4†	Schaad and White, 1974
	Pseudomonas syringae pv. *phaseolicola*	–	< 1.7	0.5, 0.8	Taylor, 1970a
Virus	Tomato mosaic virus	–	8 (l,r)	–	Hoggan and Johnson, 1936
	Tomato mosaic virus	24	< 1 (l)	–	Johnson and Ogden, 1929
	Tomato mosaic virus	–	22 (r)	–	Broadbent *et al.*, 1965

*, Summer temperatures.
†, Winter temperatures.
sr, Duration of recovery of sclerotia; vs, period of recovery of viable sclerotia; s, sclerotia; c, chlamydospores; cd, lightly cultivated affected crop debris; ps, pods and seeds; l, leaf debris; r, root debris.

do when the residues are incorporated into the soil and are subjected to more rapid decomposition (Schuster and Coyne, 1974). For example, *Xanthomonas campestris* pv. *malvacearum* (black arm of cotton) remained infective for 6 years when stored on dried plants in a laboratory (Ark, 1958). The bacterium overwintered in debris on the soil surface in Oklahoma, USA for 230 days to infect newly emerging cotton plants. Buried affected cotton debris lost infectivity in 40–100 days depending on temperature which influenced the rate of its decomposition in the soil. As free cells incorporated into soil its survival was limited to 8 days (see Table 12) (Brinkerhoff and Fink, 1964). In the arid climate of the Sudan, infected debris is a threat to the next cotton crop but in the moister climate of Tanzania, Africa this bacterium barely survives between crops in surface debris (Schuster and Coyne, 1974). The survival of bacterial cells in the soil in two other examples given in Table 12 were also strictly limited. Seedborne pseudomonad, xanthomonad and coryneform (e.g. *Curtobacterium, Clavibacter* and *Rhodococcus* spp.) bacteria of most crops, although they may overwinter on plant residues in the soil, are incapable of persisting in a free state for extended periods in natural soil (Schuster and Coyne, 1974)

Fungal and virus biotrophs of seedborne origin rarely survive long after the death of the host vegetation. For example, tomato mosaic virus (ToMV) survives for shorter periods in tomato leaf debris which disappears more quickly when buried in the soil, than in tomato stem and root debris in which the virus may overwinter to infect the following year's tomato crops (Lanter *et al.*, 1982). ToMV may persist for at least 22 months in root debris (Broadbent *et al.*, 1965) (see Table 12). Many of the nepoviruses are seed-transmitted and nematodes bearing the viruses move through the soil to infect suitable host plants. The viruses cannot survive in the soil rhizosphere in the absence of nematodes but they can and do survive in weed seeds and weed host species (Mandahar, 1981; Murant, 1983; Walkey, 1991). Nematodes move extremely slowly in the soil; the rate of migration of *Xiphinema diversicaudatum*, the vector of arabis mosaic virus, was calculated to be about 30 cm year^{-1} over a period of 75 years (Harrison and Winslow, 1961).

Survival of seedborne fungi, bacteria and viruses on debris is limited to the time taken for the crop remains to decompose and in practice the rotation of crops is used to eliminate this source of disease (see Chapter 8).

Persistence of seedborne pathogens in the soil by means of resting bodies

Plant pathogenic bacteria are asporogenous and thus have are not adapted to survival in the soil, but many of the fungi have asexual resting structures, for example chlamydospores and sclerotia, which enables a limited duration of survival in the absence of the plant host debris. *Botrytis allii* sclerotia persisted for 19 months but were non-viable after 6 months in field soil (Maude *et al.*,

1982) (see Table 12). Those organisms which invade via the seed attachment to infect the underground plant parts for example vascular wilt fungi such as *Fusarium oxysporum* f.sp. *pisi*, are suppressed or increased by the natural soil microflora (see Chapter 4, pp. 60–62). Some foot rot fungi (e.g. *Mycosphaerella pinodes*) although they invade hypogeally (Fig. 13A, 1) are disseminated aerially but also may survive in the soil by means of thick-walled resting spores, i.e. chlamydospores, as do other foot rot genera (e.g. *F. culmorum* of cereals: Cook, 1968; *F. solani* f.sp. *pisi* Dixon, 1981). Chlamydospores and sclerotia of *Mycosphaerella pinodes* overwintered and were viable when tested for 4 months in soil (Dickinson and Sheridan, 1968). This fungus was isolated from field soil in which peas had not been grown for 20 years (Wallen *et al.*, 1967); the means of its survival was not ascertained.

In general, especially with annual crops grown during the summer months, infected debris and to some extent asexual resting bodies present in the soil represent sources of diseases of limited duration which can be controlled by rotational practices (see Chapter 8). The seeds remain the major source of such diseases.

Importance of sources of secondary inoculum

In sequential cropping

Where there is a year-to-year continuity in temperate climates of a particular crop species in the same area (e.g. winter cereals followed by winter cereals, a similar cycle for horticultural brassicas for seed production, and oilseed rape following oilseed rape, or successional summer crops such as lettuce) and in tropical climates a repeated cropping of the same plant species throughout the year, the primacy of infected seed as the source of crop infection is subordinate to the field sources which ensure the carry-over and increase of disease.

Where pathogens are disseminated over considerable distances in the air the proximity of like crops and of crop refuse may be the source of inoculum from which new crops may become infected. In such circumstances different strategies are required for disease control (see Chapter 8). To appreciate the significance of the spread of disease in these circumstances it is necessary to understand how far inoculum travels and in what concentrations it constitutes a disease risk (Fitt *et al.*, 1989). There is considerable information on the distance travelled by inoculum but much less on what concentrations constitute a disease risk with increasing distance from the source. Examples have been given for distances travelled by *Alternaria brassicicola* conidia and *M. pinodes* ascospores (see p. 76). Ascospores of *Leptosphaeria maculans* released from surface debris remaining from oilseed rape harvested in the UK in July infected healthy young crops of autumn-sown rape in September, giving decreasing infection gradients with an increasing distance of 20 m (96%

plants infected), 800 m (32% plants infected) and 1500 m (3% plants infected) from the source debris (Gladders and Musa, 1980). Young plant infection was strongly correlated with plants having severe stem cankers at harvest so that some 30% of the seed crop would be lost if it was sited 800 m from a disease source. Perithecia of *Pyrenophora teres* (net blotch) in the surface debris from harvested barley crops in the UK release ascospores in the autumn which may infect volunteer plants and newly emerged barley plants of nearby winter crops (Jordan, 1981). Webster (1951) considered that the widespread occurrence of *P. teres* perithecia on barley stubble in the spring might constitute a serious source of infection for spring-sown cereals. In Canada (Piening, 1968), ascospores released from affected straw were responsible for the infection of volunteer barley plants and the spread of *P. teres* to other nearby barley crops. The role of ascospores of *Gibberella zeae* in the transfer of infection within and between crops in the disease cycles of wheat foot rot and ear scab and maize ear rot has been reviewed (Sutton, 1982).

Similarly, dry conidia of *A. brassicicola* forcibly displaced from a *Brassica oleracea* (Thousand Head kale) seed crop by the process of threshing were wind dispersed, producing infection gradients in young cabbage plants up to 200 m downwind of the source (see Fig. 19). When seed of this cabbage crop was harvested in the following summer the incidence of seedborne infection ranged from 100% in samples taken within a few metres of the original infected crop to 20% diseased at 350 m away (Humpherson-Jones and Maude, 1982). The threshing of the crop provided air-borne conidial inoculum for the infection of newly emerged crops destined for seed production in the following year.

The year-round cycle of spring-sown followed by autumn-sown followed by spring-sown onions in the UK provides an infection bridge for *B. allii* (onion neck rot) (Maude, 1977). The practice in the 1950s in the UK of planting winter followed by spring and summer lettuce crops initiated a cycle of infection by LMV which continued throughout the year when aphids were sufficiently numerous (Broadbent *et al*., 1951). There are other examples of temperate vegetables where viruses are spread by insects from one crop to the next in succession, including cauliflower mosaic virus and turnip mosiac virus in brassica crops (Broadbent, 1957), and LMV in summer lettuce (Tomlinson, 1962). Tomlinson (1987) also gives similar instances for tropical crops such as cassava, groundnut, okra and yam.

To control disease caused by crop-to-crop movement of aerial pathogens, i.e. infection bridging, requires the integration of a number of different strategies (see Chapters 6–8).

From dumps of crop waste

The preparation of harvested vegetables or stored produce in packhouses creates a considerable amount of trimmed waste which may be dumped on the surface of the soil, often close to growing crops to which it may act as a source of infection. For example, overwintering dumps of onions have been implicated in the USA (Munn, 1917; Walker, 1926) and Europe (Doorn *et al.*, 1962) as sources of infection of *B. allii* (onion neck rot) for spring-sown onions. In such instances wind-disseminated conidia of the fungus would cause infection. Sporulating conidia were found on the outer surfaces of onion dumps in the Netherlands in June (Tichelaar, 1967) at a time when the March-sown crop is young and vulnerable (Maude *et al.*, 1985).

Role of Alternate Crop and Weed Hosts in the Spread of Seedborne Diseases

Viruses

Many viruses, including those which are nematode-borne and seed-transmitted, have a wide natural host range which ensures their survival in the wild and makes them important in a wide range of annual and perennial crops (Murant, 1983). Similar or related crops can serve as main sources of virus but many of the species which are alternate hosts for viruses of commercial significance are wild plants and weeds which are located in hedgerows and headlands bordering crops, and sometimes also occurring within the cultivated crops. Perennial species provide a continuity of virus source from year to year. The importance of wild plants and weeds in virus ecology has been reviewed (Duffus, 1971; Bos, 1981). Few viruses produce conspicuous symptoms in their wild hosts which makes it difficult to recognize their presence in routine surveys of crops (Tomlinson, 1987).

The importance of wild species as sources of virus for the infection or reinfection of commercial crops varies. LMV-susceptible weeds, i.e. groundsel (*Senecio vulgaris*) and sow thistle (*Sonchus asper*), in the proximity of infected lettuce plots in the UK did not become infected (Tomlinson, 1962). Nine species of mosaic-infected weeds were found in the vicinity of fields of young lettuce in California, USA (Costa and Duffus, 1958) and Grogan (1983) notes that some 27 susceptible weed and crop species grew in proximity to lettuce crops in the Salinas Valley, California but that there was little spread of mosaic to lettuce even from large, uncontrolled weedy neighbouring areas. Unlike LMV, cucumber mosaic virus (CMV) is not seed transmitted in the vegetable crops it affects, i.e. celery, cucurbits, lettuce and spinach, but it is in chickweed (*Stellaria media*) and this alternate host probably contributes to outbreaks of disease in the commercial crops where it is a common weed (Tomlinson, 1987). Large numbers of *S. media* seeds occur in the soil which, even with low rates of seed transmission (1–2%), are probably sufficient to cause disease problems in vegetable crops (Tomlinson, 1987). In this disease the weed virus source is distributed at random throughout the crop and is therefore a direct and potent threat, whereas external plant/weed sources are less directly involved and may not contribute to the same extent to crop infection (e.g. LMV of lettuce).

Seed transmission of the nematode-borne tobraviruses and nepoviruses also occurs in several weed species, for example groundsel (*Senecio vulgaris*), and is important for their survival and spread to susceptible crops (Murant and Lister, 1967; Cooper and Harrison, 1973).

Fungi

Whereas the presence of alternate wild host species is essential to the survival of many viruses they are less important in the disease cycles of seedborne fungal pathogens. Within the Gramineae, the major temperate cereal species (e.g. wheat, barley, oats, rye, etc.) provide a limited series of alternate crop hosts for some fungi, and in addition there is a wide range of grasses whose functional role as a host species is not well defined. There is an exception in the case of the biotroph pathogen *Claviceps purpurea* (ergot) where, in Canada (Campbell, 1957), it has been shown that all indigenous and forage grasses constitute a reservoir of ergot inoculum for rye, wheat and barley, and that these cereals can be infected provided the environmental conditions are such that inoculum is disseminated at the time that the cereal crops are in blossom. However, the annual fluctuations in incidence of ergot which occurred in cereals were not found in roadside and headland grasses (Campbell, 1957) so that the status of the input of alternate host inoculum is not clear. In Europe, the sphacelial (conidial) fructifications of *C. purpurea* which occur on perennial ryegrass (*Lolium perenne*) and on annual meadow grass (*Poa annua*) are probably epidemiologically not significant for cereals. Although in southern England wheat may receive inoculum in the form of conidia in honeydew borne on precociously flowering arable weeds such as black grass (*Alopecurus myosuroides*) which has been distributed by rain-splash, insects or head-to-head contact of plants (Mantle, 1988b).

Other hosts, including weed species, may constitute an alternative source of inoculum for some downy mildew biotrophs of maize (*Sclerophthora rayssiae* var. *zeae*, brown stripe) and soyabean (*Peronosclerospora philippinensis*, Philippine downy mildew; *P. sorghi*, sorghum downy mildew) which also are reported to be seed transmitted (McGee, 1990, 1992).

Some of the seedborne graminicolous necrotrophic pathogens, such as *Leptosphaeria nodorum* and *Rhynchosporium secalis*, have a wide host range in cereals and grasses. The occurrence of the teleomorph *L. nodorum* indicates a potential for cross-infection by ascospores over a distance with that pathogen, but with both fungi the relative and respective roles of cereal and grass hosts as providers of inoculum is not well known. The pathogenicity of a wheat isolate of *L. nodorum* to the grasses *Agropyron repens*, *Lolium perenne* and *Poa annua* has been established (Shearer and Zadoks, 1972). Although *L. nodorum*-infected plants of *Agropyron repens* (couch grass) were found amongst infected wheat plants in a non-inoculated experimental plot (Shearer and Zadoks, 1972) and the same fungus occurred on *P. annua* in a severely infected wheat field (Becker, 1956) the disease contribution from the grass hosts is not known (Shearer and Zadoks, 1972). It is most probable, however, that the inoculum contribution from cereal crops to other cereal crops or to grass weed hosts would far outweigh any influx of inoculum from grass weed hosts to cereals. This condition would not obtain where grasses were grown as crop plants.

There is, however, an example from vegetables where diseased plants of a weed host, i.e. *Lactuca scariola* (wild or prickly lettuce) infected with *Septoria lactucae* (leaf spot), were the source of infection of cultivated lettuce in Australian seed production crops to the extent that the diseased lettuce seed produced was unfit for sale to the USA (Neergaard, 1977).

In other vegetables the status of the weed hosts is more complex. Gabrielson (1983) lists a number of crucifer hosts for *Phoma lingam* (teleomorph *Leptosphaeria maculans*) (black leg, canker) based on the glasshouse inoculations of many workers and a large number of crucifer weeds on which the fungus has been observed to occur in nature. To what extent weed crucifers infect crop crucifers may be open to question since McGee and Petrie (1978) demonstrated that isolates of the fungus from stinkweed (*Thlaspi arvense*) were not pathogenic to adult rape (*Brassica rapa* – the species of rape used in Canada) plants and isolates from rape were not pathogenic to stinkweed. It is apparent that although both plants are crucifers they are of different genera. With closer relationships pathogenicity has been proved and Petrie (1979), in Canada, has reported that wild mustard (*B. kaber*) is an important host of the strain of *L. maculans* which is virulent on rape (*B. rapa*). Host specificity, however, did not operate for isolates of *Alternaria brassicae* (brassica dark leaf spot) from oilseed rape (*B. napus*), swede (*B. napus*), turnip (*B. campestris*), cabbage (*B. oleracea*), field pennycress (*T. arvense*), charlock (*Sinapsis arvensis*) and hedge mustard (*Sisymbrium offinale*), all of which caused severe leaf spotting on oilseed rape, swede, turnip, and cabbage plants in the UK (Humpherson-Jones and Hocart, 1983).

Weeds may also function as symptomless carriers of inoculum of certain seed-transmitted necrotrophs. For example, pathogenic isolates of *Cercospora kikuchii* (purple seed stain) were obtained from asymptomatic leaves of six weed species collected from soyabean fields or adjacent fields in the USA (McLean and Roy, 1988) as were isolates of *Phomopsis* spp. (seed decay) obtained from three weed species common in tropical soyabean-growing areas in southern Brazil (Cercauskas *et al.*, 1983). Velvet leaf (*Abutilon theophrasti*), a common weed in soyabean crops, is reported to be a wild host for three seed-transmitted pathogens of that crop in the USA (Hepperly *et al.*, 1980). Whereas the disease potential of these alternative sources of inoculum has been recognized, their contribution to the infection of soyabean crops has not been confirmed beyond doubt. Hand weeding or herbicide treatment of weeds reduced the incidence of certain seedborne fungi in soyabean plots but whether this was because weeds when unchecked provided microclimatic conditions favourable for pod infection or because they acted as inoculum sources for infection was not established (Dhingra and da Silva, 1978).

It would appear that for fungal pathogens at least, the epidemiological significance of weed hosts needs to be addressed on a case-to-case basis.

Bacteria

Of the 18 seedborne pseudomonad, xanthomonad and coryneform bacteria listed by Smith *et al.* (1988) only *Xanthomonas campestris* pv. *tomato* is given as having a single host, tomato. The remainder have varying numbers of crop and wild hosts. In *Pseudomonas syringae* and *X. campestris* spp. there are many pathovars, most of which have a high host specificity and will only infect the genus from which they were isolated (Lelliot, 1988) which limits the range of natural hosts which can function as alternative sources of disease. For example, bacterial wilt of grass (*X. campestris* pv. *graminis*) is seedborne and infects a number of agronomically important grasses (Cleene *et al.*, 1981) but isolates from *Phleum*, *Poa*, and *Arrhenatherum* spp. will only infect those genera thus restricting the number of different grasses which might act as alternative hosts. It is important to distinguish between hosts which are good artificial hosts for test purposes and those which are the natural hosts of the organism and from which it can be disseminated under field conditions. There is a further complication where a pathovar contains cultivar-specific races which may further restrict the range of natural alternate field hosts within a single crop plant species. This is well established for *P. syringae* pv. *glycinea* (bacterial blight of soyabean) with nine races (Cross *et al.*, 1966; Thomas and Leary, 1980; Fett and Sequeira, 1981), for *P. syringae* pv. *pisi* (pea blight) with seven races (Taylor *et al.*, 1989) and more recently for *P. syringae* pv. *phaseolicola* (halo blight of bean) with an increase from two (Patel and Walker, 1965) to nine races (Teverson, 1991).

Phaseolus bean crops are grown in temperate and tropical places throughout the world and in addition to *Phaseolus vulgaris* (green beans, French beans) and *P. coccineus* (runner beans), the two main temperate hosts for *Pseudomonas syringae* pv. *phaseolicola*, tropical legumes also function as alternate hosts (Allen, 1983). These include the lima bean (*Phaseolus lunatus* var. *macrocarpus*), Mexican yam bean (*Pachyrhizus erosus*), kudzu vine (*Pueraria thunbergiana*) and tepary bean (*Phaseolus acutifolius*) (Lelliot, 1988). Further information on the host range for seedborne pathogens of legumes is given by Frison *et al.*, (1990).

In Australia *Pseudomonas syringae* pv. *phaseolicola* infects and persists on the tropical pasture siratro bean (*Macroptilium atropurpureum*) which is a potential alternate host source for infection of phaseolus beans (Johnson, 1970; Moffett, 1973). Several leguminous weeds have been identified in reciprocal inoculation studies as potential hosts for some of the bacterial pathogens which are seedborne. These include red vine (*Brunnichia cirrhosa*) weeds infected with *X. campestris* pv. *glycines* growing in a diseased soyabean crop in the USA (Jones, 1961), and the America trailing wild bean (*Strophostyles helvola*) which proved to be a host to *P. syringae* pv. *phaseolicola* (halo blight) of phaseolus beans (Gardner, 1924), as did *Neonotonia wightii* in Africa (Teverson, 1991). The latter plant, often infected with the halo blight bacterium, is a common leguminous weed present in many African bean fields; it is also a pasture legume and is of proven high susceptibility to six races of halo blight (Teverson, 1991).

In the examples given above, the crop and weed hosts were of the same plant family but it also reported that plant pathogenic bacteria, some of which are seedborne, can reside on different parts of plants, which are unrelated to the crop host, without causing disease symptoms (Leben, 1974; Schuster and Coyne, 1974). Cafati and Saettler (1980) demonstrated reciprocal spread of *X. campestris* pv. *phaseoli* from susceptible beans to *Chenopodium album* (lambs quarters) and *Amaranthus retroflexus* (pigweed) under field conditions, and they suggest that the leaves of these non-host plants, while remaining healthy, may support the residence of and epiphytic multiplication of the common blight bacterium, and that such inoculum may function as an alternative source of disease for phaseolus beans.

There is potential for transfer of bacterial inoculum from crop plant-related and unrelated weed hosts but the contribution of these secondary sources to the epidemiology of disease remains to be demonstrated.

Role of Volunteer Crop Plants in the Spread of Seedborne Diseases

Infected volunteer crop plants may also function as alternative sources of inoculum. Jordan (1981) reports that infective conidia of *Drechslera teres* were abundant in net blotch lesions on volunteer barley plants in fields of winter cereals in 1979 and 1980 in the UK. In the 1930s in North America net blotch was common on wild barley grass (*Hordeum murinum*) and this host was believed to be a constant source of infection all the year round (Butler and Jones, 1949). Volunteer crop plants growing as headland weeds might also function as disease sources for new crops, for example barley with loose smut in this situation could be a source of infection for a nearby main crop which was disease free. In northwest India, volunteer rice plants from seeds infected with bacterial blight (*X. campestris* pv. *oryzae*) may provide sufficient foci of infection to initiate outbreaks of disease in rice crops (Durgapal, 1985).

6 Disease Control: Exclusion and Reduction of Inoculum

Introduction

The previous chapters, by illustrating how seeds become infected and how inoculum survives and is transmitted, demonstrate the capacity of the seedborne phase for initiating disease outbreaks and for establishing secondary field sources of infection. Seeds are distributed internationally and within countries in commercial quantities for the production of agricultural and horticultural crops. Seeds are also distributed in smaller amounts for multiplication and as germplasm for research (Kahn, 1983) and plant-breeding purposes. All are routes which can and do function in the dissemination of seedborne pathogens. The strategies and methods by which these and field sources of infection are combated are embodied in the five general principles of disease control given by Tarr (1972).

1. Elimination of disease sources or crop sanitation.
2. Alleviation of disease by cultural practices.
3. Treatment of disease.
4. Resistance to disease.
5. Legislation against disease.

In seed pathology the application of one or more of these principles to prevent or eliminate infection of seeds can be extremely effective economically and strategically, often controlling diseases at national and international levels. Economic considerations (e.g. value of seed, volume of traded seed) also apply and may influence the choice of strategy but the general intent is to promote the production and use of healthy seeds.

The control of seedborne pathogens can be considered broadly in terms of exclusion and elimination of inoculum. In this chapter exclusion strategies which include the use of legislation, the isolation of seed production areas, the setting of minimum inoculum tolerance levels for seeds, and breeding for resistance are considered. In the succeeding chapters direct eliminatory measures including control of organisms by seed treatments (see Chapter 7) and crop treatments (see Chapter 8) are discussed (also see Dykstra, 1961). Disease control may be achieved by single or by combined strategies contained within the concepts of exclusion and elimination.

Exclusion of Seedborne Pathogens by Legislative Measures

Plant quarantine laws and regulations are based on national and international legal obligations, and have been developed in different parts of the world during the last century (Karpati, 1983). The International Plant Protection Convention (IPPC) serves as the foundation for all basic plant quarantine laws. Their aim is to secure common and effective action in preventing the spread and introduction of pests (i.e. any pathogenic agent) of plants and plant products across international boundaries and to promote measures for their control (Karpati, 1983).

The effectiveness of quarantine measures has been reviewed by Mathys and Baker (1980). Eight fundamental premises for their establishment were proposed by Morschel (1971) as: (i) measures should be based on sound biological principles; (ii) they should not be used for the furtherance of trade; (iii) quarantine must derive from adequate law and authority; (iv) as conditions change, or further facts become available, quarantines should be modified; (v) the objective of preventing introduction and spread must be considered reasonable to achieve; (vi) cooperation, both on an international scale and from the public, is required; (vii) quarantine measures can only be effective if those responsible for them are well informed; and (viii) quarantine measures are only one of the facets of domestic pest management programmes. A careful integration of measures is needed to achieve maximum effect; such measures form a basis for the concept of pest risk analysis as defined by Kahn (1979).

In the last decade, the acceleration in movement of germplasm and other genetic material around the world has been a considerable potential threat to plant health and is held as justification for a global plant quarantine system (Karpati, 1983). It is recognized that many indigenous fungal, bacterial and virus seedborne pathogens in the UK are also imported in consignments of commercial seeds. Examples include *Septoria apiicola* (celery leaf spot), *Pseudomonas syringae* pv. *phaseolicola* (halo blight of bean) and lettuce mosaic virus (LMV) in seed from the USA and similarly *Botrytis allii* (onion neck rot), *Alternaria dauci* (carrot leaf blight), *Ascochyta fabae* (teleomorph *Didymella fabae*) (leaf spot of field bean) (Maude, 1973) and *Leptosphaeria maculans* (canker of oilseed rape) from Europe (Dixon, 1981). Mandahar (1981) identifies ten seedborne viruses which have been distributed world-wide in commercial seed lots (soyabean mosaic virus (SoyMV), bean common mosaic virus (BCMV) and LMV) or introduced into one country from another (maize dwarf mosaic virus (MDMV), squash mosaic virus (SMV), strawberry latent ringspot virus (SLRSV), etc.) in samples of infected seeds. Hampton *et al.* (1982) list some six legume viruses which have been introduced in germplasm, one of which, pea seedborne mosaic virus (PSbMV), caused significant crop and seed sale losses as well as virus infection of breeding lines in institutional and commercial breeding programmes in North America. Economically important seedborne pathogens have been intercepted on seeds entering Japan (Kobayashi, 1983, 1990) as have insects, nematodes and fungal pathogens entering India in seed germplasm (Lambat *et al.*, 1983). Likewise, the range of organisms entering the USA in seed germplasm is monitored and interceptions from 1958–1964 have been reported (Kaiser, 1983). Neergaard (1977) also gives details of pathogens which have been introduced into other countries in imported seeds.

The potential for disease introduction by seedborne pathogens is often recognized before detection methods and quarantine controls have been put in place to prevent this happening. For example, the most serious disease of oil palm in West Africa is vascular wilt (*Fusarium oxysporum* f.sp. *elaeidis*) and the recent recovery of virulent isolates of the fungus from contaminated oil palm (*Elaeis guineensis*) seeds and pollen may indicate an existing and a potential problem for the oil palm industry (Flood *et al.*, 1990a,b). Both seed and pollen are widely distributed between countries in extensive breeding programmes. Current evidence indicates that plantations where wilt occurred in Brazil and Ecuador received seed from the Ivory Coast, Africa where the disease is long established (Bachy and Fehling, 1957). The isolates from all three places are vegetatively compatible, as determined by nitrate non-utilizing mutants (nits) and have similar restriction fragment length polymorphism (RFLP; see Chapter 9) DNA profiles (Flood *et al.*, 1992; Mouya, 1992). This raises the further possibility of introducing a serious vascular disease into areas from which it has not been reported, such as Malaysia where oil palm is a major crop (Flood *et al.*, 1990a,b).

Quarantine measures

Quarantine safeguards for the international exchange of plant germplasm as seeds by the International Agricultural Research Centres have been described by Kahn (1983), with recent information from the International Rice Research Institute (IRRI), Manila, Philippines (Mew *et al.*, 1990) and from the International Crops Research Institute for the Semi-Arid Tropics (ICRISAT) in India (Varma, 1990). National plant and seed quarantine procedures which apply in the UK (Dickens and Pemberton, 1983), Japan (Kobayashi, 1983, 1990), the ASEAN countries (Singh, 1983), the USA (Kaiser, 1983), China (Liang, 1983), Kenya (Olembo, 1983), Brazil (Warwick *et al.*, 1983), Australia (Morschel, 1983; Jones, 1987), Nigeria (Aluko, 1983), and India (Lambat, *et al.*, 1983) also have been described.

There are three quarantine categories in the USA: (i) restricted; (ii) post-entry; and (iii) prohibited (Waterworth and White, 1982). Seeds are represented in the first and last categories. Barley, oats, various vegetables, alfalfa (lucerne), clover and grasses are located in the restricted category and require inspection and possible chemical treatment for entry into the country. For example, a permit and chemical treatment is required for soyabean and garden bean from Africa and Australia to prevent the entry of soyabean rust (*Phakaspora pachyrhizi*) (Waterworth and White, 1982). Seeds whose entry is prohibited are those of cotton and rice (from all sources), wheat (from Asia, eastern Europe, Australia), sorghum and millets (from Africa, Asia, Brazil), corn (from Africa, Asia) and lentils (from South America) to control pests, weeds and a range of biotrophic pathogens including viruses, smuts, rusts and downy mildew.

In the UK a certificate of health (phytosanitary certificate) is not required for all seeds entering the country, but quarantine restrictions are applied to those known to have a clear disease potential (Dickens and Pemberton, 1983). There are 13 named specific seedborne organisms which have this capability.

The most serious problems for any country are those which arise when non-indigenous seedborne diseases are imported into or exported to an environment which is well suited for their establishment and spread. UK quarantine restrictions apply to these organisms, and to those indigenous organisms which are at present at low levels or are well contained but if imported seeds were not controlled problems would arise, also to those organisms with novel and damaging pathotypes or races which might be introduced with affected seeds, and finally to common pathogens which can be considered of quarantine importance where the domestic crop is wholly dependent on imported seed stocks (Dickens and Pemberton, 1983). Non-indigenous diseases may be introduced as seedborne inoculum but may fail to survive if the environment is unsuited to their establishment and spread. Thus organisms requiring high temperatures for development in the UK, for example *Curtobacterium flaccumfaciens* pv. *flaccumfaciens* and *Xanthomonas campestris* pv. *phaseoli* which cause bacterial wilt and blight of phaseolus beans in warm climates, have been found on imported seed but not as diseases in the growing crop in the UK (Dickens and Pemberton, 1983).

The statutory instrument for the implementation of quarantine regulations in the UK is The Import and Export (Plant Health) (Great Britain) Order 1980

(Anon., 1980). For a previously non-indigenous disease such as bacterial blight of pea (*Pseudomonas syringae* pv. *pisi*) conditions of entry into the country required that; (i) the consignment was free of the pathogen; and (ii) that it was accompanied by a phytosanitary certificate stating that the seeds had been taken from a region of production where the disease has not occurred for the previous 10 years, or a statement that no bacterial blight was found during at least one official inspection made since the beginning of the last complete cycle of vegetation (Anon., 1980).

Virus testing of seed-producing crops and seed mother plants of the genus *Rubus*, including regular crop inspections for virus symptoms, are necessary quarantine measures to guarantee the complete freedom of *Rubus* seeds of four viruses including black raspberry latent virus (BRLV), cherry leaf roll virus (CLRV), prunus necrotic ringspot virus (PNRSV) and tomato ringspot virus (TRSV) and allows their importation into the UK. For LMV, which is contained at low levels in UK and where the seed is wholly imported from the USA and eastern Europe, a phytosanitary certificate is required stating that seedborne infection has been indexed at a threshold value lower than that likely to cause disease (Table 13). Evidence of fumigation is required for the elimination of seedborne nematodes of lucerne, and where seed is coming to the UK from a third country a statement is required that it has not been obtained from certain other countries which have problems with bacterial wilt of lucerne (*Clavibacter michiganensis* subsp. *insidiosus*).

Other countries have similar regulations (see authors referred to above) to restrict or exclude those seedborne organisms which, if not controlled in this way, could cause considerable damage to their crops. Although they are mainly effective, quarantine procedures may not succeed indefinitely in preventing the entry of seedborne organisms for a number of different reasons; in some cases this is because suitable detection methods are not in place to intercept pathogens at the entry points or the methods lack the sensitivity required to detect the organism at a disease-causing level, or new pathotypes are introduced at low levels with imported seeds. For such reasons several non-indigenous seedborne organisms have escaped detection and have become established in Japan in the last 25 years; three of them, *P. syringae* pv. *lachrymans* (angular leaf spot of cucurbits), *C. michiganensis* subsp. *michiganensis* (bacterial canker of tomato) and cucumber green mottle mosaic virus (CGMMV), are now widespread and cause serious disease outbreaks every year (Kobayashi, 1983, 1990).

Bacterial blight of peas (*P. syringae* pv. *pisi*), formerly non-indigenous in the UK and a notifiable disease there since 1953 (Roberts *et al.*, 1991), was the cause of disease in a crop grown from one cultivar of compounding (protein) pea in 1985 (Taylor, 1986). The infection was traced to seed obtained from Holland in 1983–1984 (Taylor, 1986). Subsequently it was discovered in cultivars of combining (compounding, protein) and vining (fresh market) peas (Stead and Pemberton, 1987; Anon., 1988) and it is currently prevalent on seed

Table 13. Standards for seedborne pathogens in UK seeds regulations (from Rennie, 1993).

Regulations	Crop	Pathogen	Seed category	Standard*
Fodder	Field beans	*Ascochyta fabae*	Pre-basic	1 infected seed in 1000
			Basic	2 infected seeds in 1000
			C1	2 infected seeds in 500
			C2	1%†
	Peas	*Pseudomonas syringae* pv. *pisi*	Multiplication categories	Nil in 1 kg
Vegetables	Brassicas	*Phoma lingam*	Basic	Nil in 1000 seeds‡
	Red beet	*Phoma betae*	Basic	Nil in 200 seeds‡
	Celery	*Septoria apiicola*	Basic and certified	Nil in 400 seeds‡
	Celery	*Phoma apiicola*	Basic and certified	Nil in 400 seeds‡
	Peas	*Ascochyta* spp.	Basic	Nil in 200 seeds§
			Certified	20 seeds max. in 200§
	Lettuce	Lettuce mosaic virus	Basic and certified	Nil in 5000 seeds
	Phaseolus bean	*Colletotrichum lindemuthianum*	Basic	Nil in 600 seeds
	Phaseolus bean	*Pseudomonas syringae* pv. *phaseolicola*	Basic	Nil in 5000 seeds
	Vicia bean	*Ascochyta fabae*	Basic	Nil in 600 seeds

Table 13. *continued*

Regulations	Crop	Pathogen	Seed category	Standard*
Oil and fibre	Linseed	*Botrytis* spp.	All	5%
	Linseed	*Alternaria* spp.	All	5% (1% same fungus in flax)
	Linseed	*Phoma exigua* var. *linicola*	All	5% (1% same fungus in flax)
	Linseed	*Colletotrichum lini*	All	5% (1% same fungus in flax)
	Linseed	*Fusarium* spp.	All	5% (1% same fungus in flax)
	Turnip rape	*Sclerotinia sclerotiorum*	All	5 fragments sclerotia in 70 g
	Swede rape	*Sclerotinia sclerotiorum*	All	10 fragments sclerotia in 100 g
	Sunflower	*Sclerotinia sclerotiorum*	All	10 fragments sclerotia in 1000 g
	Sunflower	*Botrytis* spp.	All	5%
	White mustard	*Sclerotinia sclerotiorum*	All	5 fragments sclerotia in 200 g
Cereals	Wheat, barley, oats	*Claviceps purpurea*	Basic	1 piece ergot in 500 g (0 in 1 kg at HVS)
			Certified	3 pieces ergot in 500 g (1 in 1 kg at HVS)
	Barley, wheat	*Ustilago nuda*	Basic	0.5% (0.1% at HVS)†
			Certified	0.5% (0.2% at HVS)†

*Maximum permitted infection.
†Marketing standard; no official test required.
‡Where infected seeds are found, an effective treatment approved by the Minister must be applied before certification.
§Where not more than 20 infected seeds are found in basic seed or more than 20 infected seeds are found in certified seed, an effective treatment approved by the Minister must be applied before certification.

samples which have not been subjected to statutory testing (Roberts *et al.*, 1991). Up to the time of the disease outbreak the quarantine conditions given above were not breached because of the fact that most protein pea seed was produced in the UK from cultivars resistant to the dominant race (race 2) of the bacterium (Taylor, 1986). The introduction of Dutch seed (cv. Belinda) for multiplication which was infected with and was susceptible to race 2 of the bacterium brought about the consequences described and resulted in a change in regulations to include a seed test for imported protein peas (Anon., 1986), which is in keeping with Morschel's fourth premis. In June 1993, statutory controls on pea bacterial blight were reduced and the disease has been removed from the list of quarantine diseases in the European Community (EC) (Rennie, 1993; Roberts, 1993). Protein pea seed entering the UK from other EC states will no longer be subject to import controls and this may have serious disease implications (Roberts, 1993). The reasons for this change was because of the adverse effects of statutory controls on the seed trade, which were difficult to justify for a disease which is widespread in Europe and which appears to present only a limited disease risk (Rennie, 1993).

Tolerance levels for infected seeds in certification programmes and in commercial seed production

Seed certification is recognized as essential for obtaining the full benefits of new cultivars (Hewett, 1979b). Certification schemes usually follow seeds through four multiplications to provide seed for the production of commercial crops. These steps are from small amounts of pre-basic seed (breeders' seed, USA), increasing through basic seed (foundation seed, USA) and certified seed, first generation (registered seed, USA) to certified seed, second generation (certified seed, USA). The rate of seed increase may vary between crop types. Certification ensures that seeds reach set standards in quality and that the history of each seed lot is traceable (Hewett, 1979b). Laboratory and field procedures are followed to ensure that standards of purity and maximum permitted infection levels are maintained (Hewett, 1979b). Within the EC, there is free movement of seeds that meet EC standards but member states can impose additional higher standards on a voluntary basis (Rennie, 1993). The UK has imposed higher voluntary standards (HVS) for varietal and physical purity and weed seed content in cereal seed and fodder plant seed regulations, including HVS standards for loose smut and ergot in cereal seeds (Table 13 and pp. 101–102). The standards for seedborne pathogens in UK seed regulations are given in Table 13 (Rennie, 1993).

The principles on which permissable levels of infection (tolerance levels) in seed samples are based for certification programmes, for commercial seed stocks and for quarantine purposes, are described for a limited series of species in the following sections.

Inoculum threshold levels for crop infection

Gabrielson (1988) defines inoculum thresholds for seedborne fungal pathogens as 'the levels of infection on or in the seed that will significantly affect disease development and result in economic loss'. This definition embraces both superficial spore inoculum and that more deeply seated in the tissues of seeds. It describes the consequences of transmission of seedborne fungi, but it would equally apply to seedborne bacteria and viruses. Seed/seedling health testing techniques cannot be used to measure zero inoculum but, within the limits imposed by numbers and size of seeds, they can be used to determine percentages of infected seeds/seedlings in a sample which cause little or practically unimportant amounts of disease in a crop. This relationship and that of inoculum thresholds can be accommodated by Van der Plank's equation (1963) that states that the amount of initial inoculum (x_0), the average infection rate (r) and the time over which the infection occurs (t) determines the amount of disease that develops (x) and therefore $x = x_0e^{rt}$. Thus the minimum tolerance level for infected seeds is a strategy based on reducing the initial inoculum (x_0) so that the disease threshold (x_d), is not exceeded even under conditions favouring the average infection rate (r) (Hewett, 1983). To reach this point it necessary to obtain data relating the diminishing amounts of seedborne inoculum to the transmission and spread of disease. Once these have been established, minimum tolerance levels can be set and a seed/seedling health test can be used predictively to eliminate disease outbreaks of economic consequence by the exclusion of infected commercial seed stocks which exceed the permitted tolerance level. Examples where this has been done successfully include, LMV (Kimble *et al.*, 1975; Grogan, 1980, 1983), halo blight of bean (Taylor *et al.*, 1979b), brassica canker (Williams *et al.*, 1973) and leaf and pod spot of vicia beans (Hewett, 1973).

In a wider context, minimum tolerance levels for inoculum are employed legislatively in programmes to control or prevent the introduction of diseases by the use of affected basic seeds, certified seeds or imported seeds (Hewett, 1979b; Gabrielson, 1988).

Establishment of tolerance levels in infected stocks of seeds

In principle, the determination and implementation of minimum seed tolerance levels is a logical approach to disease control. In practice, however, laboratory health testing methods are suitable mainly for detecting pathogens which transmit at a relatively high incidence which is enough to cause serious disease (e.g. *Ascochyta pisi*: Anon., 1965a; Anselme *et al.*, 1970; *Botrytis allii*: Maude and Presly, 1977). In these instances, the inspection or testing of a limited number of seeds (200–1000) will give an accurate assessment of the incidence of infection. In a few cases, laboratory methods can be applied to low-incidence pathogens, for example *Phoma lingam* (teleomorph

Leptosphaeria maculans), where – because brassica seeds are small – large numbers can be handled assisted by a discriminatory test which suppresses seed germination thereby increasing the ease with which seeds can be assessed for infection (Anon., 1965b; Maguire *et al*., 1978) (see Table 11).

Formerly, and with the exception of the cereal loose smut fungi, there were no routine laboratory tests for most biotrophic pathogens (e.g. LMV) where, although the small size of seeds makes them suitable for laboratory testing, the seed tolerance levels are extremely low and detection involves 'indexing', the recognition of virus symptoms expressed on the leaves of seedlings grown from diseased seeds, and confirmatory tests of these by inoculation onto the leaves of indicator plants (Zink *et al*., 1956). Now, however, with the development of immunodiagnostic techniques (see Chapter 9), routine laboratory tests for LMV have become available for use on lettuce seed samples. There are also those necrotrophs, for example *Pseudomonas syringae* pv. *phaseolicola* (halo blight of phaseolus bean), where the large size of individual seeds has necessitated the development of a specialized extraction technique including isolation and identification methods for the detection of the bacterium (Taylor, 1970a,b,c).

Establishing predictive relationships

The data necessary to establish seed–field disease (transmission leading to rate of spread (*r*)) relationships from which tolerance levels can be set have been obtained empirically (Hewett, 1983). It is suggested (Hewett, 1983) that uncomplicated field experiments, repeated over a period of years to establish the normal range of conditions (edaphic and climatic) encountered affecting transmission and spread, are more appropriate in this respect than long-term investigations measuring the effects of varying the factors that determine spore dispersal and infection necessary for sophisticated prediction systems. In the examples which follow it will be seen that most seed tolerance values have been obtained by the first method.

Tolerance levels for Ascochyta *of pea and bean*

Simple seed health tests involving small numbers of seeds placed on agar (see Chapter 9), to allow the development of typical mycelium and fruiting bodies, have been used to establish tolerance levels for peas infected by *Ascochyta pisi* (Anon., 1965a) and for field beans infected by *Ascochyta fabae* (teleomorph *Didymella fabae*) (Hewett, 1966). In 5 years of field trials with four cultivars of peas, with seeds ranging in infection from 0.3% to 50% on agar, Anselme *et al*. (1970) found that: (i) from similar ranges of infected seeds, transmission rates and ultimately the incidence of *A. pisi* on the harvested seeds changed considerably from year to year because of varying climatic and edaphic conditions; and (ii) that emergence was reduced only at high levels of seedborne infection (13–50%). Transmission was largely unaffected but the final incidence of the fungus on the harvested seed was reduced, presumably because of the reduced density of the crop. At the lower end of the infection scale, 2–6% infected seeds produced 0.5–4% infected seedlings which gave harvested seeds with 3–9% infected seeds in the samples. These final infection incidences on the harvested seeds apparently did not exceed the minimum disease threshold levels (x_d) for the crop. It is reported that in France 5% infected seeds can be tolerated in peas intended for consumption but that a lower incidence is necessary in samples sown for seed production (Anselme *et al*., 1970). Nil infection in 200-seed tests of basic seed stocks and not more than 20 infected seeds in 200 in certified seed stocks is required in the UK (Anon., 1993a; Rennie, 1993) (Table 13). Where seedborne *Ascochyta* spp. occur in peas, an effective seed treatment must be applied to control infections of up to 10% in basic seeds and of more than 10% in certified seeds before official certification is granted (Anon., 1993a; Rennie, 1993).

During a period of expansion in the production of field beans (*Vicia faba*) as a protein source for stockfeeding in the UK (Anon., 1976) there were serious outbreaks of leaf spot caused by *Ascochyta fabae* attributable to the sowing of infected seeds (Hewett, 1966). Detailed laboratory-infection and field-transmission studies over 6 years with samples of spring beans and over 5 years with winter beans produced generally similar relationships to those described for peas, but lower tolerance limits of 1–2% seed infection were required for commercial seed intended to produce beans for feed purposes (Hewett, 1973). This was supported by the experience that no disease was found in experimental plots grown from seeds where there was no infection in 200 seeds tested on agar (Hewett, 1973). For seed production stocks of seed, there was a stricter limit of zero infection in 1000 seeds (basic seed) and up to two infected seeds in 1000 seeds (certified seed) (Hewett, 1973). The success of this field bean scheme, which was introduced on a completely voluntary basis, in the selection of healthy bean stocks was maintained by the introduction of official certification requirements for the crop (Anon., 1976). Because of the general improvement in the health of field beans and a decline in *A. fabae* leaf spot which resulted from the introduction of seed tolerance limits, the 1000 seed test was later applied to pre-basic and basic seed stocks only. In 1990 (Anon., 1990) the permitted incidence of *A. fabae* was not more than one infected seed in 1000 for pre-basic seed, two in 1000 for basic seed and two seeds in 500 for certified seed of the first generation (Anon., 1990). In the related broad bean (*V. faba*) crop the requirement for basic seed is for zero infection in 600 seeds (Anon., 1985a, 1993a).

In these examples, simple seed tests have been used to relate seed infection incidence (x_0) to disease transmission and spread (r) over several years (t), determining the amount of disease that develops ($x = x_0e^{rt}$), from which it has been possible to determine x_d, an acceptable percentage of disease. Tolerance levels have then been set for infected seeds at lower or higher levels depending on the purpose for which the seeds are to be used.

Tolerance levels for Phoma *of brassica*

Depending on seed usage (basic seed, certified seed, seed for export, imported seed), tolerance levels for seedborne *Phoma lingam* (teleomorph *Leptosphaeria maculans*) vary in different countries (Gabrielson, 1988). In 1973, severe damage to US cabbage plantings were initiated by seed lots assessed as having at maximum, 1.75% *P. lingam* infected seeds (Maguire *et al.*, 1978). Brassicas sown for transplanting form a dense canopy and produce an environment conducive to the transmission and rapid spread of a fungus such as *P. lingam* from very low levels of inoculum. Where brassicas are directly drilled as field crops, at much lower densities, similar numbers of infected seeds would be less likely to have the same effect.

In this instance, the detection of infection at a very low incidence in seed lots was considered vital to the prevention of further epidemics (Williams *et al.*, 1973) and a laboratory 2,4-D (2,4-dichlorophenoxy-acetic acid) blotter assay was modified (Maguire *et al.*, 1978) and used to assay samples of 10,000 seeds for zero infection. Statistically ($P < 0.05$) this allows for up to four infected seeds per sample, an incidence of 0.04%. Ten thousand seeds were chosen because this was the largest number of seeds which it was practical to test by existing methods (Williams, 1973). To support the detection test, a field-transmission study was done; this indicated that in six samples with increasing levels of infection of from 0.15% to 6.0%, disease transmission did not occur in samples containing 0.6% or less infected seeds (Gabrielson *et al.*, 1977) (Table 14). On this evidence, the laboratory detection test would appear to have a ten-fold safety factor (Gabrielson, 1988). However, data on the seedborne phase of this pathogen from other sources (Table 14) indicate that disease initiation may occur from much lower incidences (0.08–0.5%) of

Table 14. Seed transmission rates in *Phoma lingam* (teleomorph *Leptosphaeria maculans*).

	% infected seeds giving		
Crop	No transmission	Transmission	Reference
Cabbage	0.6*	1.50	Gabrielson *et al.*, 1977
Oilseed rape	–	0.80	Wood and Barbetti, 1977
		0.50	
		5.00	
		0.08	
Swede	0.06	0.50	Allen and Smith, 1961
	0.07	0.50	
	3.00	–	
Red cabbage	< 0.20	> 0.20	Cruger, 1979

*Laboratory test designed to test for zero tolerance in 10,000 seeds at $P < 0.05$ allows for up to four infected seeds in 10,000, i.e. an incidence of 0.04%.

seedborne inoculum, but in these cases the supporting information indicates that the amounts of crop disease which resulted were slight (Wood and Barbetti, 1977).

In this example a laboratory detection test has been used with the intention of excluding the disease as completely as possible, within the statistical limitations already stated, from the resultant crop and no acceptable threshold for disease (x_d) was defined. This may be a necessary objective in a transplant-raising environment where the rate of disease spread (r) from a few inoculum sources could be rapid and damaging to the developing crop. In that context the failure of 3% of infected seeds to transmit black leg in a dry year (Allen and Smith, 1961) is understandable but such conditions are unlikely to apply in a transplant bed.

Tolerance levels for cereal loose smuts

Microscopic examination of seed embryos to establish the numbers bearing dormant mycelium of loose smut fungus in cereal samples, form the basis of the embryo count procedure (Anon., 1985c; 1993; Rennie, 1988). This is a relatively simple laboratory test which has been used to predict the incidence of disease in field crops. The original method involved macerating wheat grains in sodium hydroxide to release the embryos which were examined after staining them in aniline blue (Skvortzoff, 1937). The method has been modified by a number of workers since then (see Marshall, 1959; Anon., 1985c, 1993b and Chapter 9).

Previously, and because of the biotrophic nature of these fungi, their incidence in seed samples could only be determined by growing the seed on to an adult state to produce loose smut in the flowering ears of plants under glasshouse or field conditions (Simmonds, 1946; Russell, 1950). The value of the embryo count system became apparent when many workers (Simmonds, 1946; Russell, 1950; Russell and Popp, 1951; Marshall, 1959) reported high correlations between infected embryo counts and the incidence of loose smut in field crops ($r = 0.83$: Marshall, 1959; $r > 0.97$: Russell, 1950) and in some glasshouse crops of susceptible barley cultivars. Although the viability of the embryo-borne mycelium cannot be determined in an embryo test, the fact that there was little difference in the viability of infected and non-infected seeds (Vanderwalle, 1942; Russell, 1950) was supportive of the highly correlated relationship between the laboratory test and loose smut in the barley crop. The embryo count method for the prediction of loose smut of barley is used in many countries as part of a strategic programme for the control of that disease (Neergaard, 1977). The method was employed extensively in the UK from 1960 when standards for loose smut permitted 0.1–0.2% infected embryos for the earliest three multiplication categories (Hewett, 1979b). Standards for wheat and barley loose smut are now 0.5% for EC minimum standard seed (basic and first and second generation seed) and 0.2% for seed certified at the HVS (Anon., 1985a). There are no standards set for farm-saved seeds which account for approximately 25% of the UK cereal area (W.J. Rennie, pers. comm., 1993).

The continued usefulness of an embryo test for predicting the incidence of smut infection in barley depends on the absence from commercial samples of cultivars with physiological resistance where there is embryo susceptibility but adult plant resistance to the pathogen (Hewett, 1972). In Canada this type of resistance was incorporated into the commercial cv. Keystone (Johnston and Metcalfe, 1961) rendering the embryo test useless for this and later varieties with similar resistance (Wallen, 1964).

The correlation between the embryo test and disease in the crop is less assured for loose smut of wheat (Popp, 1951; Batts and Jeater, 1958a). Although the hyphae of *Ustilago segetum* var. *tritici* can be seen in the scutellum of infected embryos, only 33–50% of the incidence indicated by the test appears in the field (Hewett, 1979b). This may be the result of the action of several unrelated factors including sampling variation, sowing conditions, an insufficiently refined embryo test and cultivar and loose smut race combinations at the time of infection in the field (Hewett, 1979b) (p. 62). A negative test, however, can have value in establishing that a seed lot will not exceed plot standards (Hewett, 1981) and the embryo test for wheat is applied with such a purpose in countries such as Bulgaria and India to determine the need for seed treatment (Neergaard, 1977).

Tolerance levels for LMV

The occurrence of LMV in field crops depends on the incidence of seedborne infection and on the numbers and mobility of the aphid vectors of the disease (Zink *et al.*, 1956). Early experimental work in the USA (Grogan *et al.*, 1952) demonstrated that the disease could be virtually eliminated by the use of healthy seeds.

Tolerance levels for LMV have been derived empirically by relating the incidence of infected seeds, as determined by seed (Marrou and Messiaen, 1967; Kimble *et al.*, 1975) and seedling indexing methods (Zink *et al.*, 1956; Tomlinson, 1962), to numbers of infected field plants at harvest. In California, eight repetitions of an experiment, using a mixture of infected and healthy seeds (infection gradient of 0.025–1.6%) sown into carefully arranged beds, was used to derive the incidence of LMV in mature plants at first harvest (Zink *et al.*, 1956). On average this indicated that seed transmission in excess of 0.1% caused an unacceptable threshold of disease. In the UK, Tomlinson (1962), in isolation plot trials over 2 years, demonstrated acceptable disease thresholds in harvested lettuce using seeds with less than 0.1% transmitted mosaic but not with those where transmission exceeded 2.2%. This confirmed the earlier American work and established a tolerance level of less than 0.1% which has proved effective for outdoor crisp and LMV-susceptible butterhead lettuce cultivars in the UK; similarly, a guaranteed transmission rate of not more than one in 1000 also applies in France (Marrou and Messiaen, 1967). However, there were severe outbreaks of the disease in 1960 in California from seeds reputedly indexed at 0.1% or less LMV (Grogan, 1983). Multiple factors may have combined to cause this failure; the solution was a four-point programme to eliminate LMV inoculum. The most important condition was an undertaking to plant only lettuce seed that had been indexed and found to contain zero seed-transmitted LMV in 30,000 seedlings (Kimble *et al.*, 1975; Grogan, 1980, 1983). As a result, LMV has been controlled in the Salinas valley, California (Grogan, 1983). The reasons for this are possibly twofold: (i) a seed tolerance incidence of 0 in 30,000 (actual transmission range 0–0.022%) achieved an unimportant threshold level (x_d) of the disease in field crops (see the 0.025% (2.5 seeds in 10,000) seed transmission rate of Zink *et al.*, 1956); and (ii) this stringent standard ensured that seed producers had taken all necessary precautions to eliminate LMV from their commercial seed lines.

Lettuce seeds imported into the UK in commercial quantities from the USA require an official statement that no LMV was found when not less than 3000 seedlings were grown under officially approved conditions or when not less than 5000 seeds were inoculated to suitable indicator plants or were tested by equivalent methods (Anon., 1980, 1993a, and Table 13)

Seed indexing by use of the enzyme-linked immunosorbent assay (ELISA) has been showed to be as sensitive as the use of a plant indicator (*Chenopodium quinoa*) or growing-on tests (Falk and Purcifull, 1983) (see Chapter 9).

Complete freedom from seedborne viruses is required in certification programmes for some temperate and tropical crops. Assessment is by visual inspection of standing crops which may be followed by testing the seed for the presence of a virus, usually by means of a growing-on test (Hamilton, 1983). The identity of virus in suspect plants is checked using techniques such as ELISA, sodium dodecyl sulphate (SDS) gel diffusion and other serological methods. Hamilton (1983) describes the detail of the processes involved in certification schemes for legume viruses in North and South America, Africa and Asia. Two certification schemes for barley stripe mosaic virus operate in the USA (Carroll, 1983). Both involve field inspections at various stages of seed production followed by immunodiffusion tests of barley embryos in one, or a latex flocculation test of ground seed samples in the other. Only seed which is found to be virus free is cleared for use in certification programmes (see Chapter 9).

Tolerance levels for phaseolus bean halo blight

In the examples given so far, disease incidence measured at critical stages in the growth of those crops was correlated with the numbers of infected seeds which were sown and this empirical information was used to produce seed tolerance limits. However, in setting tolerance limits for phaseolus beans infected with the bacterium *Pseudomonas syringae* pv. *phaseolicola*, all available data have been integrated to produce a working model from which the

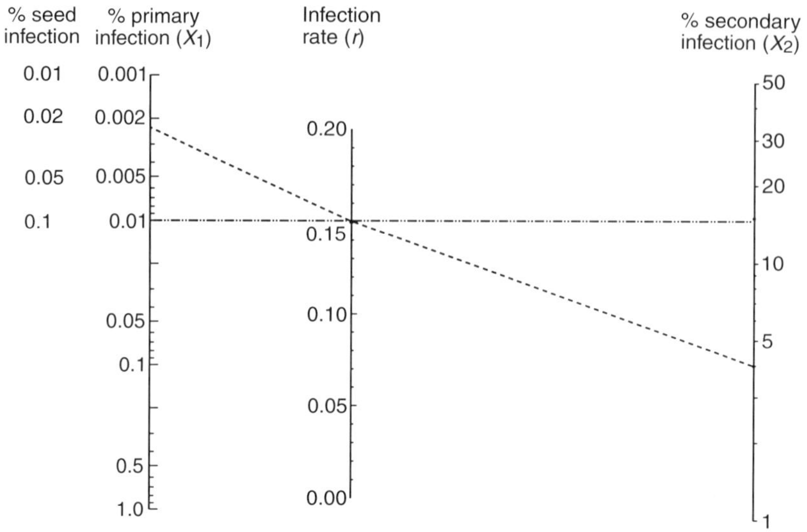

Fig. 20. Nomogram for the interpretation and production of halo blight epidemics. The nomogram is prepared by the method of Allcock and Jones (1950). To use the nomogram, place a ruler between the known values on any two of the three scales and read off an estimate of the unknown value. In the examples show here it can be predicted that, with an r value of 0.15, 0.01% primary infection (X_1) will give 15% secondary infection (— ·· — ··), whilst 0.025% primary infection (X_1) will give 4% secondary infection (X_2) (- - - -). (Redrawn from Taylor *et al.*, 1979b.)

relationship between infected seeds, disease transmission and secondary infection of crop plants can be interpreted (Taylor *et al.*, 1979b). This model is expressed in the form of a nomogram (Fig. 20) which can be used in decision making concerning acceptable seed infection limits under varying rates of infection for the crop. There are three critical elements in its construction: (i) the production of a method to give a quantitative estimate of infection in bean seed; (ii) the determination of the transmission rate from seed to seedling (primary infection); and (iii) the calculation of the infection rate for bean crops (Taylor *et al.*, 1979b).

Halo blight of bean can be transmitted by 1/10,000 to 1/16,000 bacterially infected seeds (see Table 11). To detect seed infection at those incidences required a sensitive method and also a sampling technique which could be applied accurately to quantities of large-sized seeds such as beans. Using a routine laboratory method, beans were ground to a flour and the bacteria were extracted from the flour suspended in water, grown on a selective medium and identified by phage or serological techniques (Taylor, 1970c). This method was shown to be capable of detecting the equivalent of a single infected seed in 10,000 (Taylor, 1970b). As only one infected seed in ten was shown to transmit the disease (see Chapter 4, p. 69) detection at a level of one infected seed in 1000 in the laboratory is representative of one infected seedling in 10,000 in the emergent field crop. By using a statistical technique, the most probable number method (Cochran, 1950; Taylor *et al*., 1993), which involves the testing of varying numbers of seed samples containing different numbers of seeds, it was possible to estimate the percentage infection over the range (0.8–0.05%) commonly found in commercial seed stocks in the UK at that time.

The third component of the model, the infection rate for bean crops, was derived using Van der Plank's equation $x = x_0e^{rt}$ modified to allow for the limits imposed by reduced plant tissue availability as the disease progresses in plots of limited size:

$$r = \{\log_e[x_2/(1-x_2)]-\log_e[x_1/(1-x_1)]\}/(t_2-t_1)$$

where x_1 = the initial inoculum (or disease) at time t_1, x_2 = the inoculum (or disease) present at time t_2, and r = the infection rate expressed as per unit (i.e. a single plant) per day (where t_2-t_1 is measured in days). In this equation disease is expressed as a proportion (e.g. 5% disease is equivalent to 0.05).

The model was applied to data which measured the effect of rain and wind on disease spread over a period of 3 years in the USA (Walker and Patel, 1964). Although this data was suitable for modelling purposes, on infection sites which comprised more than one infected seedling there was rapid establishment of disease which exaggerated the infection rate (r). Under UK conditions based on disease spread within small plots from infection points derived from single infected seeds, lower r values of 0.10–0.16 were obtained (Taylor *et al*., 1979b). An r value of 0.15 was considered suitable for climatic conditions in the UK.

The nomogram (Fig. 20) illustrates the relationships between primary infection (x_1) produced by the sowing of diseased seeds, the infection rate (r) and the final disease incidence in the mature crop, i.e. secondary infection (x_2). From the nomogram it can be seen that one infected seed in 1000 would create up to 15% infected plants at maturity, a commercially unacceptably high level of infection, but that 2.5 infected seeds in 10,000 would produce only 4% secondarily infected plants which is considered to be a tolerable level. This involves testing 12,000 seeds which conforms well with the previous empirically based suggestions that 10,000 seeds should form the basis of a practical seed test (Taylor, 1970b; Coleno *et al*., 1976). Although the detection test and model are used to screen phaseolus beans, the numbers of seeds tested probably depends on the seed generation (e.g. pre-basic, basic, etc.) and the quantity available. In the UK, a 5000 seed test (nil infection) is required for basic seeds (Anon., 1993a, and Table 13). In Michigan, USA, a single 10,000 seed sample forms the basis of a plant inoculation test used to exclude contaminated bean seed lots (Copeland *et al*., 1975). A modification of Taylor's (1970b) method for the detection of *P. syringae* pv. *phaseolicola* has been developed (Ball and Reeves, 1992) for routine testing for *P. syringae* pv. *pisi*

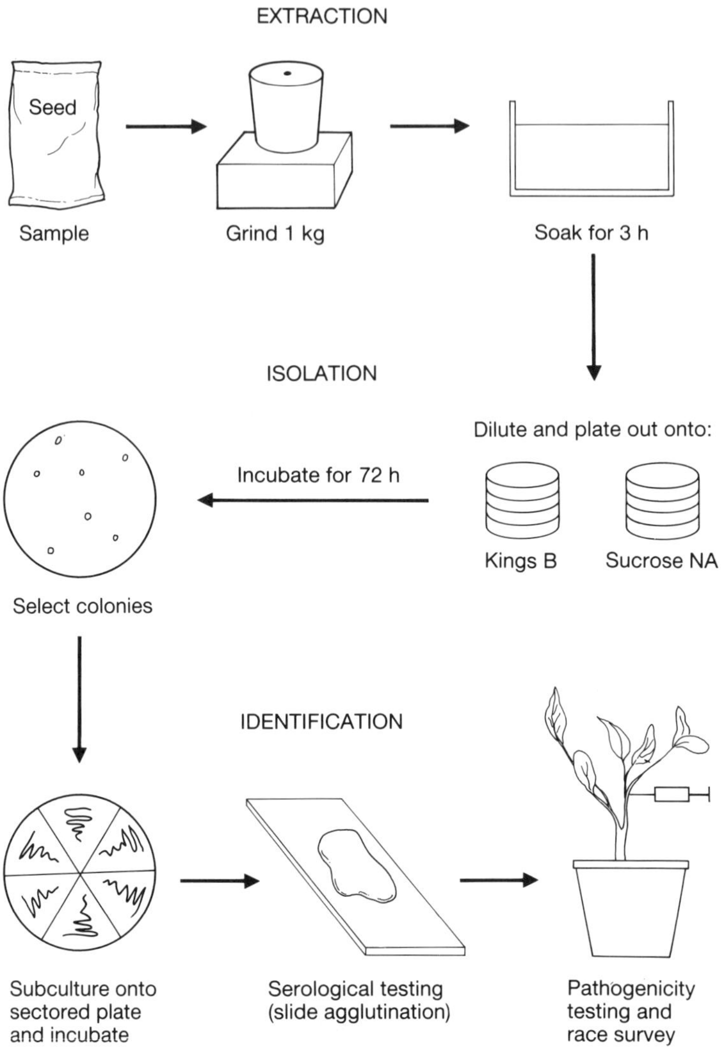

Fig. 21. Schematic representation of the protocol for testing pea seed for bacterial blight of pea (redrawn from Ball and Reeves, 1992).

in protein pea samples in early generations of seed multiplication. The protocol for this is illustrated in Fig. 21 and was based on the extraction of a milled 1 kg sample (about 3000–4000 seeds) of peas and followed similar isolation and identification procedures completed by a pathogenicity cum race survey test (Ball and Reeves, 1992).

Limitations of seed tolerance tests

In the examples described, the numbers of seeds or samples of seeds tested for infection were important in determining tolerance limits for the effective limitation of disease. Zero tolerance cannot be attained using a seed test since by definition it would involve testing complete seed stocks. When the percentage of diseased seeds is very low the probability of the occurrence of diseased seed in a sample follows a Poisson distribution (Geng *et al.*, 1983). As a consequence, both positive and negative results are possible from the same stock of seeds (Geng *et al.*, 1983; Roberts *et al.*, 1993). The limit of a test may be defined as the maximum sample size in which a single infected seed can be detected. Confidence in test results therefore requires that the size of the sample is increased or a number of samples are tested to allow statistical constraints to be applied. The mathematics of probability specifies that, for effective analysis, a test sample must have three times the number of seeds in it than the number determined by the incidence of the target organism. Thus, where the incidence of the target organism is 0.1% (one infected seed in 1000) a sample size of 3000 seeds should be used. Roberts *et al.* (1993) have combined practical considerations and mathematical studies to develop a rational approach to the design and interpretation of seed health assays for the detection of low-incidence bacterial pathogens of seeds.

Tests of single samples of seeds

Russell (1950) examined the theoretical probabilities of no loose smut appearing in embryo tests where the incidence of the pathogen was 1%. The greater the number of embryos examined the less the chance of failing to detect smut, i.e. there were only two chances in 1000 of failing to detect smut in 600 embryos but 37 chances in 100 of failing to detect it where 100 seeds were tested. Because of the Poisson distribution of infected seeds, in 38% of tests where 600 embryos were examined more than 1% infection would be found resulting in the unnecessary rejection of seed stocks (Russell, 1950). A similar problem was identified by Tucker and Foster (1958) concerning the statistics of indexing lettuce seed for mosaic content at a 0.1% level. For example, in 0.1% infective lettuce seeds about 30% of samples taken would have 11 or more infective seeds in 10,000 and about 45% of samples would have nine or fewer infective seeds in 10,000. Furthermore, there was only 69% confidence that where nine infective seeds in 10,000 were indexed that the lot contained not more than 0.1% mosaic. Thus, at least 30% of samples could have been rejected on the basis of a natural within-sample variation, and the setting of a much lower 'accept–reject' point to improve statistical confidence to 95–99% levels was suggested (Tucker and Foster, 1958). In practice the 0.1% tolerance limit failed to control the disease in commercial situations for a variety of reasons, some probably linked to the theoretical considerations stated above. The problem was fully redressed by reducing tolerance to a zero index in

30,000 seedling (Kimble *et al.*, 1975). A zero index at the 99.9% confidence level gives a probability of zero to seven infective seedlings in 30,000; field experience has proved this to be a sufficient further reduction in the incidence of seedborne virus to control LMV in California, USA (Grogan, 1983).

Tests on samples of decreasing numbers of seeds

The most probable number method is widely used for estimating bacterial densities in liquids (Cochran, 1950; Swaroop, 1951) and was applied by Taylor (1970b) to the estimation of bacterially infected phaseolus bean seed stock. Using this technique there are two assumptions: (i) that infected seeds are randomly distributed; and (ii) that it is always possible to detect at least one infected seed in the largest sample size used. This last assumption presumes the knowledge of the effective limit of the seed test (Taylor, 1970b). The test involves taking a series of samples in the ratio of 1 of x, 5 of $x/5$ and 5 of $x/50$ (where x does not exceed the sample size in which one infected seed could be detected). Interpretation tables and graphs suitable for use the estimation of bacterial and virus infections of seeds have been developed (Taylor *et al.*, 1993). False positive and false negative results from such tests may occur and can be interpreted. The mathematical validity of such approaches to seed testing for the setting of tolerance limits has been critically evaluated (Geng *et al.*, 1983; Roberts *et al.*, 1993).

Exclusion of Seedborne Pathogens by the Selection of Seed Production Areas

Seedborne organisms also can be excluded by the production of seed crops in areas environmentally unsuited for the development of disease by virtue of special climatic, edaphic or other factors. This strategy is used for the production of disease-free seeds in commercial quantities, and also in much smaller amounts as basic seeds which are then multiplied further in their natural environment in seed improvement and certification programmes. Quarantine regulations may require the production of seeds for export under such conditions. The general precepts involved are described here by reference to specific examples.

Control of fungi and bacteria

Fungi and bacteria which are transmitted under cool moist conditions and readily spread by rain splash (see Chapter 5) can be suppressed by growing seed production crops in arid or semiarid climates. For this reason, and based on observation and experiment (Walker, 1934), commercial cabbage seed production was transferred to the Skagit valley of the Puget Sound area of western Washington, USA – an area of low rainfall – to produce seeds free of the *Phoma lingam* asexual state of *Leptosphaeria maculans* (black leg) and *Xanthomonas campestris* pv. *campestris* (black rot). Although there are occasional setbacks, mainly due to fluctuations in rainfall, the Puget Sound remains one of the safest areas in the USA for the production of disease-free cabbage seed (Gabrielson, 1983).

Semiarid climates are utilized for the production of disease-free seed crops at different stages of multiplication from pre-basic seeds to commercial levels. For example, as part of bean seed improvement programmes in Michigan, USA (Copeland *et al.*, 1975) and Ontario, Canada (Sheppard, 1983) to control bacterial blights of bean (*X. campestris* pv. *phaseoli*, *Pseudomonas syringae* pv. *phaseolicola*), pre-basic seed produced under desert conditions in California and Idaho is distributed to selected growers in both areas. These growers then produce larger amounts of disease-free basic seeds (Copeland *et al.*, 1975; Sheppard, 1983) and eventually certified seed stocks for more general use.

Furrow irrigation

The use of furrow rather than overhead irrigation systems in low rainfall areas further restricts the spread of splash-dispersed pathogens including *X. campestris* pv. *phaseoli* (Sheppard, 1983), *P. syringae* pv. *phaseolicola* and *Colletotrichum lindemuthianum* on phaseolus bean, *Septoria apiicola* on celery, *X. campestris* pv. *campestris* and *Phoma lingam* on cabbage and *Ascochyta* spp. on peas (Baker, 1972).

Control of viruses

Climatic conditions, days with high temperatures (greater than 30°C) and average humidities of 36% RH (relative humidity), caused the early disappearance of the aphid vectors (*Myzus persicae*, *Macrosiphum euphorbiae*, *Aphis gossypii* and *Hyperomyzus lactucae*) of LMV in seed production crops grown in the Swan Hill area of the Murray valley, Australia. This location has been used successfully to produce lettuce seed crops practically free of the virus (Stubbs and O'Loughlin, 1962). The maintenance and reproduction of LMV-free lettuce stocks on a commercial basis in the USA is usually undertaken in two stages. Firstly the production of mosaic-free plants by roguing any diseased plants resulting from seedborne infection under aphid-free conditions in a greenhouse, and secondly by growing the remaining healthy plants to maturity in an isolated area (Shepherd, 1972).

Isolation areas

Even in semiarid climates, isolation areas of similar crop species should be separated by distances greater than the distances over which their particular seedborne pathogens or their vectors can be transported (see Chapter 5). In addition, isolation areas need to be distanced from old seed fields and from volunteer plants (see Chapter 5). For this reason nasturtium seed fields are placed as far as possible from volunteer or perennial nasturtiums in gardens or harvested seed fields to prevent the introduction of air-borne spores of *Acroconidiella tropaeoli* (syn. *Heterosporium tropaeoli*) (Baker, 1956). Lettuce seed production in the San Joaquin valley is well removed from commercial crops in the Salinas valley and elsewhere in California to prevent infection by aphids carrying LMV. Similarly, coastal planting of petunia seed fields in California is done where the beet leafhopper (*Circulifer tenellus*), which transmits the beet curly top virus (BCTV), is less frequent (Baker, 1972).

With virus pathogens, particular isolation areas must be free of weed host reservoirs. For example, wild *Lactuca* spp. and some other weeds are hosts of LMV (Shepherd, 1972). Where weed-free areas are difficult to find, eradication of the existing host weeds is a necessity. The wider the host range of a pathogen, the more difficult it is to eradicate its alterative sources. Elimination of the aphid vectors of viruses such as LMV by spraying with insecticide is of doubtful value because the virus is non-persistent.

Effects of edaphic factors in semiarid climates

The transmission rates, particularly of temperate seedborne pathogens, will be reduced under the warmer, less moist soil conditions of semiarid climates thus reducing infection in crops grown for seed (see Chapter 4, p. 56). Other factors, such as suppressive soils (see Chapter 4, pp. 60–62 and Baker 1980, 1981; Gambogi, 1983), particularly where they occur in semiarid climates, may be of value in the production of disease-free seeds, for example pea seed crops free of wilt (*Fusarium oxysporum* f.sp. *pisi*) in California.

Crop inspection and roguing

Field inspections are used to determine the incidence of affected plants in crops grown for the production of commercial seed, seeds for export and certified seed stocks. Crop inspection techniques have been developed to allow the quantitative determination of plant infection on which crop acceptance or rejection is based. Where infected plants occur they may be removed (rogued) to reduce or eliminate disease. Both methods are applied to seed crops produced under all climatic conditions.

A field inspection pattern for sweet corn and sorghum for export from Iowa, USA is shown in Fig. 22 (Groth, 1981). Five plants are collected from

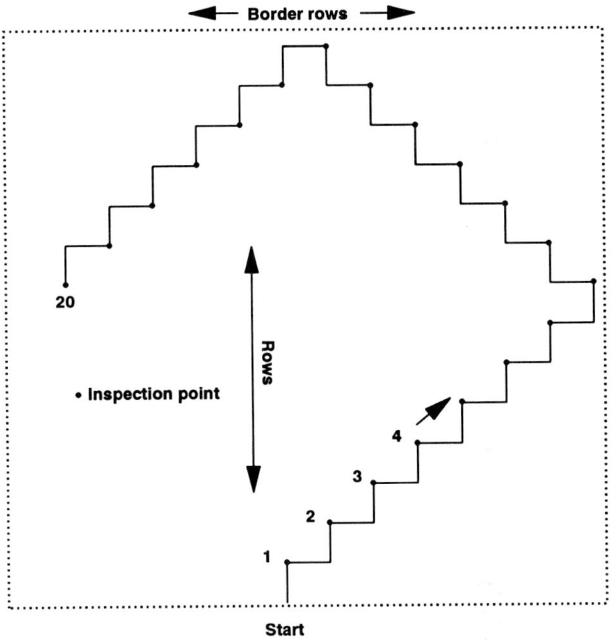

Fig. 22. Field inspection pattern for sweet corn and sorghum (redrawn from Groth, 1981).

each of 20 inspection points and tested in detail for a range of diseases; if the field meets the health requirements of the country to which the seed is to be exported then a phytosanitary certificate is issued (Groth, 1981). For example, one kale (*Brassica oleracea*) plant infected with canker (*Leptosphaeria maculans*, anamorph *Phoma lingam*) in 1000 plants was the tolerance limit for crops produced in the UK, the seed of which was for export to South Africa. However, the detection of infected plants by field inspections where there is zero tolerance can result in the rejection of that crop. Thus, to eliminate the risk of seed infection, the occurrence of any bacterial blight (*Xanthomonas campestris* pv. *phaseoli*) infection in field inspections of bean crops grown in semiarid conditions may be sufficient to warrant the immediate destruction of the crop by ploughing under (Copeland *et al.*, 1975). Similarly, phaseolus bean crops grown for certification in New South Wales, Australia are rejected if any halo blight is found in inspected crops (Anon., 1944). To improve the level of freedom from *X. campestris* pv. *phaseoli* (Sheppard, 1983), the supplementary practice of roguing (i.e removal of infected bean plants) is used, if necessary, in the crop grown from pre-basic seed under normal climatic conditions for disease expression (see also LMV, p. 109).

The use of healthy seeds for production of disease-free seed crops is good practice in certain climatic areas. For example, this is necessary for phaseolus bean seed crops grown in low rainfall (Wilson, 1947) or semiarid areas (Grogan and Kimble, 1967) where the climatic conditions suppress natural disease expression and halo blight (*Pseudomonas syringae* pv. *phaseolicola*) may not be detected by field inspections (Grogan and Kimble, 1967). As a result additional tests (e.g. pot tests of harvested seeds grown in a moister environment) may be necessary to endorse the health of the seed crop (Grogan and Kimble, 1967).

Exclusion of seedborne diseases in ware crops grown in semiarid climates

The restrictive effects of semiarid climates, such as that of California, on the transmission of several seedborne bacterial and fungal pathogens of vegetables, flowers and cereals has been suggested as an alternative to seed treatments for the control of disease in ware (commercial) crops (Baker and Snyder, 1950).

Use of 'healthy' seeds in disease control strategies

Much of the US seed industry has been located in the semiarid western states (e.g. Oregon, California and Idaho) largely because reduced transmission of seedborne pathogens resulted in fewer plant losses in seed fields (Kreitlow *et al.*, 1961; Wallen, 1964). The production of disease-free ('healthy') seeds is considered the most reliable method of controlling seedborne virus diseases and thereby of producing healthy, reasonably disease-free crops (Mandahar, 1981). Mandahar (1981) lists seedborne viruses of barley, bean, cherry, citrus, lettuce, pea, *Prunus*, strawberry and sugarbeet which can be controlled by following this strategy. Many fungal and bacterial pathogens of seed can be minimalized by such procedures. Field inspection and seed testing methodologies can set tolerance limits for seed production crops and seed, at various stages of multiplication, which will ensure economically acceptable levels of disease in the crops they ultimately produce.

However, mainly for reasons of cost, there is a diversity in seed production areas which are located in many countries and climates which makes it difficult to grow crops under conditions which are always ideal for the elimination of seedborne diseases (Maude, 1988b). Nonetheless, where there are suitable environmental conditions for the production of disease-free seeds the practical value of this approach is beyond doubt. The potential value of producing and using high quality healthy seed is recognized widely (India: Agarwal, 1983; Siddiqui, 1983; Bangladesh: Ahmed and Ahmed, 1983; Africa: Awodera *et al.*, 1983; Besri, 1983; Kabeere, 1983; Philippines: Sevilla and Guerrero, 1983). The strategies employed to achieve this objective may include all, or only some, of those described above.

Reduction in Infection by Breeding for, and Utilization of, Plant Resistance

The production of resistant plant material is a further strategy by which pathogens and pests can be reduced within, or excluded from, commercial crops. Plant-breeding methods have been used to introduce resistance to crop pathogens arising from many different sources, in some of which the main source has been infected seeds. As explained previously (see Chapter 4, pp. 62–63) some horizontal and vertical resistance mechanisms operate through the seed mother plant to reduce infection of seeds, and through the seeds to reduce or prevent transmission of disease. Many other control mechanisms also operate and resistance in the tissues of plant can be expressed in other ways, i.e. hypersensitivity reactions, resistance to virus increase, etc. In respect of seedborne pathogens, the need for resistant material is perhaps greatest where the seed is the main disease source and there is no other effective seed treatment option for disease control. This applies particularly to viruses where there are few seed treatment remedies and also to seedborne bacteria where treatments are mainly physical and cannot be applied on a large scale. For seedborne fungi, however, there are effective and relatively inexpensive chemical seed treatments available commercially. For this reason, and because of cryptic resistance (the level of field resistance present in current commercial cultivars), there has been a reduced incentive to include disease-resistant characters among the priorities of vegetable-breeding programmes (Crute, 1988). However, Crute (1988), cites 64 commercially distributed resistant cultivars in 20 different temperate vegetable types, with introduced resistance in 41 cultivars to fungi, in 19 cultivars to viruses and in four cultivars to bacteria. The greatest commercial impact of the introduction of resistance to seedborne pathogens was to the viruses, (LMV, tomato mosaic virus (ToMV) in tomato and BCMV in phaseolus beans) and to the fungus *Colletotrichum lindemuthianum* (anthracnose) in phaseolus beans (Crute, 1988). Tomlinson (1987) cites the resistance in tomato to ToMV (Pelham, 1972; Hall, 1980), controlled by two dominant genes, and the resistance to LMV in lettuce (Von der Pahlen and Crnko, 1965; Ryder, 1970), controlled by the recessive MoMo gene, as two examples of almost perfect control of vegetable viruses.

The commercial value accruing from breeding for resistance to seedborne and other pathogens in oilseed rape (Bowman, 1988) and in cereals (Priestley and Bayles, 1988) has been reviewed.

7 Disease Control: Eradication and Reduction of Inoculum by Seed Treatment

Introduction and Terminology

Seed treatment is a generic term (Scott, 1989) which does not specify the application method but indicates that seeds are subjected to a compound (chemical, nutrient, hormone, etc.), a process (such as wetting or drying) or to various energy forms (e.g. radiation, heat, magnetism, electricity). In the context of plant pathology, seed treatments may have one or several functional objectives. Fungicidal seed treatments are applied to seeds to eliminate inoculum and thereby to produce healthy seedlings and crops. This they may accomplish by killing or neutralizing seedborne pathogens, thus preventing transmission of disease. They are also applied to protect germinating seeds and emerging seedlings from soil-borne and air-borne pathogens. Very often eradicative and protective functions are necessary in a chemical seed treatment. Physical treatments involving the use of heat (wet or dry), radiation, etc., by their nature have but a single objective – to kill (i.e. eradicate) seedborne inoculum. The chemicals used in seed treatment are fungicidal when they kill fungi and fungistatic when they prevent additional growth or sporulation of an organism without killing it (McCallan and Wellman, 1942).

Seed treatment fungicides can be divided into two groups, non-systemic and systemic. Non-systemic fungicides are those which have limited penetrative activity when applied to the surface of seeds and generally are not mobile within the tissues of seeds. They provide protection against invasion by soil-borne fungi for a short period of time. They are mainly protectant in action but can have some eradicant effect when seeds are soaked in aqueous solutions or suspensions of those chemicals. Systemic fungicides, by Wain and Carter's (1977) definition, are those compounds which can prevent disease development on regions of the plant away from the site of application. Systemic compounds applied as seed treatments function mainly as eradicants of deep-seated infections of seeds, but they can be translocated via the internal tissues of seeds into the hypocotyls and the developing radicles and/or plumules of plants and there may give added protection against external soil-borne and foliar pathogens. The translocation and mode of action of systemic fungicides has been reviewed (Marsh, 1977; Edgington, 1981).

Seed treatment fungicides are classified as eradicant, protectant and systemic by Wain and Carter (1977) and as disinfestant, disinfectant and protectant by Leukel (1953). When eradicants (disinfectants) are applied to seeds they will penetrate to a limited extent and cure established infections; protectants provide local protection against infection at the site of application, for example protect the seed from infection from the soil (Wain and Carter, 1977). Systemic chemicals also may protect at a distance from the site of application (see above). Disinfestants inactivate organisms present on the surface of seeds, for example bunt spores. Since practically all effective seed treatment (eradicant and protectant) chemicals are disinfestants the distinction the term makes is slight and it is not in common use. Systemic and non-systemic fungicides, depending on the method of application, can have either or both protectant and eradicant functions.

In a wider context, the term seed treatment includes simple dusting treatments with French chalk and colouring materials to aid the flow of seeds in the drill and to improve their visibility in the furrow, pelleting which improves the uniformity of seeds for drilling (Reid, 1970), treatments which alter the physiology of seeds thereby improving their physical performance (e.g. priming (Bradford, 1986; Gray, 1994) and enzyme/protein mixtures (Soper, 1991)), the application of biological fertilizers, (e.g. rhizobial bacteria to phaseolus beans (Taylor *et al.*, 1983) and more novel techniques of application (e.g. film coating) which enable pesticides, biological agents and other systems to be added to seeds with great accuracy (Maude and Suett, 1986; Maude, 1990c). These different methods of treatment will be described and defined in the course of this chapter.

The term seed dressing, although formerly used to denote seed treatment, now refers solely to cleaning procedures for harvested seeds. The definition of seed dressing given by Scott (1989) as 'the application of finely ground solids (usually a fungicide or an insecticide) dusted onto the surface of seeds' therefore has been superseded.

This chapter is concerned principally with treatments applied to seeds to eradicate seedborne fungal, bacterial and virus pathogens, but some explanation will be given of the function of the other components which may occur in formulated mixtures of seed treatment pesticides. Information is not included on insecticides formulated singly or in mixtures as seed treatments for the control of insect pests of cereals, sugarbeet, vegetables, oilseed rape, etc. (Soper, 1992).

The use of seed treatments may be required as part of the quarantine regulations for the exchange of seed germplasm (Kahn, 1983) or for the importation of seeds in larger quantities; they can be applied at any stage in seed multiplication. For example, in 1983 more than 95% of the cereal seed sown in Scotland, of which 85% was certified, was treated with pesticides (Rennie *et al.*, 1983). Over 80 seed treatment products are listed for commercial use in the UK (Soper, 1991) for application to cereals, beet, pulses, oilseed rape, vegetables, grasses and flower seeds.

Seed treatment is the subject of a monograph by Jeffs (1986a) which contains chapters on the treatment of cereals (fungicides: Bateman *et al.*, 1986; insecticides: Griffiths, 1986), rice in Japan (Nakamura, 1986), rice in the USA and South America (Bowling, 1986), maize (Kommedahl and Windels, 1986), soyabean (McGee, 1986), cotton (Halloin, 1986), sugarbeet (Durrant *et al.*, 1986) and vegetables (Maude, 1986). Information on seed treatment is also provided by other authors including Baker (1972), Neergaard (1977), Agarwal and Sinclair (1987) and Maude (1990c). McGee (1990, 1992) includes information on seed treatment in reference works for seed technologists on maize (1990) and soyabean (1992). The control of seedborne bacteria is reviewed by Ralph (1977a). The concepts and technologies of selected seed treatments have been reviewed by Taylor and Harman (1990).

Development of Chemical Seed Treatments

The history of the development of chemical seed treatments is very much a review of the progress from inorganic to organic chemistry. At the present time, seed treatment development, though continuing to be dependent on the production of new chemicals, is closely associated with the technology necessary for efficient formulation and application of pesticides to seeds (Martin, 1988, 1994). In addition to advances in the production of new seed treatment pesticides there has been a parallel development on a smaller scale in physical treatments for seeds by the use of wet and dry heat in various forms (Baker, 1972; Neergaard, 1977; Maude, 1990c).

Periodic accounts of seed treatment provide a useful documentary of the state of development of the technology at the time. The reader is directed to the following authors for specific and general information on the topic: Leukel (1936), Horsfall (1945), Miles (1946), Buttress and Dennis (1947), Leukel (1948), Ordish (1976), Neergaard (1977), Martin and Woodcock (1983), Jeffs (1986a), Maude (1990c), McGee (1990, 1992) and Schwinn (1994).

Salt, lime, copper and formaldehyde

Seed treatments for various purposes have their origins in ancient times (Horsfall, 1945; Ordish, 1976; Jeffs, 1986b) but those for the control of seedborne pathogens are more recent. Brining (immersion of grain in salt water followed by liming to dry the grain) of wheat seed to control bunt may have been used in the UK by the middle of the 17th century (see Chapter 1). Such early seed treatments anticipated the diagnosis of disease and not until the middle of the 18th century did Tillet (1755) establish that seedborne fungi (*Tilletia tritici, T. laevis*) caused bunt of wheat and that it could be controlled by steep treatments with lime, lime and salt, or lime and nitre. At the beginning of the 19th century, Prevost's microscope examination of germinating bunt spores and of the killing of the germ tubes by exposure to copper solutions led to his recommendation that copper sulphate solutions (5 g $CuSO_4$ per litre of water) should be used to wet the wheat and control bunt (see Large, 1950; Ordish, 1976). A similar copper method was developed in parallel in the Netherlands (Ordish, 1976), but neither was widely used until Kuhn (1873) demonstrated the practical effectiveness of this seed treatment.

Both brining and steeping in copper sulphate solutions involved the immersion of large bulks of grain and these treatments were cumbersome to apply. They were eventually replaced by barn floor methods where heaps of grain were sprinkled with smaller amounts of solution and the grain and fungicide were mixed together using shovels and then spread out to dry. Although the heap method of seed treatment was widely practised it was less than satisfactory because seeds damaged during threshing were further adversely affected by imbibing copper salts through the damaged seed coats. It is reported that less damage was caused if lime was applied after the treatment of the grain as this converted the copper salt to its insoluble oxycarbonate. Copper oxycarbonate did not penetrate the grain and provided longer protection against smut (see Martin and Woodcock, 1983; Jeffs, 1986b). In Canada, treatment with 3% copper sulphate was recommended as a general preventative treatment against bunt (stinking smut – *T. laevis*) applied at 47.5 ml kg^{-1} wheat (0.3 gallons $bushel^{-1}$) to heaps of grain on the barn floor (Saunders *et al*., 1893). Steeping in formaldehyde, less toxic to seeds than copper salts, was introduced towards the end of the 19th century (Horsfall, 1945; Martin and Woodcock, 1983) for the control of bunt in North America (Bolley, 1897) and Australia (Darnell-Smith, 1917). Soaking seeds in formaldehyde, even with subsequent airing, caused some injury. Haskell (1917) utilized the vapour action of formalin to improve its anti-smut performance and reduce its toxicity by heaping the cereal seeds, spraying the seed mass with concentrated formaldehyde, covering it overnight and airing it the following morning. Formalin was used in Germany during the 1914–1918 world war as a replacement for copper sulphate which was in short supply (Jeffs, 1986b). Formaldehyde, however, although controlling seedborne smut was not a good general protectant and there is evidence that it increased foot rot infection of cereals by the soil-borne phase of fungi such as *Fusarium culmorum* and *Cochliobolus sativus*, possibly by reducing the growth of seedlings and thereby making them more vulnerable to fungal attack (Machacek and Greaney, 1935).

The first application of dry fungicide to seeds is attributed to Darnell-Smith (1915, 1917) who obtained good protection from smut by dusting seeds with basic copper carbonate powder at the rate of 2 g kg^{-1} grain. Leukel (1948) reported the continued use in the USA of basic copper sulphate and copper carbonate dusts for the control of wheat bunt and kernel smut of sorghum and the use of cuprous oxide as a seed protectant to prevent seed decay and pre-emergence damping-off of vegetables. Red and yellow cuprous oxides and zinc oxide were under test in 1942 in the UK (Croxall and Ogilvie, 1942) for this purpose on a range of species including pea, tomato, onion and cabbage. Organic coppers are still used in France as seed protectant fungicides, for example the use of copper oxyquinolate (oxine copper) for the treatment of cereal seeds (Noon and Jackson, 1992; Rennie and Cockerell, 1994). Exactly when the equipment for the application of dry powders to seeds was developed is not known but barrel dusters such as those described by Horsfall (1945) and Neergaard (1977) were probably homemade (Jeffs, 1986b).

Sulphur

Although at the beginning of the 20th century sulphur derivatives were shown to have fungicidal activity, they have been little used as seed treatments (Martin and Woodcock, 1983). Mackie and Briggs (1923) in California, USA, used flowers of sulphur applied as a dry treatment and reported the control of bunt; Uppal and Malelu (1928) in Jowar, India controlled grain smut of millet (*Sphacelotheca sorghi*) in the same way. Sulphur is simple to use and can be applied to grain before storage, where it affords protection against insect pests; Martin and Woodcock (1983) suggested that it might be suitable for use in areas of primitive agriculture.

Inorganic and organic mercury

The role of the chemicals mentioned so far was to protect seeds against organisms in the soil or to rid them of superficial inoculum of a few specific fungi such as bunt (*Tilletia tritici*, *T. laevis*). One of the major advances in seed treatment performance which came early in the 20th century resulted from the discovery of a broader spectrum of protectant activity including the limited eradicant action of the mercury fungicides. The action of mercuric chloride in controlling the dormant mycelium of *Fusarium nivale* (teleomorph *Monographella nivalis*) in the seed coats of rye (Hiltner, 1915) was significant in that it indicated eradicant action against established infections. Mercuric chloride was already known as a highly effective bactericide, but it was poisonous. The introduction of the organomercurial seed disinfectants with reduced mammalian toxicity can be associated with Riehm, who in 1913 reported on the effectiveness of chlorphenolmercury for the control of cereal bunt. Pre-war (1939–1945), the most effective organomercurials for use as cereal seed treatments were the ethyl mercury salts. Their high degree of toxicity as fungicides and bactericides and their reduced toxicity to seed germination made them ideal for the treatment of many seed types. Their discovery expanded the field of seed treatment both in the USA and in other countries (Miles, 1946). From the mid-1940s phenyl-, methyl-, methoxy- and hydroxy-mercuric preparations appeared as seed treatments. Phenyl mercury cyanimide was developed for use on seed corn in the USA and in Europe tolyl and methoxyethyl mercury were used for the treatment of seed grain (Miles, 1946). Seedborne diseases such as wheat bunt (*T. tritici*) and barley leaf stripe (*Pyrenophora graminea*), became rare in European cereal crops with the widespread adoption of mercurial seed treatments (Bent, 1978) and seedborne pathogens of cereals in the UK were very effectively controlled for more than 50 years by the routine and extensive use of organomercury fungicides (Rennie and Cockerell, 1993, 1994).

Over the past 20 years organomercury seed treatments have gradually been withdrawn. The change has been made on environmental grounds, mainly because of their toxicity and persistence (Bateman *et al.*, 1986). In a world survey made in 1983 (Bateman *et al.*, 1986) organomercurial seed treatments for cereals had ceased to be used in Australia, New Zealand, Brazil, USA and Canada. In 17 European countries nine were still using organomercurials; the UK, Eire, Belgium, Austria and Switzerland treated 85–100% of cereal seeds and the Scandinavian countries (Denmark, Norway, Sweden and Finland) treated 15–58% of cereal seeds with these compounds. In the UK in 1991 (Soper, 1991), of 17 commercial seed treatment products available for use on cereals, six were organomercurials (mainly based on phenylmercury). To comply with European Community directives, organomercury seed treatments ceased to be applied in the UK from the end of March 1992 (Noon and Jackson, 1992). The implication of this for cereal seedborne pathogens in Scotland (Rennie and Cockerell, 1993, 1994) and the UK (Yarham and Jones, 1992) has been reviewed (see pp. 126–127).

Non-mercurial organic fungicides

The poisonous qualities of organomercury fungicides was an incentive for the development of non-toxic organic alternatives. Those that were produced, however, generally affected a narrower range of fungi and were mainly protectant chemicals. Two organic sulphur fungicides, thiram (tetramethylthiuram disulphide) and captan (1,2,3,6-tetrahydro-*N*-(trichloromethylthio)phthalimide) were introduced in the mid-1930s and 1950s, respectively (Martin and Woodcock, 1983). They are powder formulations with each using a single active ingredient; they are still in use in the UK (Soper, 1991) as seed

Table 15. Classification of non-systemic and systemic organic fungicides used as seed treatments in the UK (from Hassal, 1990; Soper, 1991; Tomlin, 1994; Whitehead, 1995).

Type	Family/action	Group	Used in seed treatment
Non-systemic	Organosulphurs	Dithiocarbamates	Thiram
		Phthalimides	Captan
	Chlorinated aromatics	Chlorinated nitro compounds	(Quintozene)
		Chlorinated quinones	(Dichlone, chloranil)
		Benzotriazines	Triazoxide
Non-systemic or weakly systemic		Cyanopyrroles	Fenpiclonil
		Guanadines	Guazatine
		Azoles	Imazalil, prochloraz, bitertanol
		Dicarboximides	Iprodione, vinclozolin
Systemic		Benzimidazoles	Benomyl, thiabendazole carbendazim, fuberidazole
		Phenylamides	Carboxin
	Steroid inhibitors		
	Reduction	Morpholines	Fenpropimorph
	Demethylation	Azoles	Triadimenol, flutriafol tebuconazole
		Isoxazoles	Hymexazol
		Pyrimidines	Ethirimol
		Acylalanines	Metalaxyl
		Oxazole ketones	Oxadixyl
	Miscellaneous groups	Ethyl phosphonates	Fosetyl aluminium

Fungicides formerly in prominent seed treatment usage are shown in parentheses.

protectants, formulated singly (Table 15) or with other pesticides, for the control of pre- and post-emergence damping-off fungi in the soil. Many of the other dimethyl and ethylene bisdithiocarbamate seed treatment fungicides related to thiram and in use in the 1940s (Leukel, 1948) are no longer marketed.

The introduction of these chemicals stimulated comparative testing with copper fungicides, organomercury fungicides and non-mercurial organic compounds. Such studies, as for example by Clayton (1948) in USA and Jacks (1951) in the UK, demonstrated the effectiveness of topical applications of thiram in maintaining a high emergence level and in controlling pre-emergence damping-off fungi in a number of vegetable species. In artificially inoculated soil, thiram protected peas and lettuce from *Pythium ultimum* (causing pre-emergence damping-off) and *Rhizoctonia solani* (teleomorph *Thanatephorus cucumeris*) (causing damping-off and post-emergence wire-stem) more effectively than copper and mercury compounds but was less effective against *Fusarium solani* foot rots (Jacks, 1951). Although seed treatment with thiram gave pre- and post-emergence protection to cucumber and muskmelon in commercial seedling-raising systems, seeds (e.g. onions, spinach) were not well protected against post-emergence damping-off organisms (McKeen, 1950). The suggested explanation was that the spatial distribution of the fungicide in the soil afforded less effective protection in small seeded species than it did in large seeded species (McKeen, 1950). One effect of the addition of thiram to the soil was that the soil became more difficult to infest with *P. ultimum* and the natural population of damping-off organisms failed to increase under repeated cropping and thus protection for seedlings extended beyond the duration of persistence of the fungicide in the soil (Richardson, 1954).

The introduction of captan provided a fungicide with a very similar spectrum of activity, especially against damping-off fungi (*Pythium* and *Phytophthora* spp.) in the soil. Tests on peas and beans (deZeeuw *et al.*, 1956) and on 24 kinds of vegetable seeds (Wallen and Bell, 1956) indicated that captan was an effective general protectant fungicide often superior to thiram in achieving emergence improvement. In addition to protecting the germinating seeds from soil-borne organisms, captan and thiram have disinfectant properties and when applied as dry powders to infected pea seed both fungicides considerably reduced the transmission of *Mycosphaerella pinodes* (Wallen *et al.*, 1967) and *Ascochyta pisi* (Maude and Kyle, 1970) in field soil. Both these fungi infect the seed coat tissues, and more rarely the embryo of peas (see Chapter 2, p. 29). Captan and thiram penetrated the seed coat but rarely the embryo tissues to control fungi in the seed coats of pigeon pea (*Cajanus cajan*) (Ellis and Paschal, 1979), and it seems probable that a similar effect occurred with field peas. There is limited evidence to suggest that, from applications to the surface of peas, captan may penetrate the pea tissues to produce a fungistatic substance in the seedlings which will make them slightly less susceptible to infection by fungi (e.g. *A. pisi*) (Wallen and Hoffman, 1959).

Although the range of activity of captan and thiram may be more limited than that of the organomercurials, both fungicides have, in addition to toxicity to Oomycetes *Pythium* and *Phytophthora* spp., activity against certain Deuteromycetes (e.g. *Ascochyta* spp.), Ascomycetes (e.g. *Venturia pirina*, anamorph *Fusicladium pyrorum*) (Worthing and Hance, 1991) and Basidiomycetes (e.g. *Thanatephorus cucumeris*, anamorph *Rhizoctonia solani*). Of the 74 seed treatment products available for use in the UK on cereals, beet, pulse, oilseed, field vegetables and flower seeds (Soper, 1991), 31 contain thiram and two captan.

Other non-mercurial organic fungicides which were contemporary seed protectants and which were then superseded by thiram and captan include the chlorinated aromatic compounds, dichlone, chloranil and quintozene (Croxall and Ogilvie, 1942; Wallen, 1953; deZeeuw *et al.*, 1956). None of these fungicides are in current use in the UK (Soper, 1991) but quintozene was used in France in the late 1980s (Maude, 1986) as a seed treatment for the control of seedborne onion neck rot (*Botrytis allii*) and soil-borne onion white rot (*Sclerotium cepivorum*). Tecnazine (tetrachloronitrobenzene), more volatile than quintozene, was used to control bunt of wheat in the UK (Bateman *et al.*, 1986). Both fungicides are known to be toxic to warm-blooded mammals and fungi can become resistant to them (Bateman *et al.*, 1986).

The effectiveness of organomercurial fungicides for the control of a wide range of seedling diseases of cereals was not surpassed by the chlorinated aromatic or the organic sulphur compounds. Of the latter, thiram or captan have had particular use as seed protectants for the treatment of vegetable, oilseed rape, flower and grass seeds.

Systemic fungicides

The discovery in 1966 of systemic action in the phenylamide fungicides, carboxin and oxycarboxin (Schmeling and Kulka, 1966), completely altered many aspects of seed treatment technology from then onwards. The new systemic compounds were ideal in that, when dusted onto the surface of seeds, they penetrated the seed tissues as these imbibed water from the soil into which they were sown. Thus systemic fungicides were capable of penetrating the embryos of seeds to eliminate deep-seated infections and of further movement into the tissues of emerging seedlings providing additional protection against certain soil-borne and aerial fungal pathogens. The emphasis was shifted in seed treatment from protectant fungicides, with limited disinfectant ability, to systemic fungicides with curative, eradicant and some protectant properties (Table 16). The fact that seed treatments with systemic fungicides could be applied topically to commercial seeds of different shapes and in various quantities, generally without causing adverse effects on germination, removed the main disadvantages which can apply to physical and to physical/chemical treatments (see pp. 152–173).

Two characteristics of these novel fungicides are their selectivity for particular taxonomic groups of fungi producing a relatively narrow range of action (Table 16) and the relative ease with which fungi develop resistance to them. Those which have proved effective as seed treatments are toxic to Ascomycetes and Fungi Imperfecti (often to those which are the anamorphs of ascomycetous fungi) and some also have activity against certain Basidiomycetes. Those which are toxic to the phycomycetous fungi (*Pythium* and *Peronospora* spp.) generally do not affect fungi in the other groups (Table 16).

Table 16. Range of activity of modern fungicides against classified groups of fungi.

Date reported*	Fungicide: Group	Fungicide: Common name	Used as seed treatments against: Basidiomycetes	Ascomycetes	Deuteromycetes	Phycomycetes[†]
1966	Phenylamide	Carboxin	+	–	–	–
1974	Dicarboximide	Iprodione	+	+	+	–
1964	Benzimidazole	Thiabendazole	+R	+	+	–
1968	Benzimidazole	Benomyl	+R	+	+	–
1968	Benzimidazole	Fuberidazole	–	+	+	–
1973	Benzimidazole	Carbendazim	+R	+	+	–
1969	Pyrimidine	Ethirimol	–	+	–	–
1968	Guanadine	Guazatine	+	+	+	–
1988	Cyanopyrrole	Fenpiclonil	+	+	+	–
1973	Azole	Imazalil	–	+	+	–
1977	Azole	Prochloraz	–	+	+	–
1978	Azole	Triadimenol	+	+	+	–
1979	Azole	Bitertanol	+	+	+	–
1983	Azole	Flutriafol	+	+	+	–
1986	Azole	Tebuconazole	+	+	+	–
1977	Isoxazole	Hymexazol	+	+	+	+
1979	Morpholine	Fenpropimorph	+	+	+	–
1977	Acylalanine	Metalaxyl	–	–	–	+
1983	Oxazole ketone	Oxadixyl	–	–	–	+
1977	Ethyl phosphonate	Fosetyl aluminium	–	–	+A	+

*Authors and dates of description and the range of fungicidal activity are taken from Tomlin (1994).

[†]For convenience, the Oomycetes, the Chytridiomycetes, Plasmodiophoromycetes and the Zygomycetes are grouped together under the older name of Phycomycetes. Those fungicides which are active against this group mainly affect the Oomycetes (*Pythium*, *Phythopthora*, *Peronospora* etc. spp.).

+, activity against pathogen; –, no activity against pathogen; A, activity against *Ascochyta pisi*; R, activity against *Rhizoctonia* spp.

Seed treatments for cereals in the UK and other countries

The fungicides used singly or in various combinations to treat cereals are from the systemic or weakly systemic compounds listed in Table 15. The choice of fungicide depends on the range of target organisms to be controlled. To an extent, but not entirely, this has been influenced by the range of performance of the organomercury fungicides. Those fungicides eradicated superficial spores (e.g. covered smuts) and shallow mycelial infection of the seed coat, for example as caused by *Drechslera* spp. One of the principal reasons for the routine use of organomercury treatments was to control the damping-off and seedling establishment pathogens, particularly *Fusarium nivale* (teleomorph

Monographella nivalis) which has a great potential to cause the death of winter wheat seedlings in the UK (Rennie *et al.*, 1983, 1990); they also gave protection against soil-borne foot rot fungi such as *F. culmorum*. For many years, until mercury resistance developed (see p. 138), organomercury also controlled seedborne *Drechslera graminea* (teleomorph *Pyrenophora graminea*) (barley leaf stripe).

The organomercury fungicides had limited penetration of seed tissues and only partially controlled deep-seated infections of *F. nivale* and they had no effect on internally borne loose smut (*Ustilago segetum* var. *tritici*) mycelium (Bateman *et al.*, 1986). Many systemic fungicides are deeply penetrative of seed tissues, but in general they lack the range of fungicidal activity of the organomercury seed treatment fungicides (e.g. phenylmercury acetate) (Table 17). However, mixtures of systemic fungicides can be, and have been, targeted at a similar range of seedborne organisms (Table 17) and at an extended range including the soil-borne phases of fungi causing foot rots and seedling blights, at superficial, deep-seated and fully internal seedborne inoculum, and to give protection to the systems of young plants against early foliar infections caused by air-borne diseases such as mildews and rusts (Table 17). The published performance range (Tomlin, 1994; Whitehead, 1995) of commercially available single, double and triple formulations of these chemicals in the UK is compared with that of an organomercury for the control of barley diseases in Table 17. The experimental background to the effectiveness of some of these fungicide mixtures against many of the diseases mentioned above is given by Noon and Jackson (1992).

Activity against several seedborne pathogens of cereals has been reported for the azole (formerly triazole and conazole) fungicides myclobutanil (Green and Gooch, 1992) and metconazole (Sampson *et al.*, 1992). More recently, two azole fungicides, bitertanol and tebuconazole in separate mixtures with other fungicides, have been introduced in the UK (Table 17) for the control a range of diseases of wheat (Morris *et al.*, 1994) and of spring and winter barley (Wainwright *et al.*, 1994). A new class of non-systemic fungicides, the cyanopyrroles, have been shown to be effective in seed treatment applications against the major seedborne pathogens of wheat, rye, triticale and barley (Koch and Leadbeater, 1992; Noon and Jackson, 1992; Leadbitter *et al.*, 1994) and one of them, fenpiclonil is in use in the UK (Ivens, 1994) (Tables 15–17).

Cereal seed treatments in Europe are based on 16 systemic fungicides which are representatives of the main chemical groups, including benzimidazoles, azoles, phenylamides, guanidines, pyrimidines, and dicarboximides (Noon and Jackson, 1992). A small number of non-systemic chemicals are also used, i.e thiram, copper oxyquinolate, maneb/mancozeb and ampropylfos (Noon and Jackson, 1992). Rennie and Cockerell (1994) list the active ingredients of seed treatment formulations in use in Ireland, France, Scandinavia (Denmark, Norway, Sweden and Finland), Canada and New Zealand.

Table 17. Comparison of the range of effect of mercurial with non-mercurial commercial seed treatment formulations of fungicides against seed-, soil- and air-borne inoculum of cereal* pathogens (from Tomlin, 1994; Whiehead, 1995).

Fungicide (common names)	Seed and soil inoculum (mainly superficial)						Seed inoculum (deep seated)	Air-borne inoculum (external sources)		
	CS	FR	LS	NB	SB	B	LoS	LB	PM	R
Phenylmercury acetate	+	+	+	+	+	+	–	–	–	–
Fenpiclonil	–	+	+	–	+	+	–	–	–	–
Fenpiclonil and imazalil	–	+	+	–	+	+	–	–	–	–
Guazatine	–	+	–	–	+	+	–	–	–	–
Guazatine and imazalil	–	+	+	+	–	np	–	–	–	–
Carboxin and thiabendazole	–	+	–	–	–	+	+	–	–	–
Carboxin, thiabendazole and imazalil	+	+	+	+	+	np	+	–	–	–
Tebuconazole and triazoxide	+	C	+	+	np	+	+	–	–	–
Bitertanol and fuberidazole	+	+	–	–	+	+	+	–	–	–
Fuberidazole and triadimenol	+	+	–	+	+	+	+	+	+	+
Ethirimol, Flutriafol and Thiabendazole	+	+	+	+	+	np	+	+	+	+

*No distinction is made between different cereal types.

CS, covered smut (*Ustilago segetum* var. *segetum*); FR, foot rots (*Fusarium* spp.); LS, leaf stripe (*Pyrenophora graminea*); NB, net blotch (*Pyrenophora teres*); SB, seedling blights (*Fusarium* spp.); B, bunt (*Tilletia caries*); LoS, loose smut (*Ustilago segetum* var. *tritici*); LB, leaf blotch (*Rhynchosporium secalis*); PM, powdery mildew (*Erysiphe graminis*); R, rusts (*Puccinia hordei*, *P. striiformis*).

+, activity against pathogen; –, no activity against pathogen; C, activity against *Cochliobolus sativus* foot rot; np, not a problem for that cereal (barley, oats).

Strategies for the use of cereal seed treatments

Whereas organomercury compounds were cheap and were applied routinely to cereals in the UK, the new systemic fungicide formulations are more expensive, although many have additional disease control capacity. There is evidence to suggest that if cereal seeds were used without treatment, seedborne pathogens particularly, for example, *Tilletia tritici* (bunt) and *Pyrenophora graminea* (leaf stripe of barley) would rapidly increase and cause damage and loss in cereal crops throughout the UK (Yarham and Jones, 1992; Rennie and Cockerell, 1993).

The post-mercury fungicide seed treatments continue to be applied routinely to over 90% of winter wheat and barley crops in the UK; primarily to control the seedborne diseases bunt, leaf stripe and loose smut (Paveley and Davies, 1994). In Scandinavian countries, where the use of organomercury fungicides ceased some years ago, only a proportion of cereal seed is treated and seed treatment decisions are based on the results of diagnostic tests for seedborne pathogens (Brodal, 1993). Cereals (wheat and barley) are sown as certified and farm-saved seed in a number of countries (Rennie and Cockerell, 1994) (Table 18). The proportion of farm-saved seed which is sown untreated is not known. In some countries (Canada, France, Ireland and New Zealand) practically all certified cereal seed is treated. However, in Denmark, Norway, Sweden and Finland 5–30% of wheat and 5–35% of barley is sown untreated (Table 18). The decision to sow untreated seed is based on the results of seed health tests which indicate that threshold levels for seedborne pathogens have not been exceeded (Rennie and Cockerell, 1994). Sweden has specific standards for cereal seed health in its certification schemes. Norway operates a voluntary scheme in which all spring cereal seed is tested and seed treatment is discouraged if disease thresholds are not reached.

Table 18. Proportion of cereal seed sown as certified and farm-saved seed and the percentage of certified seed sown untreated (Rennie and Cockerell, 1994).

			% certified seed untreated	
Country	% certified seed	% farm-saved seed	Wheat	Barley
Canada	40	60	Nil	2
Denmark	90	60	10	10
Finland	35	65	30	30
France	50	50	Nil	Nil
Ireland	94	6	Nil	Nil
New Zealand	55	45	Nil	Nil
Norway	65	35	10	38
Sweden	55	45	5	5–35

An innate problem of seed health testing of winter cereal seed stocks is the delay in processing such material until a test result is available and subsequently the cost of rejection of some seed stocks. This would not apply to spring barley stocks where there would be ample time to process seed health tests and to identify suitable stocks for sowing without treatment (also see Chapter 9). In Scandinavia it is spring cereals mainly that are sown untreated (Rennie and Cockerell, 1994).

The sowing of untreated cereal seed is unlikely to have adverse effects on the crop yield in the UK (Richardson, 1986; Paveley and Davies, 1994). However, to avoid an increase in seedborne pathogens cereal seed should never be sown for more that two generations without treatment (Rennie and Cockerell, 1994) and seed stocks intended for further seed production should be treated (Rennie and Cockerell, 1993). A decision in favour of using untreated seed in the UK would need to be supported by a knowledge of the health status of the seed and of the implications of this (Rennie and Cockerell, 1994). Thus, unless the seed health of all such seed stocks was tested, an inevitable consequence would be the sowing of seeds with infection levels in excess of the threshold which would result in disease outbreaks in those crops and from them the spread of infection to neighbouring crops causing a general increase in certain diseases of cereals.

Seed treatments for pulses, oilseed rape and field vegetables in the UK

The temperate cereals represent a series of anatomically similar crop species (e.g. wheat, barley, oats, etc.), many of which are susceptible to the same or similar organisms facilitating the targeting of seed treatment fungicides against a recurring range of necrotrophic and biotrophic pathogens. The apparent greater diversity of seed species, shapes, sizes and pathogens represented by pulses, oilseeds and vegetables is not reflected in a greater range of seed treatment fungicides applied to these crops in the UK. Benzimidazole, dicarboximide, phenylamide and dithiocarbamate fungicides mainly are used singly or in mixed formulations involving one or more active ingredients.

The seedborne pathogens are mostly necrotrophic fungi located on and in the tissues of seed coats with limited deeper penetration in some cases, but, as with the cereals, the fungicides can also be targeted to protect against soil- and air-borne pathogens. Fully systemic fungicides such as the benzimidazole, benomyl, and the acylalanine, metalaxyl, when applied to seeds penetrate their tissues eradicating susceptible fungi (benomyl) and pass into the tissues of the emergent seedlings and young plants protecting them for a limited time from attack by soil-borne and air-borne pathogens (benomyl and metalaxyl). Dicarboximide fungicides (e.g. iprodione) have limited penetrant action and can eliminate pathogens located within the seed coat and cotyledonary tissues;

they do not protect seedlings grown from treated seeds from infection by foliar pathogens (Maude and Humpherson-Jones, 1980b).

Benzimidazoles are most toxic to the hyaline-spored Ascomycetes (e.g. *Leptosphaeria maculans*) and the Deuteromycetes (e.g. *Ascochyta pisi* (Fig. 23a,b) and *Phoma lingam*) they are not as active against certain dark-spored groups of Fungi Imperfecti, for example the porosporae which contain seed-borne pathogens including *Alternaria* and *Stemphylium* spp. (Bollen and

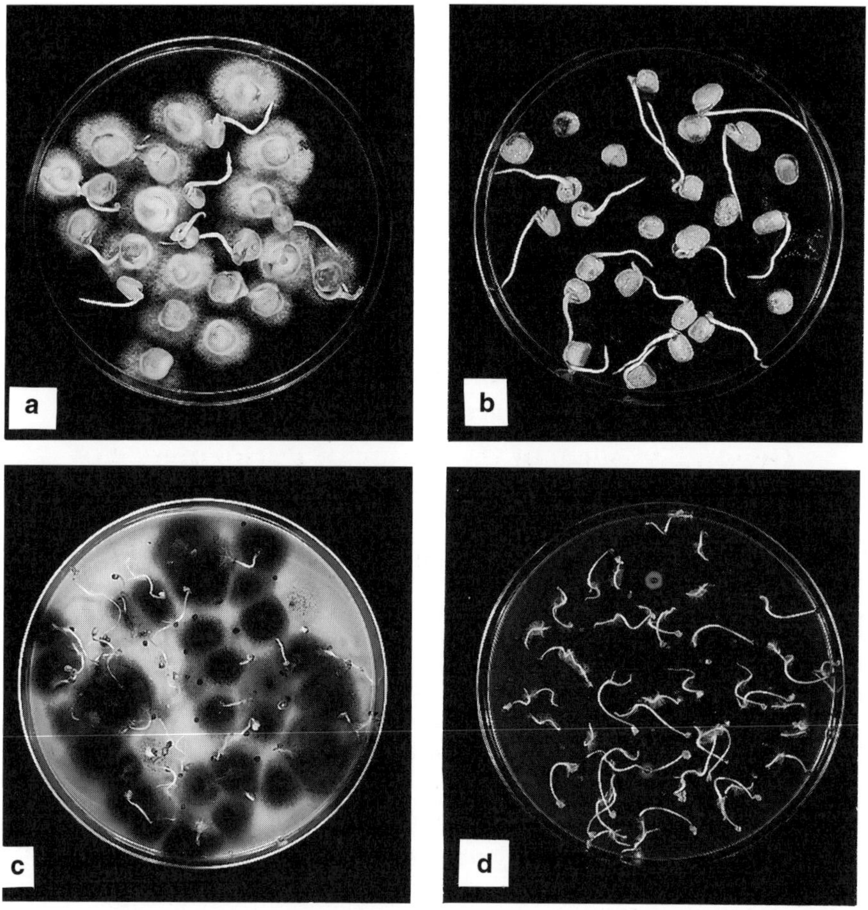

Fig. 23. Seed treatments with systemic fungicides controlling deep-seated fungal infections. (a) Untreated peas infected with *Ascochyta pisi*. (b) Peas from same sample as (a) treated with benzimidazole fungicide. (c) Untreated brassica seeds infected with *Alternaria brassicicola*. (d) Brassica seeds from the same sample as (c) treated with morpholine fungicide. Reproduced with the permission of Horticulture Research International.

Fuchs, 1970). Even greater selectivity than this may occur, for example benzimidazole fungicides may be differentially toxic to fungi within the same genus (Maude, 1986). The dicarboximides are particularly effective against dark-spored types with some activity against hyaline-spored species, and morpholine fungicides such as fenpropimorph (Fig. 23c,d) are highly active against both fungal types (Maude *et al.*, 1984; Bartlett, 1994).

Because benzimidazole and dicarboximide groups are largely ineffective against the Oomycete fungi, mainly *Pythium* spp. (Table 16), which cause damping-off under cold soil conditions, the dithiocarbamate fungicide thiram is formulated with them to extend their performance in this respect. Thiram is used in 50% of commercial seed treatment formulations (50 products) in the UK (Soper, 1991). The use of the non-systemic captan in a similar role has declined because of toxicological problems, but that of systemic acylalanine (metalaxyl), oxazole ketone (oxadixyl), phosphonate (fosetyl aluminium) and isoxazole (hymexazol) groups for the control of damping-off fungi is increasing. They are formulated with various protectant fungicides and are active against a number of soil-borne Oomycetes, including *Pythium*, *Peronospora* and *Aphanomyces* spp., attacking a wide range of crops in different countries (Tomlin, 1994). In 25% of UK seed treatment crop products (Soper, 1991), an insecticide is included to protect against attack by certain insects during emergence and the early growth of the crop.

General Aspects of the Range of Performance of Systemic Fungicides in Relation to Location of Inoculum

The foregoing account has described the target roles of systemic fungicides available for use as commercial seed treatments in the UK. A general review of their function and use which includes some information on systemic seed treatments for cereals and vegetables can be found in Marsh (1977). Their use as seed treatments for application to cereals, rice, maize, soyabean, cotton, sugarbeet and vegetable seeds has been reviewed in Jeffs (1986a) and there is more recent information on their use for the treatment of diseases of maize and soyabean (McGee, 1990, 1992). Other temperate and tropical seeds on which they have been used or tested include flower seeds, cereals, such as sorghum and millet, and the tropical food legumes, for example chickpea, pigeon pea, cowpea, black gram, mungbean, etc. There is a wealth of literature relating to their performance as seed treatments throughout the world. This cannot be reviewed in full here but a limited commentary on their functional use, on a number of temperate and tropical crop seeds, follows.

The specificity of their toxicity to certain taxonomic groups of fungi means that they can be used singly and selectively or combined with other fungicides to extend their range of activity. Fully systemic chemicals can be used to eradicate established infections in seeds and to protect the developing roots and shoots from attack by soil- and air-borne pathogens. Weakly systemic compounds have a more limited role and may be used to eliminate or reduce seedborne infections.

Eradication of seedborne pathogens

Systemic fungicides can be selected for use against biotrophic and necrotrophic pathogens located on or in the tissues of seeds (see Chapter 2, pp. 26–28) and effective elimination of these may result in the prevention of disease transmission (see Chapter 4) and thereby control of pathogens in field crops.

Superficial biotrophs

Soil-borne and seedborne teliospores of common bunt (*Tilletia tritici* and *T. laevis*) in the USA were killed by seed treatments with carboxin and thiabendazole (Hoffman, 1971). The seedborne phase of this disease is controlled by a number of chemicals, some non-systemic and some systemic (Hoffman and Waldher, 1981). In Pakistan, though seed treatment (carboxin, triadimenol, etc.) reduced disease and increased the 1000-grain weight, infection was not completely eradicated (Akhtar and Khokhar, 1987). Seed treatments containing carboxin killed the majority of seedborne teliospores of *T. controversa* (dwarf bunt) in tests in Montana, USA (Mathre *et al.*, 1990) and seed treatment is recommended as a method for controlling the movement of the disease from place to place by the dissemination of untreated contaminated seeds. In Sweden, seed treatment formulations containing bitertanol and fuberidazole gave over 90% control of seedborne inoculum of this pathogen and 93% control of soil-borne inoculum (Johnsson, 1991). By comparison, teliospores of *T. indica* (Karnal bunt of wheat) proved extremely resistant to chemical surface sterilants, fumigants and physical seed treatments (Smilanick *et al.*, 1988) and to 47 non-systemic and systemic fungicides of which, with the possible exception of mercurial compounds, none were capable of killing the teliospores when applied to infected seeds (Warham *et al.*, 1989). Very similar results were obtained in India (Aujla *et al.*, 1986) where maximum delay in germination of *T. indica* teliospores was achieved with certain benzimidazole, azole and organomercurial fungicides applied dry to contaminated seeds. Aujla *et al.* (1989) have reported more effective inhibition of teliospore germination by the application of fungicides as slurries to contaminated seeds.

Corn head smut (*Sphacelotheca reiliana*), of importance in maize crops in North America, was not controlled by seed treatment with carboxin when teliospores of the fungus were added to the soil (Stienstra *et al.*, 1985). In this situation, band application of a fungicide to the location of the inoculum, i.e. the soil, was necessary to combat the disease.

Internal biotrophs

The systemic movement and eradicant action of the phenylamide fungicide, carboxin, applied to the surface of cereal seeds against loose smut (*Ustilago segetum* var. *tritici*) inoculum located deep in the embryonic axes of barley and wheat embryos heralded the era of systemic fungicides (see p. 122). One hundred percent disease control of barley loose smut with applications of carboxin was reported (Pederson, 1982) but with some reduction in seed performance in this case; treated smutted seed germinated and grew more slowly and yielded less than healthy seeds (Pederson, 1982). However, this fungicide has been highly effective against seedborne loose smut globally (Moseman, 1968; Brooks and Buckley, 1977; Bateman *et al.*, 1986; Richardson, 1990) as has the azole fungicide, triadimenol, introduced some 12 years later in 1978, which is additionally effective against a number of other seedborne smut fungi (Moore, 1979; Bateman *et al.*, 1986; Richardson, 1990) (Table 16). However, the effect of triadimenol is temperature dependent, 100% activity was obtained at 10°C but only 66% at 25°C (Martin and Edgington, 1980). To preserve the general protectant activity of this azole fungicide and to ensure smut control under warm growing conditions, seed treatment with a mixture of triadimenol and a phenylamide fungicide is suggested (Martin and Edgington, 1980). Similar effects of soil temperature on the efficacy of seed treatment with triadimenol (plus fuberidazole, see Table 17) and other sterol-inhibiting fungicides, nuarimol and triadimefon, for the control of head smut (*U. bullata*) of prairie grass (*Bromus willdenowii*) have been reported from New Zealand (Falloon, 1988). The fungicides named controlled infection at 17.5°C and below but were less effective at temperatures above 20°C, and such fungicides may fail to control head smut where treated seeds are sown into warm soils (Falloon, 1988).

The mycelium of endophytic fungi, which is deeply located in the tissues of grass seeds (see Chapter 2, p. 17,30) is also susceptible to systemic fungicidal action. For example, *Acremonium coenophialum* in the seeds of tall fescue (*Festuca arundinacea*) was controlled by applications of the azole fungicides (triadimenol and/or triadimefon) in experiments in Argentina (Maddaloni *et al.*, 1989) and the USA (Williams *et al.*, 1982; Siegel *et al.*, 1984).

Superficial and internal necrotrophic pathogens

Most necrotrophic pathogens are located on the surface of seeds (see Chapter 2, pp. 26–28) and in the seed coat/pericarp tissues (see Chapter 2, pp. 28–31). Seed treatment of fungi in these locations using non-systemic protectant fungicides and even organomercurials reduces but does not eliminate seedborne inoculum. Such infections can be virtually eliminated by the use of weakly systemic or fully systemic fungicides.

Pathogens of field vegetables

Benzimidazole fungicides are particularly effective eradicants of the seed infections of, *Ascochyta pisi* of pea in Europe (Anselme *et al.*, 1970; Maude and Kyle, 1970; Biddle, 1981; Salter and Smith, 1986; Biddle, 1994); *A. fabae* f.sp. *lentis* (syn. *A. lentis*) of lentil (*Lens culinaris*) in germplasm accessions in the USA (Kaiser and Hannan, 1988); *A. rabiei* of chickpea (*Cicer arietinum*) (India: Grewal, 1982; Turkey: Maden, 1983; the USA: Kaiser and Hannan, 1987); *Botrytis allii* of onion in Europe (Maude and Presly, 1977; Bochow and Bottcher, 1978; Janyska and Rod, 1980; Maude, 1983a; Bottcher, 1988); and *Phoma lingam* infections of horticultural brassicas (*Brassica oleracea*) in the UK and USA (Maude *et al.*, 1973; Gabrielson *et al.*, 1977; Maude *et al.*, 1984). Seed treatments with benzimidazole fungicides have also been applied for the control of purple stain of soyabean (*Cercospora kikuchii*) in India (Agarwal *et al.*, 1974) and Japan (Fujita, 1990).

Dicarboximide fungicides, particularly iprodione (weakly systemic), have sufficient penetrant action to control seed infections caused by the seedborne dark-spored fungi *Alternaria brassicicola* (on horticultural brassicas (*B. oleracea*) (Maude and Humpherson-Jones, 1980b; Maude *et al.*, 1984)) and *A. brassicae* (on horticultural and agricultural brassicas (*B. napus*) in the UK (Maude *et al.*, 1984)) and also *A. dauci* infections of carrot seeds (Maude and Bambridge, 1991; Maude *et al.*, 1992).

Pathogens of cereals

The necrotrophic cereal fungi, which respond to seed treatment, cause seedling diseases resulting in establishment losses which are of seed- and/or soil-borne origin. Most *Drechslera* spp. (teleomorphs *Pyrenophora* spp.) which cause seedling blight are seedborne but they may also infect from soil-borne inoculum, as can the *Septoria nodorum* state of *Leptosphaeria nodorum* (seedling blight) and *Rhynchosporium secalis* (scald). In the past such diseases were controlled or reduced by the use of organomercurials. In the UK, these have been replaced by the fungicides shown in Table 17 and the range of products available in other countries is commented on by Bateman *et al.* (1986). Blast (the *Pyricularia oryzae* state of *Magnaporthe grisea*) is the most important seedborne disease of rice in temperate and tropical regions of Asia, North and South America, and West and East Africa (Nakamura, 1986). In Japan, rice seed is treated for control of seedborne inoculum of *P. oryzae* (rice blast), *Fusarium moniliforme* (anamorph of *Gibberella fujikuroi* – Bakanae disease) and *Bipolaris oryzae* (anamorph of *Cochliobolus miyabeanus* – leaf spot) using benomyl plus thiram and thiophanate methyl plus thiram seed treatment formulations (Nakamura, 1986). The formulations which are applied topically to seeds also give some control of the white tip nematode (Nakamura, 1986). These seed treatment mixtures and benomyl on its own have also been applied as seed soaks for the control of the same organisms (Fujii, 1983). The systemic fungicide pyroquilon (Schwinn *et al.*, 1979) is reported to give more effective eradication of rice blast than benomyl when tested as a seed treatment (Prabhu, 1985). This fungicide's activity, however, is limited to *P. oryzae* (Schwinn *et al.*, 1979). Outbreaks of seedling blight of sorghum, caused by *Gloeocercospora sorghi*, can be severe in India and Japan (Mathur *et al.*, 1987). Inoculum occurs as minute sclerotia (100–200 μm in size) in the endosperm of seeds; it is controlled by seed treatment with benzimidazole fungicides (carbendazim, thiabendazole) (Mathur *et al.*, 1987).

Control of soil-borne pathogens

Non-systemic fungicides, such as thiram, are applied to seeds to protect them from soil-borne fungi which cause pre- and post-emergence damping-off and the death of seedlings of many crops during crop establishment. In such cases the seeds act as vehicles which convey fungicide into the soil where it is toxic to organisms in the rhizosphere immediately surrounding them. Doses of 1.5–5.0 g a.i. (active ingredient) kg^{-1} seed are used to obtain short-term protection for vegetable seeds against these organisms (Maude and Thompson, 1989). For longer term protection against soil-borne pathogens capable of attacking throughout the growth of the field crop, it is necessary to increase the rate of application of chemical to seeds; in the past, 250–500 g calomel kg^{-1} seed was applied to onion seeds to protect the green crop against onion white rot (*Sclerotium cepivorum*) (Maude, 1986). Systemic fungicides applied to control organisms adversely affecting establishment may act from the seed surface against soil-borne pathogens but they also penetrate the tissues of germinating seeds and emerging seedlings where they may also function against such invaders.

The choice of systemic fungicides depends very much on the taxonomy of the pathogen (Table 16).

Pre-emergence, post-emergence and crop pathogens

Pythium, *Phytophthora* and *Aphanomyces* spp. are soil-borne Oomycete fungi causing pre- and post-emergence losses in many crops. *Pythium* spp., *A. euteiches* and *Fusarium solani* f.sp. *pisi* singly or in combination cause root rot of pea (Harper, 1968). Early in the study of systemic fungicides it was discovered that seed treatment with benzimidazole fungicides did not control *Pythium* spp. causing pre-emergence damping-off of pea (Harper, 1968; Crosier *et al.*, 1970). The acylalanine and phosphonate groups of chemicals are selective for these fungi (see Table 16).

Seed treatment with metalaxyl effectively controls *Pythium* damping-off of pea (Paulus and Nelson, 1977; Locke *et al.*, 1979; Kraft, 1982). Seed treatment mixtures of metalaxyl with chloroneb (a systemic fungicide highly fungistatic to *Rhizoctonia solani* (teleomorph *Thanatephorus cucumeris*) (Worthing and Hance, 1991)) have been used successfully to control a damping-off disease complex of cotton caused by *P. ultimum* and *R. solani* (Schulteis, 1989)

In addition to controlling *Pythium* spp. which attack during germination, seed treatment with metalaxyl is effective against *Pythium* spp. (cavity spot) infections of carrot roots especially during the early stages of crop growth (White, 1986; Petch *et al.*, 1991a). A seed treatment formulation of mancozeb and metalaxyl is in commercial use in the UK for that purpose (Whitehead, 1995).

The acylalanine and phosphonate fungicides are active against a wider range of soil-borne Oomycetes than thiram and species not affected by that fungicide such as the *Aphanomyces* spp., for example *A. euteiches* which causes root rot of pea and *A. cochlioides* which causes black root of sugarbeet are respectively controlled by seed treatment with fosetyl aluminium (Oyarzun *et al.*, 1990) and the isoxazole fungicide hymexazol (Tsvetkova and Guseva, 1980; Payne and Williams, 1990). *A. euteiches* f.sp. *phaseoli* causing root rot of bean (*Phaseolus vulgaris*) in New York State was not controlled using metalaxyl, and seed and soil treatment with the non-systemic fungicide fenaminosulf was needed to achieve control (Dillard and Abawi, 1988). The lack of effect of metalaxyl is hardly surprising since this fungicide can be used in selective media for the isolation of *A. euteiches* (Tu, 1988).

Rhizoctonia spp. are soil-borne Basidiomycetes which cause damping-off and stem base rotting (wirestem) of numerous plant species in many different climatic regions of the world. *R. solani* (teleomorph *Thanatephorus cucumeris*, Basidiomycetes) attacks a wide range of crop plants including bean, beet, brassicas, potato and tomato as well as many herbaceous ornamental plants (Hide, 1988). In peas it produces a stem base rot in conjunction with the root rot fungi (*Pythium* and *Aphanomyces* spp.) causing a disease complex during crop establishment (Harper, 1968). *Macrophomina phaseolina* (syn. *R. bataticola*) attacks many plant species, especially in hot countries favoured by temperatures over 28°C (Smith, 1988). These fungi are mainly soil-borne and can be controlled by seed treatment with phenylamide, dicarboximide or benzimidazole fungicides.

Seed treatment with carboxin was more effective than similar treatments with conventional fungicides in preventing seedling infection by *R. solani* of phaseolus beans, radish (Beyries, 1969) and peas (Amin, 1981). However, Kataria and Grover (1978) found that seed treatments using the benzimidazole fungicides were more effective than carboxin or chloroneb in protecting mung bean from attack by *R. solani* in glasshouse experiments. Similar results were obtained in India in the control of *R. solani* seedling infection of cotton (Alagarsamy and Jeyarajan, 1989), but in Zimbabwe field infection of that crop was more effectively controlled by seed treatments with the non-systemic fungicides tolclofos-methyl and pencycuron plus captan than by treatments with the systemic compounds carboxin and benodanil (Hillocks *et al.*, 1988). Pyrimidine and azole fungicides were not toxic to *R. solani* when applied as seed treatments to rape seed (*Brassica campestris* and *B. napus* cultivars) grown in an inoculated soil-free medium in a constant environment (Katari and Verma, 1990). Azole fungicides, though fungistatic, were not sufficiently long lasting to protect roots against infection and phenylamide fungicides also were quickly broken down in the host tissues. Benzimidazole fungicides were not particularly effective in this test and the best control of pre- and post-emergence damping-off was obtained with the dicarboximide fungicide, iprodione, and with tolclofos-methyl (an aromatic hydrocarbon), neither of which were systemically translocated to the foliage but were capable of accumulating in the hypocotyl and roots thus providing long-term protection (Kataria and Verma, 1990).

Benzimidazole-based fungicides did prove effective in India and Senegal, West Africa as seed treatments for the control of *Macrophomina phaseolina* root and damping-off infections, of tomato, French bean and egg plant (Vyas *et al.*, 1981), cowpea (Gaikward and Sokhi, 1987), chilli (Satija and Hooda, 1987), and black gram (*Vigna munga*) and green gram (*V. radiata*) (Agrawal and Nema, 1989).

The efficiency of seed treatment fungicides may be modified by environmental factors. Thus thiophanate methyl and carbendazim seed applications were more effective at 20°C than at higher temperatures in controlling *M. phaseolina* infections of mungbean (Hooda and Grover, 1990). Similarly the clay content of soil reduced the efficacy of carbendazim used as a seed treatment for the control of this pathogen causing dry root rot of chickpea (Taya *et al.*, 1990).

Fusarium solani and *F. oxysporum* are soil-borne Hyphomycetes, some of which have specialized races. *F. solani* causes damping-off and foot rots in temperate parts of the world. Less specialized forms of *F. oxysporum* cause seedling blights and rots but the specialized forms are more important causing vascular wilts in temperate and tropical climates. Generally, seed treatments with benzimidazole-based systemic fungicides protect against infection from soil-borne *F. solani* sp. and *F. oxysporum* sp. (Russell and Mussa, 1977; Utikar *et al.*, 1978; Shuckla *et al.*, 1979; Gambogi, 1983; Maude, 1986). The benzimidazoles, in addition to being effective in seed treatment applications against individual soil-borne *Fusaria* spp. (Kotasthane *et al.*, 1987; Goncalves *et al.*, 1991), are also toxic, as is the phenylamide fungicide carboxin, to some fungal complexes in the soil (Gupta *et al.*, 1989). For example, field infections of cotton caused by *F. oxysporum* f.sp. *vasinfectum*, *M. phaseolina*, *R. solani* and *F. solani* are controlled by seed treatment with carbendazim or carboxin followed by a soil drench of either fungicide (Chauhan *et al.*, 1988).

Similarly, benzimidazole-based fungicides are effective as seed treatments for the control of infections of pea and soyabean caused by *R. solani*, *F. solani* and *Sclerotium rolfsii* (teleomorph *Corticium rolfsii*) (Neweigy *et al.*, 1985).

They were less successful against *S. rolfsii* alone, which was controlled by seed treatment with carboxin (Neweigy *et al.*, 1985).

Seed treatment with systemic fungicides such as chloroneb, which is effective against a number of damping-off organisms (Worthing and Hance, 1991), has protected against root rots caused by *Fusarium* and *Rhizoctonia* spp. affecting cucumber and tomato in protected cropping systems (Saifutdinova, 1985).

Crop diseases of winter wheat in cold latitudes (43° and 46° north) caused by soil-borne inoculum of *Typhula ishikariensis* (typhula blight) and *F. nivale* (teleomorph *Monographella nivalis*) (pink snow mould), which produce considerable plant losses in some years, can be controlled by seed treatment mixtures of triadimenol and thiabendazole (Lawton and Burpee, 1990). Likewise, seed treatment with benomyl (3.0 g a.i. kg^{-1}) or with triadimenol (0.3 g a.i. kg^{-1}) reduced the incidence of *Gaeumannomyces graminis* var. *tritici* (take-all) and improved the yield of soft white spring wheat (*Triticum aestivum*) in southern Alberta, Canada (Conner and Kuzyk, 1990).

Control of foliar pathogens

The physiological and chemical requirements for the movement of systemic fungicides within plant tissues has been reviewed by Edgington (1981). Many move upwards with the transpiration stream in the dead tissues of the xylem (apoplastic movement). Some may move within the living cells of the phloem (symplastic movement) with the photosynthates to accumulate in roots or shoots, or they may move in both directions (ambimobile) (Edgington, 1981). When applied to the surface of seeds, systemic fungicides in addition to acting directly and externally against soil-borne fungi may also be translocated internally in the seedling tissues via the hypocotyl to root or shoot to control soil and foliar pathogens. Most systemic fungicides when applied to seeds move apoplastically to accumulate in the cotyledons and leaves with lesser quantities remaining in the roots. Some, for example the dicarboximides, are ambimobile (Edgington, 1981). However, iprodione when applied as a seed treatment shows little systemic movement to aerial parts to counteract foliar pathogens, but it accumulates and provides protection in the hypocotyl and roots of plants (Kataria and Verma, 1990).

Depending on the dose rate, systemic fungicides applied to seeds can be recovered from the tissues of plants over several weeks or longer periods of time. From seed treatment applications (rate *c.* 1 g a.i. kg^{-1}), carbendazim was detected for up to 21 days in the above ground tissues of rice plants (Parida *et al.*, 1990) and for 30 days in those of the urd bean (*Vigna mungo*) (Chatrath *et al.*, 1984) and chickpea (*Cicer arietinum*) (Ramakrishna and Chatrath, 1987). From high dose rates (250 g a.i. kg^{-1} seed) it was possible to detect carbendazim in the leaves of overwintered onions for at least 8 months (Presly and Maude, 1980). Likewise, metalaxyl was detected for up to 20 days in the tissues of pea seedlings grown from treated seeds (Singh *et al.*, 1985), but in cowpea metalaxyl (less than 1%) was detected 50 days after seed treatment (Singh, 1989). In cotyledonous seeds such as the pea (metalaxyl) and chickpea (carbendazim), most fungicide was recovered from the cotyledons and roots and least from the stems and leaves (Singh *et al.*, 1985; Ramakrishna and Chatrath, 1987). In cowpea, however, most metalaxyl was recovered from the cotyledons and leaves and not from the roots (Singh, 1989).

Not all systemic fungicides retain their fungitoxicity in plant tissues, the phenylamide carboxin is relatively quickly oxidized to the non-toxic sulphoxide (Snel and Edgington, 1970; Briggs *et al.*, 1974).

Those which accumulate and remain toxic in the above ground tissues of plants grown from treated seeds provide protection against foliar pathogens for different lengths of time. Field levels of inoculum and artificially applied inoculum may produce quite different constraints on dose rate and duration of the performance of seed treatments with systemic fungicides.

In some cases long-term protection may be achieved with high doses of fungicide provided they are not toxic to the seeds. Protection of green onions from the *Botrytis cinerea* (collar rot, grey mould) was obtained for at least 8 months in overwintered crops by treating the seeds with carbendazim at 250 g a.i kg^{-1} seed (Presly and Maude, 1980). In pot tests, powdery mildew (*Erysiphe polygoni*) was controlled on the foliage of pea for 9 weeks by seed treatment with benomyl at concentrations of 0.3–0.5% (Jhooty and Behar, 1972). At seed treatment rates of 62.5–250 g metalaxyl kg^{-1} seed, Crute (1980) reported control of lettuce downy mildew (*Bremia lactucae*) in inoculation tests for up to 4 weeks but not beyond.

Against field inoculum pressure, much lower rates of application of metalaxyl (0.5 g a.i. kg^{-1} seed) applied to pea seeds halved the severity of leaf infection due to downy mildew (*Peronospora viciae*) over a 12-week period (Brokenshire, 1980) and low levels of primary foliar infection were virtually eliminated by seed treatment (0.7 g a.i. kg^{-1} seed) (Miller and de Whalley, 1981). At similar rates, seed treatment with metalaxyl + ziram (zinc ethylenebis dithiocarbamate) was sufficient to control sorghum downy mildew (*Pereno-sclerospora sorghi*) at low disease pressure (Anahosur and Lakshman, 1989) but with some adverse effects on seed germination (Lakshman and Mohan, 1989).

Powdery mildews, particularly those infecting the foliage of cereals, for example *Erysiphe graminis* f.sp. *hordei* of barley and *E. graminis* f.sp. *tritici* of wheat, have been controlled by seed treatment with systemic fungicides. The pyrimidine fungicide ethirimol (see Tables 15 and 16) was one of the first successfully used in the 1970s to control powdery mildew of barley by a single application to seeds for the duration of a growing season with consequent improvements in plant vigour and yield (Brooks, 1970, 1972; Hall, 1971; Finney and Hall, 1972; Jenkyn, 1974). In 1981, superior control of barley powdery mildew by seed treatment with the systemic azole fungicide triadimenol, at an extremely low rate of active ingredient (0.375 g a.i. kg^{-1}), was reported in the UK (Martin, *et al.*, 1981). In the USA, triadimenol seed treatment successfully controlled powdery mildew of winter wheat with yield increases in two out of three years (Frank and Ayres, 1986), but in subsequent work greater improvements in yield and control of the same disease are claimed for the use of a single foliar application (Lipps and Madden, 1988) or for the combination of seed treatment followed by a foliar spray (Christ and Frank, 1989). Seed treatment with triadimenol is reported to reduce the incidence of brown and yellow rusts on the foliage of cereals (Martin *et al.*, 1981; Christ and Frank, 1989).

Resistance to Fungicides Applied as Seed Treatments

There have been many reviews on the mode of action of systemic fungicides and of the development of resistance to them (Dekker, 1977; Edgington, 1981; Langcake *et al.*, 1983; Corbett *et al.*, 1984; Trinci and Ryley, 1984; Hassall, 1990) including a detailed account in these respects of the benzimidazole group (Davidse, 1986). Reviews of the development of resistance in non-systemic compounds have been published for inorganic fungicides (Ashida, 1965) and for organic fungicides (Georgopoulus and Zaracovitis, 1967). Most non-systemic chemicals used in seed treatment are multisite inhibitors, i.e. total cell poisons, and for that reason resistance has developed rarely to these conventional fungicides. Even over a long period of commercial use the number of cases in which the emergence of resistance has been of practical significance is small (Dekker, 1977).

One notable exception was the development of resistance in *Pyrenophora avenae*(oat leaf spot) to organomercury applied as a seed treatment in Scottish seed oats in the 1960s (Noble *et al.*, 1966). Mercury resistance was found to be widespread in seed oats and was reported from Northern Ireland (Malone, 1968), in samples from England and Wales and the Netherlands (Dickens and Sharp, 1970), also from New Zealand (Sheridan *et al.*, 1968; Sheridan, 1971) and other places (Richardson, 1990). More recently, the emergence of resistance to organomercury (seed treatment application) in *P. graminea* (leaf stripe) of barley in Scotland and England (Jones *et al.*, 1989) led to a significant disease outbreak in Scotland in 1990, causing losses estimated at approximately UK £2 million (W. J. Rennie, pers. comm.), which resulted in the introduction of a code of practice by seed merchants and farmers for the control of the seedborne phase (Anon., 1991).

Most systemic fungicides affect only a single site or process within a target pathogen to achieve selective toxicity (Edgington, 1981). For example, carbendazim, the active principle of benomyl, possesses selective activity for the microtubulin protein of fungi but not for plant microtubulin protein (Davidse, 1973). It is this site- or process-specific mode of action which distinguishes the systemic fungicides from conventional protectant fungicides and which makes systemic chemicals more vulnerable to the development of fungicide resistance (Dekker, 1977).

There are exceptions, however, with those systemic chemicals which are not directly toxic to fungi but which alter the host plant physiology to counteract infection, e.g. fosetyl aluminium (aluminium triethyl phosphonate) (Edgington, 1981).

Fungicide-insensitive forms of fungi may occur naturally or arise by mutation within a field population of that particular organism. In either case, they are present initially at extremely low frequencies within the population. They can increase to dominant proportions as a result of selection pressure resulting from repeated applications of the same fungicide over a period of time (Dekker, 1977).

Seed application of a systemic fungicide to eradicate a seedborne pathogen is direct and final; the fungicide is unlikely to select quickly for resistance but where the use is to control soil-borne and particularly foliar pathogens then fungicides within the tissues of plants may be present at sublethal or lethal levels, providing continuous selection pressure over a sufficient period of time for more rapid elicitation of low or high level resistance in these pathogens.

A case in point was the use of a systemic fungicide in seed treatment applications to control powdery mildew (*Erysiphe graminis* f.sp. *hordei*) on the foliage of barley. Ethirimol, introduced at the beginning of the 1970s, was used extensively in the UK for this purpose. By 1973 Wolfe and Dinoor (1973) had established an increase in frequency in insensitive forms of the fungus by monitoring field populations of *E. graminis* f.sp. *hordei* on a large scale. This was confirmed by Shephard *et al.* (1975) in samples taken from 300 barley fields in England and Scotland. Although loss of disease control was not apparent, the greatest levels of insensitivity occurred where seed treatments with ethirimol had been used for the longest time and the use of these on winter barley was suspended to provide a break to allow restitution of a balanced population of isolates (Shephard *et al.*, 1975). With the introduction of other ergosterol biosynthesis inhibitors, for example the azole (formerly conazole, triazole) fungicides triadimefon and triadimenol, population shifts towards increased insensitivity of the fungus occurred (Fletcher and Wolfe, 1981); insensitivity to triadimenol continued whilst it was being used as a seed treatment from 1981 to 1984 (Wolfe *et al.*, 1984). This and related work emphasized the need for growers to diversify between fungicides having different modes of action both in space and time (Wolfe *et al.*, 1984; Bolton and Smith, 1988). The increase in insensitivity to azole fungicides continued between 1984 and 1986 (Heaney *et al.*, 1986) although mildew control was not reduced using triadimenol seed treatment. However, the introduction in 1985 of ethirimol formulated with flutriafol (azole group) in a seed treatment mixture apparently introduced sufficient diversity in the mode of action to give extremely effective control of powdery mildew of barley without measurable shifts in sensitivity at that time (Heaney *et al.*, 1986). The dual mode of action of the azole, flutriafol and the pyrimidine, ethirimol (to which thiabendazole was added to control *Fusarium* spp.) has continued to give effective control of powdery mildew in spring and winter barley cultivars without a detectable change in the sensitivity of the fungus to the fungicide mixture (Noon, 1986; Stott *et al.*, 1990).

Where the action of systemic fungicides is directed at seedborne inoculum, and this is virtually eradicated, then the selection pressure put by the toxicant on the pathogen is largely discontinuous. As a result, the integrity of the fungicide can be maintained for a considerable time without the development of insensitive forms. For example, seed treatment control of seedborne *Ustilago segetum* var. *tritici* with the phenylamide fungicide carboxin remained effective for some 20 years before resistance to the fungicide was discovered by French workers (Leroux, 1986; Leroux and Berthier, 1988). Strains of phenylamide (carboxin and fenfuram)-resistant *U. segetum* var. *tritici* occurred in the winter barley cvs Panda, Viva, Gerbel and Barberusse (Leroux, 1986; Leroux and Bertier, 1988) in France. Disease control was restored by using seed treatment fungicides with a different mode of action (e.g. the ergosterol biosynthesis inhibitors, flutriafol and triadimenol) (Leroux and Berthier, 1988) but in the meantime there have been associated problems elsewhere arising from the loss of disease control with carboxin-treated barley, cv Panda, imported from France into the UK (Locke, 1986) and in stocks of cvs Panda and Gerbel imported into Ireland (Dhitaphichit and Jones, 1991).

A gene which confers resistance to carboxin has been isolated from the maize smut pathogen, *U. zeae* using gene transfer techniques (Keon *et al.*, 1992).

Once treated seeds are sown, systemic fungicide on the seed surface and within the underground parts of the developing seedling can exert a selection pressure on soil-borne fungal inoculum over a period of time. Such exposure has resulted in the emergence of resistance in soil-borne pathogens. The occurrence of insensitive forms of *Monographella nivalis* resulting from seed treatments with benzimidazole fungicides is reported widely from Europe (Olvang, 1984; Harkte and Buchenauer, 1985; Locke, 1986).

Accelerated Degradation of Seed Treatment Chemicals in the Soil

Accelerated microbiological degradation of soil-applied pesticides, fungicides (Woodcock, 1978), herbicides and insecticides, is a well-documented phenomenon (Felsot, 1989; Chapman and Harris, 1990; Racke and Coats, 1990; Suett, 1990, 1991). It results from repeated applications of the same pesticide to soil which selects for microorganisms capable of rapidly metabolizing the chemicals, causing a decline in persistence of the chemical with a concomitant loss in biological efficacy. The decline in biological efficacy occurs usually over several years depending on the number of applications and the structure of the chemical that was used (Suett, 1991).

Dose rate comparisons between soil- and seed-applied treatments might suggest that the small amounts of chemicals routinely applied to seeds (< 5.0 g a.i. kg^{-1} seed) would seem unlikely to apply sufficient pressure on the soil microflora to select for microorganisms which would induce their accelerated degradation. However, where seed treatments are used at relatively high application rates on densely sown crops coupled with field applications of the same pesticide, loss of disease control has been reported. Thus seed treatment (of 50–62.5 g a.i. kg^{-1} seed) followed by one or two stem base sprays (0.15 g a.i. m^{-1} row) of the dicarboximide fungicide iprodione used repeatedly on the same land (the equivalent of 6–10.5 kg iprodione ha^{-1} at each application) resulted in failure to control soil-borne onion white rot (*Sclerotium cepivorum*) in green onions after 8 successful years of commercial use in the UK (Walker *et al.*, 1984; Entwistle, 1986). Enhanced degradation of iprodione, and of the related dicarboximide, vinclozolin, also used in the UK for the control of onion white rot, was confirmed (Walker *et al.*, 1986; Walker and Welch, 1990). Rapid degradation of iprodione can be induced with a single pretreatment of the soil at a very low concentration of fungicide and the degrading ability is easily spread from one soil to another; soil bacteria are involved in the accelerated degradation of these dicarboximide fungicides (Walker *et al.*, 1986; Walker and Welch, 1990). Loss of control of onion white rot has resulted in the withdrawal of both fungicides in the UK for this seed treatment use.

Topical Application of Fungicides to Seeds

Formulation of chemicals for seed treatment

Seed treatment fungicides, both systemic and non-systemic, are formulated in several ways for application to seeds. Five main groups are recognized in the UK (Ivens, 1994). The abbreviations are standard codes derived from the Catalogue of Pesticide Formulation Types (GIFAP, 1989).

1. Powders for dry seed treatment (DS), i.e. they contain the active ingredient plus additives to prevent the cohesion of particles and/or stickers to improve the adhesion to seeds.
2. Wettable powders (WP), i.e. they contain the active ingredient, fillers and a wetting agent for use in slurry (water and fungicide applied together or one after the other) applications to seeds.

3. Water dispersable powders (WS), i.e. fine powders (particles up to 90 μm) to be dispersed at high concentration in water for application as slurries to seeds.
4. Solutions for seed treatment (LS), i.e. fungicide solutions (non-aqueous) for application directly to seeds.
5. Flowable concentrates (FS), i.e. stable fluid dispersions (solids in liquid) for application directly or after dilution to seeds.

The advantages and disadvantages of the type of formulation for seed treatment have been presented by Godwin *et al.* (1988) (Table 19) which indicate that flowable concentrates have superior overall qualities by comparison with the other types. Fifty three percent of pesticide products listed for seed treatment use in the UK, excluding those for potatoes (Ivens, 1994), are formulated as flowable concentrates with 18% formulated as dry powders, 13% as solutions and 11% as wettable/dispersable powders; other minor formulations make up the remainder. Some formulations containing single active ingredients are available but to broaden their range of action, and possibly to

Table 19. Advantages and disadvantages of the major types of seed treatment formulation (from Godwin *et al.*, 1988).

Formulation	Advantage	Disadvantage
DS (powder for dry seed treatment)	'On farm' application possible	Reduced adhesion Reduced efficacy
WS (water dispersable powder for slurry treatment)	'On farm' application possible	Efficacy not maximized
FS (flowable concentrate)	Accurate loading Good adhesion Excellent efficacy Application machinery can be washed out No organic solvents used	Specialist machinery needed for application
LS (solution)	Accurate loading Excellent adhesion Efficacy maximized	Organic solvents present can be phytotoxic Specialist machinery needed for application
FC (formulated products applied in polymer film coats)	Accurate loading Excellent adhesion Efficacy maximized No phytotoxicity No hazards to health or environment	Specialist machinery needed for application

reduce the likelihood of selection for resistant forms of fungi, systemic fungicides have been formulated with other systemic compounds which are non-related chemically and/or with non-systemic protectant fungicides. In addition, the change in formulation technology to liquid pesticides noted by Jeffs and Tuppen (1986) has continued (Elsworth, 1988).

Formulated fungicides can also be used in alternative methods of application to seeds, for example in pelleting and film-coating (FC) treatments.

Application of seed treatments

The methods of, and the machinery for seed treatment throughout the world are reviewed in Jeffs (1986a) and advances in this area of technology in the UK are discussed by Elsworth (1988) and in Hungary by Gyurk (1988). Although similar seed treatment chemicals have been applied to both agricultural and horticultural seeds, the scale of operation is different. The large volume of agricultural seeds (e.g. cereals, pulses, oilseed rape) has necessitated machinery with a high and continuous throughput of treated seeds (up to 25 t h^{-1}) whereas horticultural seeds, which are diverse in size and shape and often of low volume but high value, have generally been treated in single quantities using batch treaters.

Seed treatment machinery

Independent machinery manufacturers are few in the UK at present and the majority of seed treaters are engineered and produced by the agrochemical industry and supplied to the seed trade, generally with the agrochemicals for particular uses (Elsworth, 1988). Seed treatment machinery may be static and located at seed houses or it may be mobile and transported onto farms by contractors who treat on site. Cereal seeds in the UK are treated using both systems. In Australia, 79% of all cereal seed is treated using mobile seed treaters (Jeffs and Tuppen, 1986).

Jeffs and Tuppen (1986) divide the types of machine used throughout the world for treating seeds into eight groups based on the method of applying the formulation to the seed and the mixing process. The main types of machine recently available in the UK and to an extent elsewhere in the world (Jeffs and Tuppen, 1986) apply pesticides by: (i) drum mixing; (ii) auger mixing; and (iii) spinning disc distribution.

Drum mixing

Not many drum treaters are still used for the treatment of cereals in the UK (Elsworth, 1988). Their main use was for the application of liquid formulations of organomercurials to seeds. These chemicals have been banned for seed treatment use in Europe. A measured dose of liquid chemical was applied to a batch of seed as it passed along a slightly inclined revolving drum. The

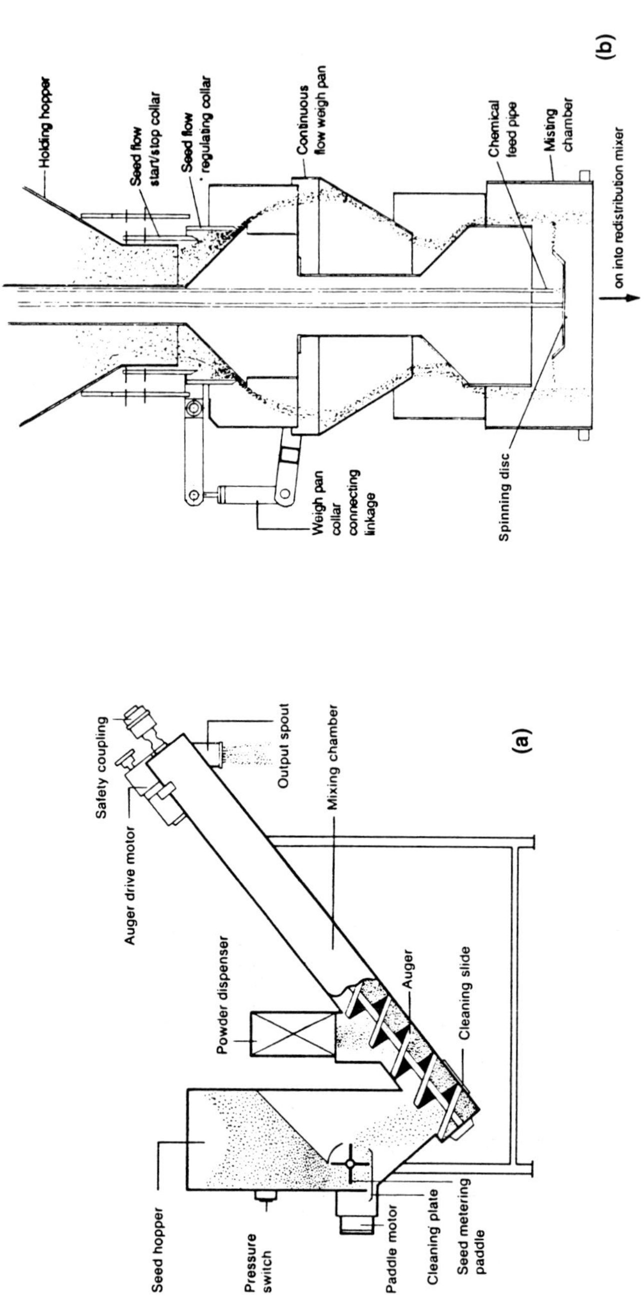
Holding hopper
Seed flow start/stop collar
Seed flow regulating collar
Continuous flow weigh pan
Chemical feed pipe
Misting chamber
(b)
on into redistribution mixer
Spinning disc
Weigh pan collar connecting linkage
Safety coupling
Auger drive motor
Output spout
Mixing chamber
(a)
Powder dispenser
Auger
Cleaning slide
Seed hopper
Pressure switch
Paddle motor
Cleaning plate
Seed metering paddle

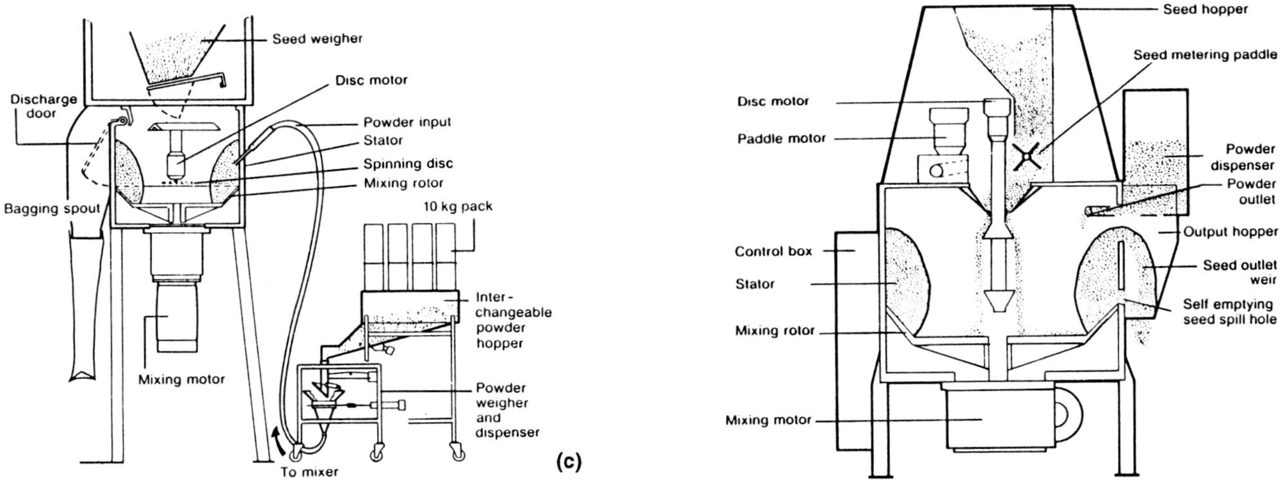

Fig. 24 Sectional diagrams of agricultural seed treaters. (a) Plantector; (b) Mist-O-Matic; (c) Rotostat; (d) Centaur. (a, c and d are reproduced from Pflanzenschutz-Nachrichten Bayer 34/1981, 3, with permission of Bayer PLC; b is reproduced by courtesy of DowElanco.)

treated seeds passed into bins or bags. Drum treaters are used for the application of pesticides in powder form to vegetable seeds (Maude, 1986) and for treating soyabean seeds (McGee, 1986).

Auger mixing

Grain is discharged from a hopper metered by paddles passed along an inclined auger conveyor. A powder dispenser is activated when grain is released and mixing of the pesticide and seed occurs as the seed moves along the auger to be discharged into bins or bags. Auger treaters (Fig. 24a) can be modified to apply slurry and flowable formulations of pesticides (Jeffs and Tuppen, 1986; Elsworth, 1988). Drill box treaters sometimes make use of the auger principle for treating and loading the treated grain into drill boxes for sowing (Jeffs and Tuppen, 1986).

Spinning disc distribution

In principle liquid formulations of pesticide are delivered to a spinning disc from the edge of which they are dispersed by centrifugal force as filaments which break up into fine droplets and are uniformly distributed onto seeds. The method by which the seeds are presented to the pesticide 'spray' is part of engineering technology and serves to identify the main treaters available for commercial use in the UK (Jeffs and Tuppen, 1986; Elsworth, 1988) and elsewhere (Jeffs and Tuppen, 1986). For example, with continuous treaters in the Misto-O-Matic range (Fig. 24b), cereal, vegetable (Jeffs and Tuppen, 1986) and soyabean seeds (McGee, 1986) are treated by falling through a peripheral curtain of pesticide spray. Otherwise, seeds constantly circulated in the format of a 'toroidal doughnut' on the sides of the treatment chamber receive a coarse spray of pesticide from a rotary atomizing disc (Jeffs and Tuppen, 1986). This system of application was introduced with the Rotostat-type of batch treater (Fig. 24c – up to 50 kg seed treated per batch) for cereals and has been developed further as a machine (Centaur, Fig. 24d) for continuous treatment (Jeffs and Tuppen, 1986; Elsworth, 1988).

Film coating

The technology of film coating seeds has developed over the past decade in the UK (Maude and Suett, 1986; Martin, 1988) and in Europe and North America (Halmer, 1988, 1994). Coating is a general term which in its broadest sense includes any process for the addition of materials to seeds (Taylor and Harman, 1990). Film coating, in its most developed form, is the application to seeds of a thin durable water-permeable coat (usually a polymer binder) which can be used to contain pesticides and other products securely so that they can have maximum effect when the seeds are sown (Maude, 1990c).

Seed shape is not changed by film coating but seed weight can increase 1–10% (Halmer, 1988) and, in terms of added material, film-coated seeds lie between dust- or slurry-treated seeds and pelleted seeds (Clarke, 1988). Liquid seed treatments applied by conventional seed treatment machinery, as described on pp. 143–146, rarely increase the seed weight by more than 1% and the liquid carrier evaporates rapidly making drying of the seeds unnecessary. Larger amounts of liquid are used in film coating and the polymer film has to be dried onto the seeds to complete the process. Film coatings may be applied to seeds in fluidized beds (Fig. 25A) or in revolving perforated bowls or drums (Fig. 25B) (Clarke, 1988, Halmer, 1988); common to both systems is the use of warm air (*c.* 30°C) to evaporate the aqueous phase of the binder, causing it to adhere to the seeds.

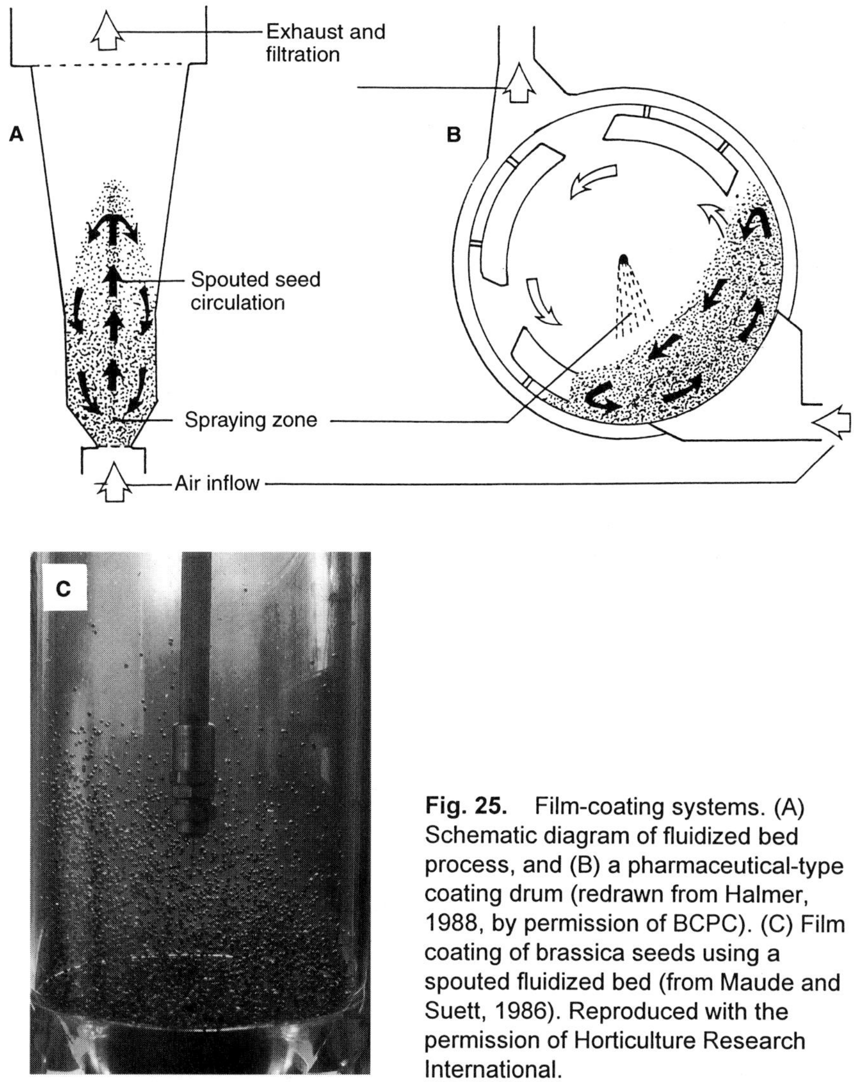

Fig. 25. Film-coating systems. (A) Schematic diagram of fluidized bed process, and (B) a pharmaceutical-type coating drum (redrawn from Halmer, 1988, by permission of BCPC). (C) Film coating of brassica seeds using a spouted fluidized bed (from Maude and Suett, 1986). Reproduced with the permission of Horticulture Research International.

Film-coating methods have been adopted and adapted for commercial use by seed companies particularly for the treatment of horticultural and agricultural (e.g. vegetable and oilseed rape) seeds (Soper, 1991). Formulated pesticides can be applied at low doses (up to 5 g kg^{-1} seed) and high doses (up to 160 g kg^{-1}) in these enclosed systems that present no hazards to the operator, and because pesticides are bound onto seeds virtually none falls off after treatment to cause environmental contamination; treated seeds also flow easily in drills. Different pesticides can be applied in a single film to seeds or potentially in separate layers of polymer (Bacon and Clayton, 1986). However, emergence problems can be caused where successive layers carry high rates of fungicide (Petch *et al.*, 1991a). The general advantages and disadvantages of film-coating methods of seed treatment have been discussed (Bacon and Clayton, 1986; Maude, 1990c).

Fluidized beds. Fluidized bed techniques are based on the Wurster process (Wurster, 1959; Hall and Pondell, 1980) where seeds are circulated vertically and continuously in moving columns of warm air in cylindrical chambers and at the same time are sprayed from below (Halmer, 1988; Horner, 1988) or above (Maude, 1990c) with aqueous binder, which dries rapidly onto the seed (Fig. 25A). Different types of distribution plates in the base of treatment chambers create different fluidization patterns, for example where air is channelled through a conical distributer, seeds are cascaded upwards in the form of a spout and this called a spouted bed (Fig. 25A,C) (Nienow *et al.*, 1991), whereas where a fully perforated plate is used, a more regular distribution of air and seeds is achieved.

Fluidized beds are versatile, treating 0.5–50 kg of various seed species in single batches (Halmer, 1988). The system has also been used for semi-continuous seed treatment using the 'SHR' (spraying, homogenizing, redrying) method (Horner, 1988), based on a carousel mechanism where seeds are moved along to receive coatings at successive fluidization stations with seed treatment capacity of up to several hundred kilograms per hour (Horner, 1988).

Perforated bowls or drums. Seeds are sprayed with liquid binder in rotating perforated bowls or drums (Fig. 25B) through which warmed air is drawn in the course of treatment. This is the principle on which pharmaceutical equipment is based for the coating of tablets, etc. (Halmer, 1988). Quantities of up to 100 kg can be batch-treated at a time. Recently, a drum system with a throughput of 4–6 t has been developed by a seed company in the UK (Maude, 1990c).

Film-coating adhesive. The chemistry of the adhesives (binders) used commercially in film coating is often a closely guarded trade secret. There are many products available (Hall and Pondell, 1980) including water-soluble or dispersible polysaccharides and synthetic polymers which are used in industrial coating processes (Davidson, 1980). It is likely that those used in commercial practice for film coating seeds are from the latter two groups (Halmer, 1988). Water-miscible polyvinylacetate-based adhesives are examples of the polymeric binders which have been used for the application of fungicides (Sauve and Shiel, 1980; Maude and Suett, 1986) and insecticides (Nevill and Burkard, 1988) to seeds.

In low dose applications of pesticides, 0.5% polymer (w/w seeds) is sufficient to create a durable film which is not phytotoxic (Bacon and Clayton, 1986; Maude and Suett, 1986). Polymer (polyvinyl acetate) alone applied in excess of 16% by weight of seeds can reduce oilseed rape germination (Suett, 1988) and high dose applications of fungicide (50–100 g metalaxyl kg^{-1} seed) applied with 5% binder (w/w seed) in single and multilayers reduced the emergence of carrots (Petch *et al.*, 1991a).

Pelleting of seeds

The technique of pelleting seeds was developed in the USA in the 1940s and introduced into Europe some two decades later (Halmer, 1988). The main object was to build small irregularly shaped seeds into spheres facilitating precision drilling in order to achieve optimum plant stands and thereby reduce the need for thinning. It was recognized that such a system also had potential for the incorporation of pesticides (Longden, 1975). The International Rules for Seed Testing (Anon., 1985b) define seed pellets as 'More or less spherical units, developed for precision sowing, usually incorporating a single seed with the size and shape of the seed no longer readily evident. The pellet in addition to the pelleting material, may contain pesticides, dyes or other additives.'

The pelleting process involves the rolling of seeds together with fillers and binders with the gradual addition of water followed by drying to add incremental layers to the seeds until the correct size/grade of pellet is reached. Pesticides can be added discretely to different layers of the pellet or can be mixed throughout the matrix depending on the compatibility of the chemicals being used. Rotating mills, cylindrical drums or pans, conveyor belts, etc. are used to apply the coating materials which include fillers (e.g. cellulose powder, diatomaceous earth, chalk, non-ionic synthetic polymers, peat and many others (see Halmer, 1988)) and binders (e.g. calcium sulphate, starch, cellulose derivatives, polyvinyl polymers, etc.). Weight increase ratios range from 2:1 (sugarbeet, sweet corn) to 150:1 (petunia, lobelia) (Halmer, 1988); the process is a batch treatment capable of treating up to 100 kg at a time.

Sugarbeet is the main seed species treated worldwide but pesticides also are applied in pellets to a proportion of seed types in most countries. In the UK, the use of a diethyl mercuric phosphate for treatment of sugarbeet to control seedborne *Phoma betae* (teleomorph *Pleospora betae*) (black leg) has been replaced by a modified form of the thiram soak treatment (Maude *et al.*, 1969) developed by Durrant *et al.* (1988) which is applied as a steep during pelleting. The fungicide hymexazol is added to pellets in the UK and most of Western Europe for the control of damping-off caused by *Pythium* and *Aphanomyces* spp. (P. Halmer, pers. comm.). Sugarbeet pellets may also contain insecticides such as methiocarb and tefluthrin for the control of soil pests including wireworms (*Agriotes* spp.) (Durrant *et al.*, 1986; P. Halmer, pers. comm.). The incorporation of a novel systemic insecticide, imidacloprid, in sugarbeet pellets has resulted in the effective control of leaf aphids and a complex of soil pests at crop establishment (Elbert *et al.*, 1990; Schmeer *et al.*, 1990; Asher and Dewar, 1994). Additionally, as a cereal seed treatment, particularly in barley yellow dwarf virus (BYDV) high risk areas having mild winters, imidacloprid is a potential alternative to autumn insecticidal sprays for virus vector control (Schmeer *et al.*, 1990).

Certain vegetable seeds are also pelleted (Maude, 1986) and fungicides and/or insecticides are added during the bonding process (Halmer, 1988) or to the raw seed by various means before pelleting (Maude, 1986). In earlier estimates (Maude, 1986), 1–2% of all treated vegetable seeds in the UK were pelleted; in 1990 (MAFF, 1991) this had risen to 19% (by weight) with some 50% or more of lettuce, carrot, leek and onion seeds (by weight) receiving this treatment.

Seed pelleting has been reviewed in general and specific terms by a number of authors (Reid, 1970; Longden, 1975; Tonkin, 1984; Durrant *et al.*, 1986; Halmer, 1988; Scott, 1989; Maude, 1990c; Taylor and Harman, 1990).

Requirements for efficient treatment of seeds using topically applied pesticides

The main requirements for the efficient treatment of seeds are defined (Elsworth and Harris, 1973) as: (i) loading – the correct ratio of chemical to seed must be maintained within narrow limits; (ii) distribution – active chemicals must be uniformly divided between the seeds; (iii) retention – the chemicals must adhere strongly enough to seeds to avoid losses during handling following treatment; (iv) hazard – no risk to operators must be created; and (v) contamination – there should be no environmental pollution.

The advantages and disadvantages of the main seed treatment formulations are given in Table 19. A general problem with powder formulations is lack of adhesion so that the loading achieved on seeds is small although there is uniform distribution of pesticide between seeds (Lord *et al.*, 1971). The result is poor efficacy of the treatment and an environmental problem of loose pesticide in bags of seed and causing obstruction of drills (Bacon and Clayton, 1986; Koshiek and Jeffs, 1986). Liquid formulations (insecticides) applied to cereals are better retained and the average loadings of seeds are closer to target but seed-to-seed distribution of the chemical is irregular, most seeds carrying too little pesticide and some carrying enough to be phytotoxic (Jeffs *et al.*,

1968; Lord *et al.*, 1971). Where comparisons of equivalent doses of active ingredient and different formulations have been made (Maude *et al.*, 1986) the retention of thiram by pea seeds applied as a powder was 38% and as a slurry 80%. A problem of both types of formulation is the presence of inert materials. This problem is overcome in flowable concentrates where the amount of inert materials in the formulation is reduced and the accuracy of dosing increased (Godwin *et al.*, 1988; Roberts *et al.*, 1988). Accuracy of dosing and efficacy also obtains with modern liquid pesticides, but where organic solvents are used in their formulation phytotoxicity may be a problem in some cases (Godwin *et al.*, 1988). There is little published information in the public domain on the performance of these newer formulations.

Recovery of pesticides from film coat applications is high and up to 90% of the target dose (the amount of active ingredient applied per weight of seeds) has been achieved with low rate seed treatment fungicides (up to 5 g a.i. kg^{-1} seed) (Suett *et al.*, 1985) and recoveries of from 86% to 108% of the target dose of fungicide/insecticide mixtures applied by this method to peas (Suett *et al.*, 1985; Salter and Smith, 1986). In comparative application studies (Fig. 26a), dust and slurry loadings of the fungicide iprodione onto brassica seeds ranged from 63% to 81% and 71% to 81% of target, respectively, compared with 93% of target with a film coat application (Maude and Suett, 1986). Similarly, recovery of pesticide from film-coated seeds subjected to simulated rough handling by means of retention tests (Jeffs, 1974) was 86% of the target dose compared with 48% (powder) and 30% (slurry) (Suett and Maude, 1988). The film coating method gives an even distribution of pesticide between individual seeds with low coefficients of variation (c.v.) as reported by several authors (Maude and Suett, 1986; Bacon *et al.*, 1988; Halmer, 1988; Horner, 1988).

In principle, pelleting, because all materials are securely attached to the seeds, should, and does with some treatment methods, provide high recovery rates (e.g. 98%) of the active ingredient (Halmer, 1988). Seed-to-seed distribution of pesticide may not be as uniform (c.v. 23%) as it is with film coating. Other methods of pelleting may be even more variable. The main problem with this well-established commercial technology is that there is insufficient published information to gauge the reproducibility of pesticide retention in the pellets.

One direct effect of optimization of dosing by techniques such as film coating is that seed treatments are more uniformly applied (Fig. 26a) and are thereby more effective against seedborne inoculum (Fig. 26b,c) and foliar inoculum (Bacon and Clayton, 1986) so that it may be possible to reduce the rates of application (Maude and Suett, 1986). This principle may not hold, or need to hold for soil-borne pests and diseases where treated seeds act as a vehicle for the placement of chemical in the soil and the uniformity of seed-to-seed dosing is likely to be quickly negated by their exposure to fluctuating and unpredictable environmental factors although biological efficacy is unaffected (Suett and Maude, 1988).

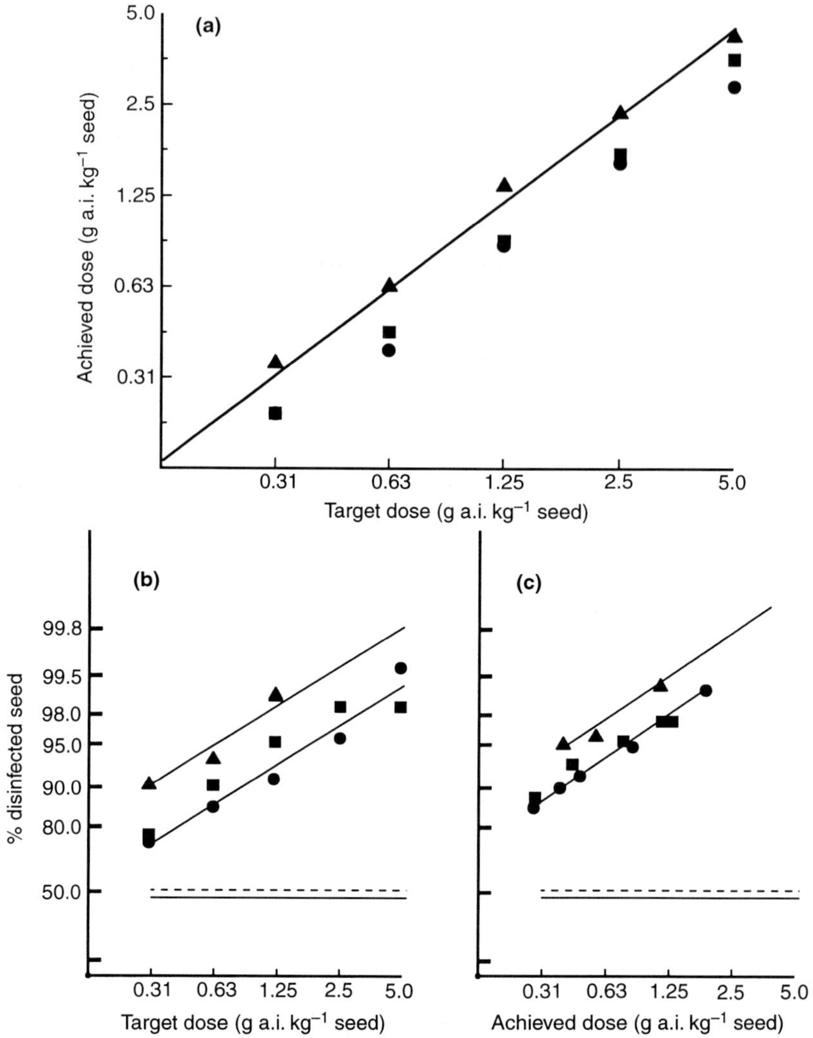

Fig. 26. (a) Comparison of target and achieved seed loadings of iprodione by different application methods; (b, c) Effect of target (b) and achieved (c) doses of iprodione on the eradication of *Alternaria brassicae* infections of cabbage seeds. (Redrawn from Maude and Suett, 1986.) Iprodione treatments: ▲, film coat; ■, slurry; ●, dust; - - - -, nil treatment; ——, film coat only.

Seed Treatment using Immersion Techniques

Seed immersion methods are those where seeds are steeped for varying periods of time in aqueous or solvent-based liquids at ambient or raised temperatures with or without the addition of chemicals to eradicate seedborne organisms. Their use has declined in practice because of the development of modern fungicides and advances in application technology. These have produced for the seed industry pesticide formulations which can be applied directly and topically for the control of seed, soil and some foliar pathogens in many crops. Seed immersion methods now may have a decreased use for the treatment of fungi, but they remain important for the treatment of seedborne bacteria and to a lesser extent seedborne viruses.

Seed soaks in aqueous fungicides

The treatment of seeds by steeping them in aqueous chemicals to control pests and diseases is of ancient origin (Martin and Woodcock, 1983). The 17th century practice of brining (see Chapter 1, p. 2) and the use of seed steeps in inorganic and organic fungicides to control seedborne pathogens have been mentioned (see p. 117). The principle underlying such techniques is that immersion of seeds in aqueous solutions or suspensions of pesticides results in partial or full hydration of the host (seed) and pathogen (bacterium, fungus) tissues making both more susceptible to the penetration of chemicals than they would have been in the dry state. Such penetration was necessary if deep-seated pathogens were to be eliminated. This approach to seed treatment was particularly relevant during the era before systemic fungicides became available for commercial use.

With the older more toxic chemicals (copper sulphate, formalin, etc.), presoaking seeds in water before submission to fungicide treatment was a method of reducing the phytotoxicity of these compounds (Braun, 1920). Otherwise, steeps in fungicides such as diethyl mercuric phosphate (EMP) for control of black leg (*Phoma betae*) of sugarbeet (Gates and Hull, 1954) were applied at a low concentration (40 $\mu g\ l^{-1}$) and for a limited time (20 min). Commercial treatment of sugarbeet by EMP steeping was unique to the UK (Durrant *et al.*, 1986) and it was practised until 1992 when the use of plant protection products based on mercury ceased (Soper, 1992).

Prolonged seed soaks in less toxic protectant fungicides have been used with some success to control internal fungal pathogens of seeds, in particular those of vegetables. Many such pathogens were virtually eliminated by soaking seeds for 24 h at 30°C in 0.2% aqueous suspensions of the dithiocarbamate fungicide, thiram (Maude *et al.*, 1969). In principle this method depended on the low solubility of the fungicide thiram (about 30 $\mu g\ ml^{-1}$ water at 30°C) which was imbibed in solution by seed tissues over the 24 h period of the soak in sufficient quantities to eradicate deep-seated fungi but without harming the seed tissues (Maude *et al.*, 1969; Maude, 1983b, 1986). At that temperature incremental increases in the water imbibed by seeds, over periods of up to 24 h, were accompanied by similar increases in disease control (Fig. 27A–C). Dissimilarities in the sensitivity of different pathogens to the treatment occur; deeper seated infections of the embryonic axes of barley by the biotrophic

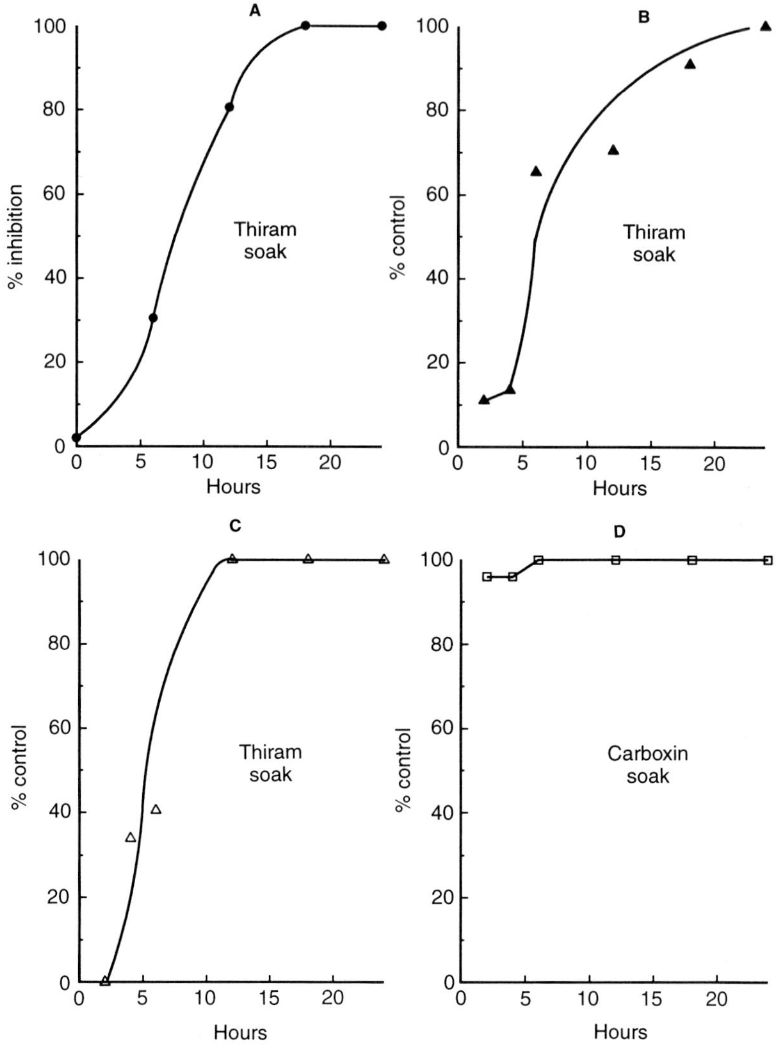

Fig. 27. Seed immersion treatments: influence of treatment, pathogen and mode of action of fungicide upon disease control. (A) Percent imbibition of seeds in aqueous thiram with time. (B) Percent control of seedborne *Drechslera avenacea* with increasing duration of immersion of infected oats in 0.2% thiram at 30°C. (C) Percent control of seedborne *Ustilago segetum* var. *tritici* with increasing duration of immersion of infected barley seeds in 0.2% thiram at 30°C. (D) Percent control of seedborne *U. segetum* var. *tritici* with increasing duration of immersion of infected barley seeds in 0.2% carboxin at 30°C.

loose smut fungus *Ustilago segetum* var. *tritici* were eliminated by 12 h immersion in aqueous thiram, those of the necrotroph *Drechslera avenacea* (teleomorph *Pyrenophora avenae*) located in the seed coats of oats took a full

24 h to control (Fig. 27B,C). A reduction in the temperature of soaking by 5°C gave reduced disease control (Maude, 1967). The rate of eradication using thiram was slow and was linked with the gradual uptake of the non-systemic chemical by the seeds; the difference in mode of action and uptake of the systemic fungicide (carboxin) applied in a similar fashion resulted in high levels of eradication of seedborne *U. segetum* var. *tritici* shortly (2 h) after application (Fig. 27C,D).

The soak treatment was not fully effective against seedborne *Alternaria brassicicola* of brassicas (Maude *et al.*, 1969; Chirco and Harman, 1979; Maude and Humpherson-Jones, 1980b), but improved control of that fungus and *Phoma lingam* was obtained by applying the treatment under partial vacuum to seeds (Harman and Nash, 1978). More recently, a modified form of the thiram soak treatment (0.2% thiram for 12 h at 25°C; Payne and Williams, 1990) has replaced the EMP steep for the control of *P. betae* in monogerm sugarbeet (Durrant *et al.*, 1988) and this modification has been incorporated into the pelleting process for that crop in the UK (Halmer, 1988). In New Zealand, the full treatment gave effective control of seedborne *A. radicina* (black rot) of carrots (Soteros, 1979b); in the USA 24 h soaks in thiram and in iprodione at 30°C gave equivalent levels of control of *A. dauci* of carrots in seedling assays based on the treatment of 13,000 seeds (Strandberg, 1984). In Finland, Tahvonen (1978) reported that thiram soaking applied for 6, 12 and 24 h controlled *A. radicina* and *A. dauci* infections of carrot seeds but that the 12 and 24 h treatments reduced the emergence of seedlings. In the UK, the thiram soak treatment is in commercial use for the treatment of red beet seed to control black leg (*P. betae*) and to improve emergence (Maude, 1973) and for the treatment of celery to control leaf spot (*Septoria apiicola*).

The advantages of the treatment are: (i) it is a non-selective treatment controlling many seedborne fungi and in addition giving protection against some soil-borne pathogens; (ii) it is a batch treatment where up to 50 kg seed can be treated at a time, compared with smaller amounts (up to 5 kg) using the hot water treatment (see pp. 162–163). Its disadvantages are: (i) it is not effective against seedborne bacteria; and (ii) after 24 h of soaking seeds need to be dried for 6–12 h. Thiram soaking and drying equipment for commercial use is available (Maude, 1986).

The aim of thiram soaking was to achieve penetration of seed tissues to eradicate internal pathogens. Except for the UK uses specified above, this treatment has been superseded by the advent of systemic chemicals which have greater inherent mobility of action and when applied to the surface of seeds can penetrate their tissues to achieve the same effects.

Seed soaks in antibiotics

Dekker (1957) tested the effect of imbibing pea seeds in antibiotics (rimocidin and pimaricin) to eradicate deep-seated fungal infections of pea seeds, and antibiotics such as kasugamycin are toxic to certain fungi and bacteria (Worthing and Hance, 1991; Tomlin, 1994). However, the main purpose of immersing seeds in antibiotics has been to control seedborne bacteria. Antibiotics applied to the surface of seeds have not been sufficiently penetrative to be effective (Ralph, 1977a) against bacteria mainly located within the seed coat tissues (see Chapter 2, pp. 28–29). As a result bacterially infected seeds have been immersed in aqueous solutions/suspensions of antibiotics or in water at high temperatures with or without the addition of various chemicals to kill or neutralize the pathogens. Because of the priority for their use in medicine, the availability of antibiotics to agriculture and veterinary science is now proscribed in many countries. Formerly, many antiobiotics (Sutton and Bell, 1954; Klisiewicz and Pound, 1961; Humaydan *et al.*, 1980) were tested against seedborne bacterial inoculum. Of these, streptomycin probably was the most successful (Ralph, 1977a).

Dose rate and duration of treatment are loosely related. Thirty minute steeps in 3000 $\mu g\ ml^{-1}$ of antibiotics, including streptomycin were used to control black rot *Xanthomonas campestris* pv. *campestris* of crucifers (Klisiewicz and Pound, 1961); 2 h soaking in 500 $\mu g\ ml^{-1}$ solutions of streptomycin was proposed by Humaydan *et al.* (1980) for the control of seedborne inoculum of this pathogen. A 2 h soak in streptomycin or kasugamycin at 1000 $\mu g\ ml^{-1}$ was suggested by Taylor and Dudley (1977) for the control of halo blight (*Pseudomonas syringae* pv. *phaseolicola*) of bean; Keyworth and Howell (1961) eradicated seedborne *Curtobacterium flaccumfaciens* pv. *betae* (syn. *Corynebacterium betae*) (silvering disease of red beet) by soaking seeds for 18 h in a 400 $\mu g\ ml^{-1}$ solution of streptomycin sulphate. A 30 min soak in 250 $\mu g\ ml^{-1}$ of streptomycin sulphate was sufficient to control *P. syringae* pv. *sesami* on sesame seed (Thomas, 1959). In Japan, streptomycin and kasugamycin soaks have been used to control *P. glumae*, the cause of grain rot of rice (Nakamura, 1986). Such treatments, although not achieving complete eradication of seedborne inoculum in all cases, gave very high levels of control.

Phytotoxicity is a major problem of soak treatments in antibiotics, particularly with the use of streptomycin. Streptomycin exerts a number of suppressive effects at cell level (Brian, 1957), including suppression of chloroplast development, which partly may be prevented or antidoted by chemical means (Gray, 1955; Ark and Thompson, 1958; Venis, 1969; McMeekin, 1973). Phytotoxicity commonly results causing bleaching of the cotyledons and stunting of seedlings; potassium, sodium and calcium salts gave some reversal of toxicity from antibiotics to brassica seeds but also reduced the antibacterial effects (Klisiewicz and Pound, 1961). Similarly, Humaydan *et al.* (1980) found that the phytotoxic effects of soaking brassica seeds for 2 h in 500 $\mu g\ ml^{-1}$ streptomycin were counteracted by rinsing the soaked seeds in water followed by a 30 min steep in 0.5% sodium hypochlorite (NaOCl), and that this did not reduce the eradicative control of seedborne *X. campestris* pv. *campestris*. More recently, this surface sterilant was tested directly against the pathogen by soaking bacterially infected brassica seeds for 30 min in mixtures of sodium hypochlorite, lactic acid and water which eradicated *X. campestris* pv. *campestris* from a number of different brassica seed types but reduced emergence in some (Harman *et al.*, 1987).

The testing of antibiotics for the control of seedborne bacteria has continued in some centres. Phytotoxicity and incomplete eradication of *X. campestris* pv. *campestris* from naturally infected brassica seeds occurred in tests of a range of antibiotics in South Africa (Aveling and Robbertse, 1990). However, in India, 30 min soaks in a 100 μg ml^{-1} mixture of streptomycin sulphate + tetracycline + vitamin B_{12} reduced the incidence of this pathogen and increased yield (Kishun, 1984).

Seed soaks in inorganic chemicals

Many seedborne viruses are embryo located and treatments which inactivate them in the embryo may be phytotoxic (see pp. 171–172). Where, however, the pathogen contaminates or is only slightly invasive of seeds, as is tomato mosaic virus (ToMV) for example, then remedial treatments are possible (Broadbent, 1976; Mandahar, 1981). Most tomato seed is extracted from the fruit by fermentation or by chemical means. Extraction of the seeds in 20% hydrochloric acid for 30 min or with 10% or higher concentrations of trisodium orthophosphate for 10 min inactivates the virus in many instances and these treatments are used by seedsmen (Mandahar, 1981). They are more successful when the virus is wholly superficial and are less effective when the virus has penetrated the seed coat (Broadbent, 1976). Also, where the virus content (capsicum mosaic virus) of pepper seeds was high it was necessary to increase the immersion time in 10% trisodium orthophosphate to 2 h to achieve control, but not complete elimination, of the virus (Rast and Stijger, 1987; Stijger and Rast, 1988).

Seed infusion in organic solvents

Most aqueous seed soak or steep systems require that the treated seeds should be dried afterwards, which after 24 h of immersion can amount to 6–12 h of additional drying. Solvent infusion methods are based on the concept that pesticides, some of them relatively insoluble in water, can be dissolved in organic solvents which facilitate their entry into seeds which have been immersed in such mixtures for relatively short periods of time. No drying is required after treatment because the solvents evaporate. Solvents function as carriers which aid the penetration of the chemical either by lowering the surface tension of the seeds or by introducing it in the lipophilic phase. Fungicide penetration appears to be limited to the seed coat tissues of intact seeds but if seed coats are cracked or broken fungicides may penetrate as far as the cotyledons (Dhingra and Muchovej, 1982a).

Acetone was one of the first solvents to be tested in this way (Milborrow, 1963); it was used to introduce the non-systemic contact insecticide DDT (dichlorodiphenyltrichloroethane) into peas and cereals. Subsequently, dichloromethane (Anderson, 1973; Meyer and Mayer, 1973; Triplett and Haber, 1973) was used to introduce the hormone coumarin into dry lettuce seeds. Some solvents are not suitable for the introduction of non-systemic protectant fungicides (e.g. captan, thiram, quintozene, chlorothalonil) into dry seeds but are effective in transferring systemic fungicides (e.g. benomyl, thiabendazole, iprodione) in this way (Ellis *et al.*, 1976; Muchovej and Dhingra, 1979). The role of organic solvents in the infusion of non-systemic and systemic fungicides into seeds is discussed by Dhingra and Muchovej (1982a).

Although solvent treatment of seeds is usually considered to be non-injurious this is not the case for all seed types and all durations of application (Coolbear *et al.*, 1991). Acetone infusion of peas, for example, can damage the seeds and reduce their storage capacity depending on the initial moisture content of the seeds (Coolbear *et al.*, 1991).

In the main, seed treatments with fungicides using the solvent infusion technique have been tested against soil-borne pathogens (Tao *et al.*, 1974; Papavizas and Lewis, 1976; Locke *et al.*, 1979; Phillips, 1992). Examples where fungicides have been applied in solvents to control seedborne pathogens include internal infection of soyabean seeds by *Phomopsis* spp. (Ellis *et al.*, 1976), infection of asparagus seeds by *Fusarium moniliforme* and *F. oxysporum* (Damicone and Cooley, 1981) and contamination of sunflower seeds by *Sclerotinia sclerotiorum* (Herd and Phillips, 1988).

Systemic fungicides applied by solvent infusion or by conventional methods (e.g. dusts, slurries) can achieve similar levels of disease control but those applied in solvent can be used at lower rates of active ingredient and the same batch of solvent can be used on more than one occasion (Kraft, 1982; Herd and Phillips, 1988; Phillips, 1992). A disadvantage is that at a commercial level solvent infusion is a batch treatment which limits the amount and type of seed that can be treated, and the technique may present a health hazard to the operator. However, Dhingra and Muchovej (1982b) have described a prototype organic solvent seed treater with a 40 kg (soyabean or bean) capacity which is sealed, protecting the operator from exposure to harmful vapours. The need for this method of seed treatment is debatable at a time when modern systemic fungicides and technologically advanced methods for their application to seeds at low dose rates are available.

Seed priming methods

Osmotic priming and process engineering of seeds

Seed priming is a method of controlling imbibition, in this case by osmotic means, to allow uptake of water by the seed sufficient to initiate the germination process but insufficient to cause emergence of the radicle. Because most seeds reach the same stage of advancement during priming, when they are sown subsequently germination is more rapid and also more synchronous (Heydecker *et al.*, 1973; Heydecker and Coolbear, 1977; Bradford, 1986). Priming involves the immersion of seeds in a priming fluid or osmoticum, for example polyethylene glycol (PEG) for 10 days at 15°C. Seeds can be primed in small quantities by placing them on filter paper that has been irrigated with PEG or in larger quantities by using PEG in bubble columns or stirred bioreactors (Bujalski *et al.*, 1992) or in bioreactors capable of priming commercial quantities of seed (Bujalski and Nienow, 1991; Bujalski *et al.*, 1991; Talavera-Williams *et al.*, 1991; Gray *et al.*, 1992; Gray, 1994). Priming of large quantities of seeds, followed by drying and then coating the primed and dried seeds with a polymer film containing pesticides or other agents to produce fully advanced seeds protected against pests and diseases, is an integrated process achieved by engineering methods that has been termed 'process engineering of seeds' (Bujalski *et al.*, 1992).

Some of the options available and the steps involved in the process are illustrated in a flow diagram for vegetable seeds (Bujalski *et al.*, 1992) (Fig. 28). Steps 1, 2 and 3 comprise priming, at the beginning of which there are various treatments which may or may not be applied to the osmoticum. At the end of priming seeds are separated by sieving (step 2) from the osmoticum (which may be re-cycled or disposed of), washed and centrifuged to achieve surface dryness (step 3). The 'wet' primed seeds may then be used or they may be dried (step 4) and used later. There are various methods of drying primed seeds, including the use of warm air in fluidized bed systems (Nienow *et al.*, 1994). Primed and dried seeds may then be sown directly or they may be treated and sown. Seed treatment pesticides may be applied using conventional machinery or these may be applied by film coating utilizing fluidized bed or other application methods (step 5).

Fungicides, particularly thiram, have been added to priming osmotica (Khan *et al.*, 1978, 1980–81; Knypl *et al.*, 1980) or applied to seeds (benomyl + thiram; Dearman *et al.*, 1986) for priming, presumably as general palliatives to reduce the seed-contaminating microfungi in PEG osmotica. The value of such treatments is questionable, particularly as the general microbial inoculum of osmotica which is derived from the seeds which are being primed, such as that from carrots, may have little or no influence on the subsequent performance of the fully primed and dried seed (Petch *et al.*, 1991b).

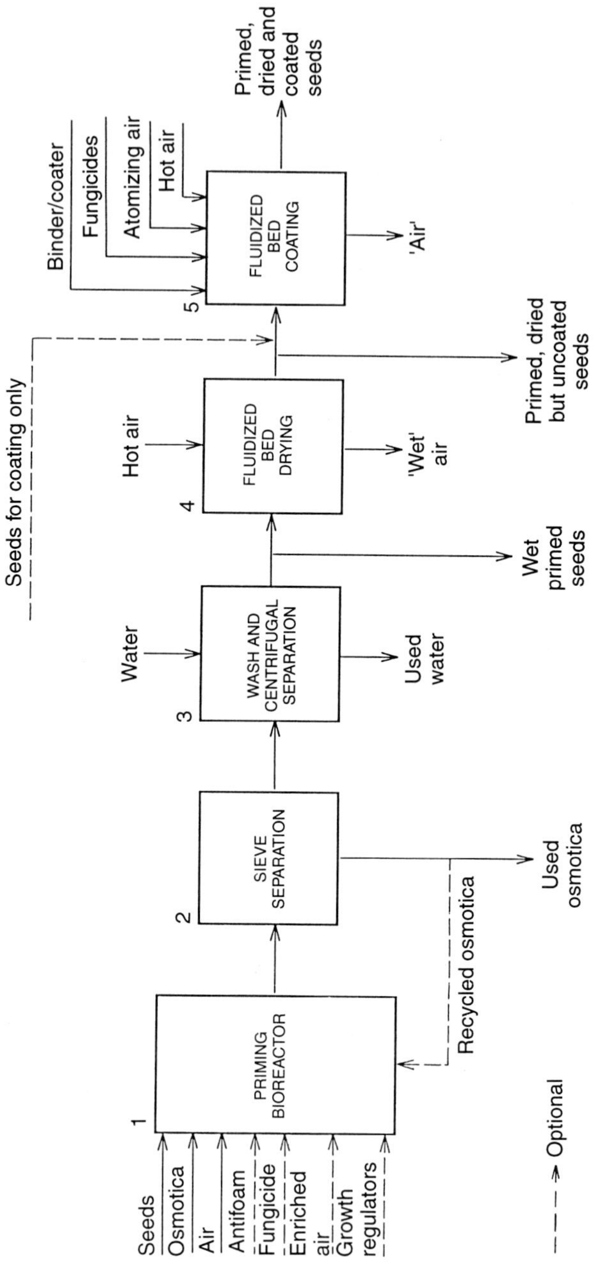

Fig. 28. Flow sheet for the process engineering of vegetable seeds.

Specific seedborne pathogens may present a different problem in a liquid osmoticum where their regrowth conceivably could increase contamination or even infection of the seed mass. In respect of seed infection of carrots by *Alternaria dauci* (leaf blight), priming in PEG did not alter the incidence of the pathogen in the seeds and although seedling infection was apparent earlier because of the quicker emergence of primed seeds, the rate of infection of seedlings from primed and unprimed seeds was similar (Maude *et al.*, 1992). The addition of fungicides (thiram or iprodione) to priming osmotica as part of a process engineering programme (Fig. 28) only partially controlled seed-borne infection and a further application of the fungicide was necessary after drying or during film coating the seeds to achieve maximum control (Maude *et al.*, 1992). In effect, a single application of fungicide made after priming and drying was sufficient and the addition of fungicides at the priming stage may have little advantage for controlling seedborne fungi but could be useful to protect the primed and dried seeds against soil-borne pre-emergence damping-off organisms, thereby improving field performance.

Lettuce mosaic virus was inactivated in PEG-imbibed lettuce seed incubated continuously at 40°C for 6–10 days but not by dry heat treatments (Walkey and Dance, 1979) (also see pp. 171–172).

Drum priming

Drum priming (Rowse, 1991, 1992) involves the hydration of seeds over a period of 24–48 h in a drum revolving at 1–2 cm s^{-1}. Seed mixing is extremely uniform and after this period controlled hydration has occurred but the seed surface is dry. The degree of hydration is calculated and the rate of wetting and the water supply is computer controlled. Seeds are then tumbled for 5–15 days. The technique is operated commercially under licence in the UK (Gray, 1994).

Solid matrix priming (SMP)

Technical and logistical difficulties occur with osmotic priming using PEG solutions. Osmotic solutions may require continuous aeration and a large volume of priming solution is needed per quantity of seed (Taylor *et al.*, 1988); with large seeds it has not been possible to achieve the high PEG:seed ratios necessary to make the system economic (Gray, 1994). SMP is a process developed in the USA by which seeds are mixed with a solid material and water in known proportions. This mixture allows the seeds to imbibe and attain a threshold moisture content, but prevents radicle emergence (Taylor *et al.*, 1988). Seeds may take 6 days at 15°C to become fully primed after which the seeds are dried back to their original moisture content. To increase aeration to the seeds, solid materials have been sought which are capable of generating low water potential, produce few solutes and have a high 'flowability'. Certain shales, calcined clay, diatomaceous silica, etc. are used for various methods of solid matrix priming (Taylor *et al.*, 1988; Khan, 1992).

SMP methods have improved field emergence of some large-seeded vegetable species (Khan *et al.*, 1992). SMP also provides a delivery system for biological control agents (Harman and Taylor, 1988; Harman *et al.*, 1989) and fungicidal seed treatments (Khan *et al.*, 1992). Provided comparisons are made for seeds achieving the same water potential during treatment, the effects of PEG priming, drum priming and SMP on germination and seedling emergence are identical (Gray, 1994).

Seed Treatment using Physical Methods

Thermotherapy is the term used by Baker (1962a) to define the various forms of heat treatment which can be applied to control the infection of planting material, including seeds. This comprises the use of heat in a wet form (e.g. hot water, aerated steam) or in a dry form (e.g. elevated temperatures, solar heat). Water is twice as efficient in heat transference as saturated water vapour (steam) and five times as effective as dry heat. Therefore temperatures and/or exposure times have to be increased when aerated steam or dry heat is used to achieve equivalent levels of disease control to hot water. The death of organisms by the application of moist heat is due to coagulation of some proteins in cells; death by dry heat is principally an oxidation process (Baker, 1962a; Sykes, 1965). Energy has also been applied as microwaves, radiation, etc. for the treatment of seeds.

Hot water

The basis for effective hot water treatment is that the temperature of the seed mass should be raised quickly to the lethal level for the infecting organism; this temperature should be maintained for a sufficient time to kill the pathogen but not the seed and then the process should be stopped quickly (e.g. by plunging the seed mass into cold water) and the seeds dried. The margin of difference between the effective destruction of the pathogen and the risk of damaging the seeds or other material is small, and precise temperature control is needed (Tarr, 1972). Baker (1962a,b) notes that microorganisms, including viruses, are often killed by heat treatment at temperatures and immersion times only slightly injurious to the host.

Hot water is probably most effective against superficial organisms but it has penetrative properties and can reduce the incidence of internal pathogens located within the seed coats and possibly deeper in the storage tissues of seeds. The treatment probably denatures the external tissues of seeds in addition to killing the organisms but does not (within the elapsed treatment time) substantially affect the storage tissues which provide a food base for the germinating seeds. Presoaking of seeds in cold water to eliminate air between the surrounding dead tissues (e.g. glumes, fruit coats, etc.) and the seed proper is sometimes practised (Baker, 1950, 1962a, 1972) to aid the conduction of heat applied as hot water.

The different methods of hot water treatment are described by Baker (1962a,b). The majority of seeds may be treated for a 30 min period at temperatures close to 50°C (Baker, 1972). Treatment damage, when it occurs, may delay germination, retard or slightly stunt seedlings or weaken their growth or kill the seeds (Baker, 1972). Such effects occur more frequently with older seeds of reduced vigour.

Fungi

Hot water was first used by Jensen in 1883 to treat potato rot caused by *Phytophthora infestans*. The method was later adapted (Jensen, 1888) to a 5 min seed immersion at 52.8°C to control loose smut (*Ustilago segetum* var. *avenae*) of oats (*Avena sativa*) and *U. segetum* var. *tritici* of barley (*Hordeum vulgare*). Since that time, hot water has been used in the treatment of a large number of fungal seedborne diseases, some of which are listed in Table 20. Its advantage is that it is a relatively simple, cheap and non-chemical treatment. Its disadvantages include incomplete eradication of seedborne fungi (many authors, see Maude, 1983b), seed damage (Baker, 1972) and the fact that only small amounts of seed can be treated at one time. Although highly penetrative against deep-seated infections of small seeds (e.g. endophytes of grass seeds (Siegel *et al*., 1987)), hot water treatment is ineffective against internal infections of larger seeds (e.g. *Ascochyta pisi* of pea). As a commercial treatment for seedborne fungi, hot water has been largely replaced by treatments using systemic fungicides.

Bacteria

With the withdrawal of mercury-based agrochemicals (Ralph, 1977a; Schaad *et al*., 1980a), some of which were used to control certain seedborne bacteria (Ralph, 1977a), and greater restrictions on the use of antibiotics, physical treatments remain an available and perhaps a necessary technology for the control of these organisms. Because of the generally higher thermal death point of bacteria, wet heat is likely to be more injurious to the seeds and less effective against seedborne bacteria than it is against seedborne fungi (Ralph, 1977a). Notwithstanding, hot water has been used in the treatment of many seedborne bacteria (Table 20).

Complete disinfection can be achieved when bacteria are on the seed surface (*Erwinia carotovora* subsp. *carotovora* of tobacco seed: McIntyre *et al*., 1978; *Pseudomonas syringae* pv. *pisi* of pea: Grondeau *et al*., 1992) but hot water is less effective when bacteria are internally located and the exten-

Table 20. The use of hot water in the treatment of seedborne diseases.

Crop, disease and pathogen	Treatment	Reference
FUNGI		
Brassica canker (*Leptosphaeria maculans*)	30 min at 50°C	Walker, 1969
Brassica canker (*Leptosphaeria maculans*)	25 min at 50°C	Millard, 1945
Brassica dark leaf spot (*Alternaria brassicae*)	20 min at 50°C	Randhawa and Aulakh, 1984
Brassica dark leaf spot (*Alternaria brassicicola*)	18 min at 50°C	Schimmer, 1953
Celery leaf blight (*Septoria apiicola*)	30 min at 48–49°C	Krout, 1921
Celery leaf blight (*Septoria apiicola*)	25 min at 50°C	Bant and Storey, 1952
Cereal loose smut (*Ustilago segetum* var. *tritici*)	5 min at 25.8°C	Jensen, 1888
Cereal loose smut (*Ustilago segetum* var. *tritici*)	1.5–2 h at 49°C or 5–6 h at 41°C	Doling, 1965b
Cereal loose smut (*Ustilago segetum* var. *tritici*)	5 h at 21°C presoak + 1 min at 49°C + 11 min at 52°C	Walker, 1969
Millet downy mildew (*Sclerospora graminicola*)	10 min at 55°C	Thakur and Kanwar, 1977
Nasturtium leaf spot (*Acroconidiella tropaeoli*)	1 h in water + 30 min at 51.7°C	Baker and Davis, 1950b; Baker, 1956
Rice blast (*Magnaporthe grisea*)	6–12 h in cool water + 1–2 min at 50°C	Nakamura, 1986
Rice Bakanae disease (*Gibberella fujikuroi*)	7 min at 57°C	Nakamura, 1986
Rice leaf spot (*Cochliobolus miyabeanus*)	7 min at 51°C	Nakamura, 1986
Safflower leaf spot (*Alternaria alternata, A. carthami*)	30 min at 50°C	Zazzerini *et al.*, 1985
Squash foot rot (*Fusarium solani* f.sp. *cucurbitae*)	15 min at 55°C	Gries, 1946

Table 20. *continued*

Crop, disease and pathogen	Treatment	Reference
BACTERIA		
Brassica black rot (*Xanthomonas campestris* pv. *campestris*)	30 min at 50°C	Walker, 1923
Brassica black rot (*Xanthomonas campestris* pv. *campestris*)	30 min at 50°C	Clayton, 1925
Brassica black rot (*Xanthomonas campestris* pv. *campestris*)	20 min at 40°C in ACA	Schaad *et al.*, 1980a
Brassica black rot (*Xanthomonas campestris* pv. *campestris*)	30 min at 52°C	Lin, 1981
Brassica black rot (*Xanthomonas campestris* pv. *campestris*)	20 min at 35–40°C in ACA	Lin, 1981
Brassica black rot (*Xanthomonas campestris* pv. *campestris*)	20 min at 38–40°C in AZS	Huang and Lee, 1988
Mungbean black rot (*Xanthomonas campestris* pv. *phaseoli*)	20 min at 52°C	Gaur *et al.*, 1984
Pea blight (*Pseudomonas syringae* pv. *pisi*)	15 min at 55 or 60°C	Grondeau *et al.*, 1992
Stock black rot (*Xanthomonas campestris* pv. *incanae*)	10 min at 54–55°C	Baker, 1972
Tobacco hollow stalk (*Erwinia carotovora* pv. *carotovora*)	12 min at 50°C	McIntyre *et al.*, 1978
Tomato canker (*Clavibacter michiganensis* subsp. *michiganensis*)	60 min at 53°C	Marinescu, 1975
Tomato canker (*Clavibacter michiganensis* subsp. *michiganensis*)	20 min at 52°C in ACA	Fatmi and Schaad, 1991
Wheat black chaff (*Xanthomonas campestris* pv. *translucens*)	20 min at 45°C in ACA	Forster and Schaad, 1988

ACA, 0.25% or 0.5% cupric acetate acidified with acetic acid; AZS, 0.1 M acidified zinc sulphate.

sion of duration of immersion or an increase in treatment temperature can reduce seed germination (McIntyre *et al.*, 1978; Grondeau *et al.*, 1992).

Particular use has been made of hot water for the control of seedborne *Xanthomonas campestris* pv. *campestris* (black rot) of crucifers. This organism can cause disease outbreaks of economic importance from very small numbers of infected seeds (Schaad *et al.*, 1980b). The treatment was introduced by Walker (1923) utilizing a temperature (50°C) and immersion time (30 min) similar to those applied for the control of seedborne fungi. However, though reducing black rot, hot water did not eradicate it (Clayton, 1925) and deep-seated infections by this bacterium may be less amenable to control (Srinivasan *et al.*, 1973; Schaad *et al.*, 1980a). Nonetheless, control of other internally borne bacteria such as *X. campestris* pv. *incanae* in stock seeds by hot water treatment for 10 min at 54–55°C is reported (Baker, 1972), as is that of *X. campestris* pv. *phaseoli* in mung bean (*Vigna radiata*) by treatment at 52°C for 20 min (Gaur *et al.*, 1984).

Attempts at improving efficacy and reducing phytotoxicity of hot water have resulted in a number of pretreatment and treatment modifications, some of which have been mentioned above. Hot water treatment is usually applied directly to seeds in a dry state. However, nasturtium seeds infected with *Rhodococcus fasciens* (bacterial fasciation) are soaked in cold water to eliminate the air space between the fruit and seed coat before hot water treatment (51.7°C for 30 min) (Baker, 1950). The application of heat (50°C) to seeds stored at a high moisture content (75%) for 3 days was a modification which successfully eliminated *P. syringae* pv. *lachrymans* in vacuum-inoculated cucumber seeds but which had adverse effects on seed germination (Leben and Sleesman, 1981). Moist air storage of tall fescue seed at 49°C for 7 days eliminated the fungal endophyte *Acremonium coenophialum* (with only a minor reduction in seed viability) (Siegel *et al.*, 1984). Successful treatment of *X. campestris* pv. *phaseoli* of phaseolus bean at lower humidities (45–55% relative humidity) for shorter periods (22–32 h) at temperatures ranging up to 60°C has been reported (Kolev, 1984).

Because *X. campestris* pv. *campestris* of crucifer seeds was not eradicated by hot water treatment at 50°C for 20 min nor by steeping in acidified cupric acetate at 25°C, Schaad *et al.* (1980a) combined the treatments and eradicated the pathogen with a 20 min soak in hot (40°C) acidified cupric acetate. However, Lin (1981) found hot water treatment (30 min at 52°C) and hot acidified cupric acetate were equally effective against the seedborne *X. campestris* pv. *campestris* of brassicas.

The latter treatment with minor modifications has been effective in the eradication of other seedborne bacteria including black chaff of wheat (*X. campestris* pv. *translucens*: Forster and Schaad, 1988) and bacterial canker of tomato (*Clavibacter michiganensis* subsp. *michiganensis*: Fatmi and Schaad, 1991). However, some depression of germination results from the use of hot acidified cupric acetate seed soaks and it is suggested that the treatment should be applied only to foundation (breeders') seed from which healthy seed crops may be produced (Schaad *et al.*, 1980a; Forster and Schaad, 1988).

The most likely explanation for the synergistic effect of combined heat and copper is an increased membrane permeability of seeds or bacterium or both, allowing copper ions to come into closer contact with the membrane proteins of cells of *X. campestris* pv. *campestris* to which they are toxic (Schaad *et al.*, 1980a). Hot acidified zinc sulphate seed soaks (for 20 min at 38–40°C) are reported to be as, or more, effective than the hot copper acetate treatment for the eradication of *X. campestris* pv. *campestris* from crucifer seeds (Huang and Lee, 1988).

Aerated steam

The principles involved in the use of aerated steam treatment as applied to soils and planting materials, including seeds, have been explained by Baker (1962a, 1962b, 1969). The heat capacity of aerated steam is about half that of water and 2.5 times that of air (Baker, 1969) and, as a result, higher temperatures and longer exposures are needed to achieve similar levels of control of seedborne pathogens. However, seeds are drier when removed from aerated steam, seed coats are not damaged and germination is not reduced by as much as it is when treated with hot water.

Aerated steam treatment is a compromise between hot water and dry heat and it is applied by the introduction of air into a steam line, thus slightly drying the steam and reducing its temperature to a predetermined operating level. Usually, seeds are treated at 56–57°C for 30 min; in some cases it is necessary to raise the moisture level of the seeds before treatment to achieve the maximum heat penetration (Baker, 1962a).

This method of seed treatment has been applied to the control of fungal and bacterial pathogens (Table 21). Most of the fungi listed in Table 21 can be eradicated or significantly reduced by aerated steam treatment (Miller and McWhorter, 1948; Maude, 1966b; Smith, 1966; Baker, 1969; Bertus, 1972; McGee and Kellock, 1974; Prichard, 1974; Navaratnam *et al.*, 1980; Hall and Taylor, 1983). One fungus, *Ascochyta fabae* f.sp. *lentis*, was not controlled (Kaiser and Hannan, 1987). In many cases, however, the treatment reduced the rate of germination (Hall and Taylor, 1983) or the total germination of seeds (Maude, 1966b; Baker, 1969; McGee and Kellock, 1974; Navaratnam *et al.*, 1980), but in others germination (Miller and McWhorter, 1948; Bertus, 1972) or emergence (Pritchard, 1974) was unaffected.

Aerated steam has proved less successful against seedborne bacteria. *Xanthomonas campestris* pv. *malvacearum* of cotton was not controlled (Navaratnam *et al.*, 1980) and others, for example *X. campestris* pv. *campestris* of brassicas, *X. campestris* pv. *phaseoli* of phaseolus bean, *X. campestris* pv. *carotae* of carrot (Miller and McWhorter, 1948), *Clavibacter michiganensis* subsp. *michiganensis* of tomato (Navaratnam *et al.*, 1980) and *Pseudomonas*

Table 21. The use of aerated steam for the control of seedborne diseases.

Crop and Pathogen	Treatment	Reference
FUNGI		
Celery (*Septoria apiicola*)	56°C for 30 min	Navaratnam *et al.*, 1980
Clover (*Fusarium avenaceum*)	49–60°C for 5–30 min	McGee and Kellock, 1974
Corn (*Drechslera maydis*)	54–55°C for 17 min	Prichard, 1974
Crucifers (*Leptosphaeria maculans*)	56°C for 30 min (H)	Baker, 1969
Crucifers (*Alternaria brassicae*)	56°C for 30 min (±H)	Baker, 1969
Lentil (*Ascochyta lentis*)	45–75°C for 30 min	Kaiser and Hannan, 1987
Lettuce (*Septoria lactucae*)	54.4°C for 20–25 min	Bertus, 1972
Lobelia (*Alternaria tenuis*)	50–51°C for 15–20 min	Hall and Taylor, 1983
Parsnip (*Itersonilia pastinacae*)	45.5°C for 30 min	Smith, 1966
Pea (*Ascochyta pisi*)	55–75°C for 20–80 min	Maude, 1966b
Pea (*Mycosphaerella pinodes*)	55–75°C for 20–80 min	Maude, 1966b
Red beet (*Pleospora betae*)	56°C for 30 min (H)	Baker, 1969
Red beet (*Pleospora betae*)	57.2°C for 20 min	Miller and McWhorter, 1948
Sweet corn (*Fusarium moniliforme*)	60–64°C for 30 min	Navaratnam *et al.*, 1980
Wheat (*Septoria nodorum*)	52–62°C for 30 min	Navaratnam *et al.*, 1980
Zinnia (*Ascochyta zinniae*)	57°C for 30 min	Baker, 1969
BACTERIA		
Beans (*Pseudomonas syringae* pv. *phaseolicola*)	55–60°C for 30–60 min	Ralph, 1977b
Crucifers (*Xanthomonas campestris* pv. *campestris*)	54°C for 30 min	Navaratnam *et al.*, 1980
Nasturtium (*Rhodococcus fascians*)	51.7°C for 30 min	Baker, 1950
Tomato (*Clavibacter michiganensis* subsp. *michiganensis*)	56°C for 30 min	Navaratnam *et al.*, 1980

H, moisture level of seeds raised before treatment.

syringae pv. *phaseolicola* of phaseolus bean (Ralph, 1977b), were not controlled by treatments which were safe for the germination of seeds.

Many of the fungal seedborne pathogens against which aerated steam was tested are now controlled commercially by seed treatments with systemic fungicides. There remains a niche for the use of aerated steam, as a batch treatment, particularly for fungal infections of some small-seeded species and the method is applied for the control of *Alternaria alternata* on lobelia seeds in the UK (Soper, 1991).

Dry heat

Dry heat in the form of hot air is five times less effective in heat transference than hot water (Baker, 1962a,b). Thus dry heat applied to control seedborne fungi is less injurious to both the seeds and the pathogen than wet heat applied over the same temperature range (Maude, 1966b) (Table 22). However, the long exposures, the possibility of injuries at high temperatures and the contingent fire hazard have discouraged the use of dry heat by seedsmen (Miller and McWhorter, 1948).

Dry heat is usually oven applied. It has been shown to kill fungal spores on the surface of seeds, for example uredospores of *Puccinia antirrhini* on snapdragon seeds (Baker, 1972). Dry heat has been used to eliminate or reduce artificial or natural bacterial contamination of seeds often with little impairment of seed germination (Table 23). Treatments include exposure for 3 days at 65°C to reduce *Pseudomonas syringae* pv. *pisi* contamination of pea seed (Grondeau *et al.*, 1992); 3 days at 70°C to control *P. syringae* pv. *lachrymans* contamination of cucumber seeds (Umekawa and Watanabe, 1978; Umekawa, 1987); control of bean seed contamination with *P. syringae*

Table 22. The comparative effects of wet and dry heat* upon germination and *Mycosphaerella pinodes* infection of pea seeds (from Maude, 1966b).

		Wet heat method		Dry heat method	
Pea type	Temperature (°C) used	Peas germinated (%)	Peas infected (%)	Peas germinated (%)	Peas infected (%)
Round seed	Control	97	41	97	41
	55	95	7	97	31
	65	0	0	93	29
	75	0	0	98	38
Wrinkled seed	Control	82	19	83	26
	55	78	6	86	20
	65	41	0	76	15
	75	0	0	73	17

*All heat treatments applied for 20 min.

Table 23. The use of dry heat for the control of seedborne bacteria.

Crop and pathogen	Treatment	Effect on Infection	Effect on Germination	Reference
Barley *Xanthomonas campestris* pv. *translucens*	4 days at 72°C	E	SR	Fourest *et al.*, 1990
	7 days at 72°C	E	GR	Fourest *et al.*, 1990
Bean *Pseudomonas syringae* pv. *phaseolicola*	1 day at 60°C	E	ok	Naumann and Karl, 1988
	3 days at 50°C	E	ok	Naumann and Karl, 1988
Bean *Pseudomonas syringae* pv. *phaseolicola*	3 h at 50°C	R	ok	Tamietti & Garabaldi, 1984
Bean *Pseudomonas syringae* pv. *phaseolicola*	2 h at 70°C	R	SR	Belletti & Tamietti, 1982
Cucumber *Pseudomonas syringae* pv. *lachrymans*	3 days at 70°C	R	SR	Umekawa and Watanabe, 1978; Umekawa, 1987
Pea *Pseudomonas syringae* pv. *pisi*	1 day at 65°C	R	ok	Grondeau *et al.*, 1992
Rice *Pseudomonas glumae*	2 days at 65°C	E	ok	Ziegler and Alvarez, 1989
Rice *Pseudomonas fuscovaginae*	6 days at 65°C	E	ok	Ziegler and Alvarez, 1987
Tomato *Clavibacter michiganense* subsp. *michiganense*	1 h at 80°C	E	ok	Marinescu, 1975

E, infection eradicated; R, infection reduced; GR, germination greatly reduced; SR, germination slightly reduced; ok, germination unaffected.

pv. *phaseolicola* by exposure for 2 h at 70°C (Belletti and Tamietti, 1982) or for 3 days at 50°C or for 1 day at 60°C (Naumann and Karl, 1988). An increase in duration of exposure to 11 days at 71, 75 or 84°C was necessary to eradicate *X. campestris* pv. *translucens* (black chaff) from heavily contaminated barley seeds (Fourest *et al.*, 1990). The authors suggest that oven-applied dry heat may be sufficiently penetrative to rid the seed of more deep-seated infections and advocate such treatments for use in international breeding and selection programmes. Grondeau *et al.* (1992) also propose the use of thermotherapy for pre-basic and basic quantities of lightly affected peas which, after treatment, could be grown on for multiplication. *P. syringae* pv. *pisi* was not eliminated from heavily infected samples of peas with dry heat applied for 3 days at 65°C and treatment at an elevated temperature (80°C) for 24 h was also unsuccessful in this respect. Similarly *P. fuscovaginae* (sheath brown rot) was not eradicated by heat treatment (6 days at 65°C) from all samples of badly affected rice; a milder treatment (3 days at 60°C) is suggested for less heavily infected seed samples (Zeigler and Alvarez, 1987, 1989). Similar treatments were applied with some success to small quantities of rice affected by *P. glumae* which causes grain discoloration (Zeigler and Alvarez, 1989).

Dry heat treatment may affect the storage potential of seeds. The germination of heat-treated cucumber seed was seriously impaired after 11 months in store (Umekawa, 1987).

Immersion in hot liquid carbon tetrachloride, which has the same specific heat (0.2) as air and with little or no penetration of seeds, was suggested by Watson *et al.* (1951) as less damaging than hot water for the dry heat treatment of peas and beans. Cruickshank (1954) used a range of temperatures and exposures in carbon tetrachloride and demonstrated heat transference by controlling two seedborne diseases (pasmo, *Mycosphaerella linicola*, and browning, *Aureobasidium lini*) of flax without adversely affecting germination with treatment for 15 min at 70°C or 20 min at 65°C. Immersion in carbon tetrachloride of larger seeds (e.g. peas internally infected with *M. pinodes*) for 3 h at 70°C reduced, but failed to eradicate, the fungus without causing phytotoxicity (Maude, 1960). Zinnen and Sinclair (1982) demonstrated that refined soyabean oil also could be used as a heat transference medium by controlling of internal infections (*Phomopsis* spp.) of soyabean seeds by immersing them in hot oil for periods of 5 min at 70°C to 10 s at 140°C.

Although thermotherapy has been widely practised in the production of virus-free plants (Walkey, 1991), the treatment of seeds carrying a virus internally, by heat and/or chemicals, has not been completely successful because treatments which inactivate the virus generally impair seed viability (Shepherd, 1972; Mandahar, 1981). Most attempts have been made to eradicate ToMV from tomato seeds (Broadbent, 1976). Dry heat treatments do not eliminate the virus in the endosperm of the seeds and they impair germination; more successful control of less deeply infected seeds may be obtained with treatments at 70°C for 3 days or at 80°C for 1 day and delays in germination may be eliminated by treating freshly extracted seeds (Broadbent, 1976). Elimination of ToMV without impairment of germination is reported from Yugoslavia with heat treatment for 24 h at 85°C (Noor-Hassan *et al.*, 1988). Dry heat or hot air treatments have been tested or applied for the control of other mosaic viruses, mainly external viruses on seed coats, including cucumber green mottle mosaic virus (Fletcher *et al.*, 1969), cowpea mosaic virus (Verma, 1971), lettuce mosaic virus (Rohloff, 1963; Pelet, 1971; Howles, 1978; Walkey and Dance, 1979) and capsicum mosaic virus (probably pepper mild mottle virus) (Rast and Stijger, 1987; Stijger and Rast, 1988). Hot water, dry heat and hot air treatments have reduced or controlled cowpea banding mosaic virus in cowpea (Sharma and Varma, 1975) and cucumber mosaic virus in vegetable marrow seeds (Sharma and Chohan, 1971).

Although there have been some considerable successes using thermotherapeutic methods for the treatment of viruses borne externally on the coats of seeds, phytotoxicity remains a major impediment to commercial exploitation.

Radiation

Radiation in the form of solar heat is applied in hot climates mainly for the control of soil-borne organisms and weeds (Katan, 1981; Katan and DeVay, 1991). Solar heat has also been employed in India for the treatment of loose smut (*Ustilago segetum* var. *tritici*) by soaking wheat seeds for 4 h in water and then drying them in the afternoon sun for 4 h at temperatures in excess of 35°C (Luthra and Sattar, 1934; Luthra, 1953). Presumably the action of moistening the seeds increases the efficacy of the heat treatment. A similar treatment is reported to have controlled *X. campestris* pv. *vignicola* in artificially inoculated cowpea seeds (Jindal *et al.*, 1989).

Ionizing radiation, particularly the use of gamma rays, has been used extensively for the elimination of bacteria in meat, poultry and sea fish (Josephson and Peterson, 1983) and for prolonging the shelf life of fruit and vegetables by delaying ripening and senescence (Eckert and Sommer, 1967; Urbain, 1986). Generally, fruit and vegetables will not tolerate irradiation at doses which will control the organisms which cause postharvest decay (Eckert and Sommer, 1967; Urbain, 1986). Gamma irradiation, however, is reported to eliminate surface fungi from maize seed without adversely affecting germination (Cuero *et al.*, 1986) and similarly to give a substantial reduction of the endophytic fungus *Acremonium coenophialum* in tall red fescue seeds at non-injurious doses (Bagegni *et al.*, 1990).

Reduction of barley stripe mosaic virus (BSMV) was obtained in plants grown from dry infected seeds (24% diseased) exposed to up to 10,000 roentgens. The virus, however, was not reduced in similarly irradiated seeds when these were pre-imbibed in water (Halliwell and Langston, 1965). Exposure of infected *Prunus* seeds to irradiation at doses (of 20,000 rad) which reduced the transmission of prunus necrotic ringspot virus (PNRSV) and prunus dwarf virus (PDV) impaired germination (Megahed and Moore, 1969).

Other producers of radiation, including lasers and microwave ovens, have been tested for seed treatment. Laser treatment of infected barley seeds reduced the transmission of four fungi (*U. segetum* var. *segetum*, *Pyrenophora graminea*, *Cochliobolus sativus* and *Fusarium* spp.) in the field (Bel'skii and Mazulenko, 1984) and microwave treatment (625 W for 20 min) of tobacco seed artificially contaminated with cells of *Erwinia carotovora* subsp. *carotovora* eliminated the bacterium without loss of germination (Hankin and Sands, 1977). However, the germination of larger seeds (cabbage and beans) was adversely affected by microwaves possibly because they were unable to radiate heat away from the seed during treatment (Hankin and Sands, 1977). Similarly, high frequency radiowaves failed to eliminate external (*Drechslera avenacea* on oats: Grainger and Simpson, 1950) and internal (*Ascochyta pisi* and *Mycosphaerella pinodes* on pea: Seaman and Wallen, 1967) fungal seed infections at time/temperature combinations which did not damage seed germination.

Use and Application of Microbial Seed Treatments

The treatment of seeds with microorganisms which are beneficial to plant growth and/or which achieve disease control has been the subject of considerable investigation for many years, often with mixed results (Weller, 1988). Aspects of the subject have been reviewed by Weller (1988), Taylor and Harman (1990), Harman (1991), Paulitz (1992) and Harman and Nelson (1994).

Seed inoculants, particularly with nitrogen-fixing rhizobial bacteria, have been used with some success to produce nitrogen gains in legumes and cereals (Alexander, 1985). Others, for example *Bacillus subtilis*, applied as a seed treatment substantially increased the yields of carrots, oats and peanuts (Weller, 1988).

Control of soil-borne organisms

Biological control of plant diseases by seed treatment with microorganisms has been directed mainly against soil-borne organisms (Harman and Taylor, 1990; Harman, 1991). Both fungi and bacteria have been tested for this purpose (Weller, 1988; Taylor and Harman, 1990; Harman, 1991). Generally, biological control (biocontrol) is more variable and less effective than control obtained with manufactured pesticides (Harman and Lumsden, 1990) often due to the poor ecological competence of the biocontrol microorganism, i.e. the ability of the organism to survive and compete in nature. The ability of such organisms to colonize root surfaces is known as rhizosphere competence and their intrinsic ability to grow on and colonize seed surfaces is called spermosphere competence (Harman and Nelson, 1994).

Treatments which may be effective under sterile protected conditions are less effective when exposed to the greater variability of the soil environment; however, in some cases biocontrol agents used as seed treatments have been shown to be as effective as similarly applied pesticides in achieving disease control (McQuilken *et al.*, 1990; Parke, 1990).

Many fungi and bacteria have been tested as seed treatments to provide short-term protection against seed rots and damping-off fungi (e.g. *Pythium* spp.) in the soil which attack during the germination and emergence of crop seeds. In this function biological seed treatments largely have been effective because the pathogens are limited in time and space and the area of host tissue available for infection is relatively small and can be effectively covered with antagonists (Paulitz, 1992). Such microorganisms have been collectively called bio-protectants (Harman, 1991). Fungi (e.g. several *Trichoderma* spp.) and bacteria (e.g. *Enterobacter* and *Pseudomonas* spp.) have been tested for this purpose (Harman *et al.*, 1981, 1989; Hadar *et al.*, 1983, 1984; Parke, 1990). Bacteria of the genus *Pseudomonas* (Osburn *et al.*, 1989) and fungal oospores of the mycoparasite *Pythium oligandrum* have been used to treat sugarbeet seeds to control damping-off caused by *Pythium* spp. (Lutchmeah and Cooke, 1985; Walther and Gindrat, 1987; McQuilken *et al.*, 1990) and *Aphanomyces cochlioides* (McQuilken *et al.*, 1990).

Longer term protection is required for soil-borne pathogens which can attack throughout the growth of a crop, for example *Sclerotium cepivorum* (onion white rot). Seed treatment applications of biologically active organisms have been shown to be effective against the fungus for considerable periods (12–16 weeks) in naturally infested soil in some situations (Ahmed and Tribe, 1977; Utkhede and Rahe, 1980), but in others seed treatment was a less effective method than the application of biocontrol organisms as soil amendments or in pellet form (Kay and Stewart, 1994).

Control of seedborne organisms

The use of microorganisms to control seedborne pathogens has received less attention. An additional effect of seed treatment with *P. oligandrum* oospores was the control of seedborne *Pleospora betae* (black leg of sugarbeet: Walther and Gindrat, 1987). The seedborne phase of this fungus on sugarbeet has also been reduced by applications of ascospores of *Chaetomium globosum* and bacterial cells of a *Pseudomonas* species (Gordon-Lennox *et al.*, 1987). Randhawa *et al.* (1987) demonstrated that strains of the bacterium *Erwinia herbicola* applied in suspension to cotton (*Gossypium hirsutum*) seeds naturally infected with *X. campestris* pv. *malvacearum* considerably reduced black arm infection in the emerged seedlings. Seed treatment with antagonistic fungi or bacteria increased the emergence of brassicas and reduced the transmission of seedborne *Alternaria brassicicola* (Wu and Lu, 1984). Recently, Maude (1990b) and Peach *et al.* (1994) reported control of the seedborne fungus *Botrytis allii* (onion neck rot) by seed treatment with bacterial cells of an *Enterobacter* species.

Effects of edaphic factors on performance

Edaphic factors (e.g. soil pH, soil temperature, etc.) have a bearing on strategies for the use of biocontrol organisms. For example, low soil pH may favour the growth of biologically active fungi whereas alkaline conditions may favour bacteria, potentially influencing the choice of agent (Harman and Lumsden, 1990). Mixtures of microorganisms and pesticides can be used to extend the activity of a biocontrol agent to a wider range of environmental conditions (Smith *et al.*, 1990) or to control pests against which the organism is not effective (Alexander, 1985).

Mode of action of biologically active organisms

The mode of action of biologically active organisms is complex and extremely varied; bacteria and fungi can suppress pathogens by substrate competition, niche exclusion and antibiosis (Brian and McGowan, 1945; Stanghellini and Hancock, 1971; Weller, 1988; Roberts and Lumsden, 1990; Whipps and Lumsden, 1991, Paulitz, 1992; Harman and Nelson, 1994); fungi in addition also control by mycoparasitism (Deacon and Henry, 1978; Ayers and Adams, 1981; Wu and Lu, 1984; Whipps and Lumsden, 1991, Harman and Nelson, 1994). The involvement of mycoparasitism in protecting seeds against fast-growing pathogens, such as *Pythium ultimum* or the *Rhizoctonia solani* state of *Thanatephorus cucumeris*, is doubtful (Paulitz, 1992). For example, *P. ultimum* is capable of colonizing pea seeds within 10 h, whereas mycoparasitism occurs only after 24 h of mycelial contact (Lifshitz *et al.*, 1986). Other mechanisms may apply and these include lysis of fungal cells by bacterially produced extracelluar enzymes, production of ammonia by bacteria, attachment of bacteria to fungi interfering with fungal growth, etc. (Paulitz, 1992; Harman and Nelson, 1994).

Harman and Nelson (1994) compared the mechanism of biological action of three bacteria (strains of *Enterobacter*, *Pseudomonas* and *Serratia*) and two fungi (from the genera *Trichoderma* and *Gliocladium*). They concluded, from the scientific information available, that the biocontrol effects of the bacteria resulted from: (i) the production of antibiotics and toxicants; (ii) the production of cell wall degrading enzymes; (iii) competition through inactivation of molecules in seed exudates necessary for germination of fungi and fungal propagules and, in a few examples, from siderophore competition for iron; and (iv) adherence of bacterial cells to hyphae affecting their viability. Mechanisms of action of the fungi included: (i) the production of antibiotics and toxicants; and (ii) mycoparasitism and the production of cell wall degrading enzymes.

Performance improvements and requirements

Biologically active microorganisms differ fundamentally from manufactured pesticides in that they are alive and must grow well if they are to be successful. Strategies for improving the efficiency and reliability of biological pesticides in the field include: (i) the use of genetically superior strains; (ii) development of seed treatment systems that produce a conducive environment for the biological control agent and minimize the competition from soil-borne microflora; and (iii) the production of the microorganism in liquid fermentation to a high quality and viability (Harman, 1991; Harman and Nelson, 1994; Rhodes and Powell, 1994).

Ideally, biological seed treatments should meet the following criteria: (i) they should provide excellent and reliable control of pathogens on a wide range of crop seeds; (ii) they should be inexpensive and simple to apply; (iii) they should have no deleterious effect during storage; and (iv) the biocontrol agent must retain viability and efficacy for over 1 year at room temperature (Harman and Nelson, 1994).

Methods of application

Biologically active bacteria are applied as cells to seeds; fungi can be applied as mycelial fragments, sclerotia and asexual or sexual spores to seeds. Application methods are varied and extend from bacterially inoculated peats for the introduction of *Rhizobia* spp. onto seeds (Vincent, 1970), to simple slurry applications containing biocontrol inoculum, to methods where an aqueous binder containing fungal spores is sprayed onto seeds followed by the deposition of solid particulate material to give a double layer, allowing a slower rate of release of bioprotectant (Taylor *et al*., 1991). Bacterial and fungal biocontrol agents also have been applied to seeds using a film-coating technique (Maude, 1990b; McQuilken *et al*., 1990; Peach *et al*., 1994) and fungi have been pelleted onto seeds by commercial methods (Lutchmeah and Cooke, 1985; McQuilken *et al*., 1990). Systems such as SMP have been developed to increase the numbers of biocontrol organisms per seed and to enhance seed performance (Harman and Taylor, 1988; Harman and Nelson, 1994; also see pp. 161–162).

The formulation technology for the application of biological control agents to seeds has been subdivided by Rhodes and Powell (1994) into dry formulations, aqueous formulations and coatings; examples of some of these are given above. In general, microorganisms applied by these techniques survive relatively well on stored seeds.

Commercial exploitation of microbial seed treatments

The efficient use of microorganisms to combat plant diseases by their application in seed treatments requires the development of biological control systems that are effective, reliable and economical (Taylor and Harman, 1990; Harman, 1991). To a certain extent, these requirements have been achieved for some biocontrol agents which are in the final stages of large-scale production, field testing and registration (Harman and Taylor, 1990). Others are beginning to be used commercially (Harman and Nelson, 1994). A seed inoculant containing *Bacillus subtilis* (Quantum 4000 ™) is on sale in the USA (Connick *et al*., 1990) and a seed treatment biofungicide (Mycostop) containing mycelium and spores of *Streptomyces griseoviridis* is commercially available in Finland, Hungary and Bulgaria and registration procedures for it have begun elsewhere (Mohammadi, 1992). Gliogard (active ingredient *Gliocladium virens*) is a commercial product for the control of *Pythium* damping-off of cotton (Howell, 1991) and is now available in the USA (Scheffer, 1994). The bacteria *Pseudomonas cepacia* and *Agrobacter radiobacter* are also being used in the commercial production of biological seed treatments (Rhodes and Powell, 1994).

Registration of microbial products

The environmental implications of the release of biocontrol agents has been reviewed by Lynch (1992) who argues that they should be subjected to the same regulatory scrutiny as any pesticide, but that this may be modified depending on their mode of action. Microbial products, in fact, are regulated in accordance with pesticide regulations and also by regulations which govern the handling and release of microorganisms (Rhodes and Powell, 1994). There is no universal scheme for their registration; most countries have not as yet formulated specific guidelines for this purpose but deal with biological products on a case by case basis (Rhodes and Powell, 1994). The combination of efficacy, economic constraints and legislative restrictions will determine which biological products are developed (Scheffer, 1994).

Efficacy of Eradicative Seed Treatments

Complete eradication of seedborne pathogens by physical or chemical means is not possible. However, what is required of eradicative seed treatments is that the number of infective seeds which remain after treatment is below the threshold necessary to cause disease outbreaks of commercial significance. In this respect, eradicative seed treatments have been highly successful and, in many cases, have reduced levels of transmission to within the set or accepted tolerance limits for particular organisms thereby achieving disease control.

They have been used on seeds in commercial quantities and also on seeds in small amounts, for example in germplasm exchanges and in certification schemes (Rennie, 1993).

Eradicative performance has been linked to tolerance requirements for some seedborne pathogens of vegetables in examples given by Maude (1983b,1988b). From these it is apparent that the performance of an individual seed treatment may vary. Many factors influence this. These include the number of times a test is repeated, the number of seeds that are tested and the severity of infection of the samples. Laboratory tests assess the control of seedborne inoculum; glasshouse and field trials in consequence assess the control of transmission of infection. In the latter environments other factors, for example edaphic, climatic and host effects, obtain and these in turn affect transmission (see Chapter 4) modifying the outcome of such tests.

In describing the efficacy and physical/mechanical data requirements for the approval of seed treatments in the UK, Slawson and Gillespie (1994) state that treatments should be tested against seed stocks carrying moderate to high levels of infection, five to six field trials held in each of 2 years should achieve similar disease control results, and field evidence may be supported by experiments in controlled or glasshouse conditions if it is possible to contrive challenging conditions. To support a claim for 'control' on a pesticide product label in respect of seedborne diseases of cereals, which have the potential to multiply rapidly from one generation to the next (i.e. bunt on wheat, covered smut on barley and oats, loose smut on wheat and barley and leaf stripe on barley), evidence of consistent control of greater than 98% is required (Slawson and Gillespie, 1994). A lesser claim for 'partial control' may be accepted in certain cases (e.g. leaf stripe of barley) where consistent control of 85–98% is demonstrated, with the restriction that this level of control is not sufficient for crops grown for multiplication.

For other pests and diseases, effectiveness of the test pesticide must be judged by comparison with an untreated control and an approved standard. The level, duration and consistency of control of the test product must be similar to that provided by the standard and where no suitable standard exists there must be a definable benefit in disease control (Slawson and Gillespie, 1994).

8 Disease Control by Cultural Measures and Sanitation Practices

Introduction

Crop rotations were devised to improve soil structure and fertility and to reduce soil-borne weeds, pests and diseases (Anon., 1954). The growing of unrelated plant species in sequence on a particular area of land has the effect of diluting soil-borne inoculum causing specific pathogens and pests to decline with time. For example, a four course rotation of winter wheat, roots, spring barley and clover led to the control of take-all of wheat (*Gaeumannomyces graminis*) in the UK (Yarham, 1988).

Rotation of Crops

The majority of seedborne necrotrophs (fungi and bacteria) and some biotrophs (particularly certain viruses) are not capable of survival as free organisms in the soil; necrotrophs persist as saprophytes on the debris of the crops they infect (see Chapter 5, pp. 79–82 and Table 12). In moist climates decay of debris is rapid but this may be more prolonged under arid or semiarid conditions. Surface-borne litter takes longer to decay than buried crop debris (see Chapter 5, pp. 79–81 and Table 12). Limited climatic change may also affect the rate of decay. For example, in the recent series of dry years in the eastern counties of England, soil-borne inoculum has been an important source of wheat bunt but few cases have been reported from the wetter west or from Scotland (Yarham and Jones, 1992).

Where infected seed is the main source of the disease, cultural control is achieved by ploughing under the affected crop remains as soon as possible and by practising a rotation of non-host species for a set number of years. A typical example is *Ascochyta fabae* on *Vicia faba* beans in the UK (Hewett, 1973). It has no alternative hosts and, as the crop is typically a break-crop in a sequence of cereals, this rotation minimizes the amount of inoculum from debris and diseased volunteer plants (Hewett, 1978). Similarly, the rotation of soyabean fields with corn (maize) reduces the infection of soyabean seeds by *Phomopsis* spp. (McGee, 1981). Autumn ploughing and a 1-year rotation was suggested by Lipps (1983) to rid the soil of crop residues and thereby control anthracnose (*Colletotrichum graminicola*) of corn in Ohio, USA. For most of the organisms listed in Table 12, a 2-year break would be sufficient to cleanse the land (Maude and Shuring, 1970; Maude *et al*., 1982); in practice, rather longer rotations are used (Anon., 1978). For example, it is suggested that onions are grown on a 4–6-year rotation in the UK, partly to ensure breakdown of crop debris, but mainly to reduce the chances of infection of new crops by soil-borne fungi and nematodes associated with onions (Anon., 1978).

Cultivation of tillage land is also important in controlling those fungi with monocyclic disease cycles which produce apothecia on the soil surface from which ascospores are released to infect the flowers of plants. Shallow ploughing (40–75 mm deep) to bury the ergots of *Claviceps purpurea* removes this source of infection and prevents the reappearance of this disease in grasses (O'Rourke, 1976) and cereals (Yarham, 1988). Ploughing to a similar depth in the autumn and early spring prevents ascocarp development of *Gloeotinia granigena* (blind seed disease) of ryegrass and is part of a disease control programme devised by Hardison (1963). Deep ploughing is suggested for the eradication of sclerotia of *Claviceps fusiformis* which causes ergot of pearl millet in India (Thakur, 1983).

The success of rotational practices is less assured with polycyclic diseases of seedborne origin, especially where the field sources of pathogens have primacy in the disease cycle. There are particular problems where crops are contiguous in time and space, for example where winter cereals follow winter cereals on the same land (Prew, 1981) or oilseed rape follows oilseed rape in the same area, etc. and where the pathogens, particularly fungi, are disseminated over considerable distances as wind-borne conidia or ascospores (see Chapter 5, pp. 75–76). In general, the functioning of trash-borne inoculum as a disease source is reduced where cultivations are used and the type of cultivation is important (Yarham and Norton, 1981); the use of resistant cultivars and agrochemical sprays are strategic options in these situations.

In addition to the use of resistant cultivars, deep ploughing is practised in the UK for the control of the canker *Leptosphaeria maculans* of oilseed rape (*Brassica napus*) where stem debris bearing the pseudothecia of the fungus is ploughed under in late summer to prevent ascospore release and thereby reduce the threat of infection to nearby autumn-sown crops (Yarham, 1988). This practice also reduces the risk to overwintering horticultural brassicas (*B. oleracea* types, e.g. Brussels sprouts, winter cabbage, broccoli, etc.) and to forage brassicas (*B. campestris* types, e.g. swede) which are attacked and are particularly susceptible to the virulent form of the fungus from oilseed rape debris (Humpherson-Jones, 1984). Ploughing is also used to reduce inoculum levels of debris-borne diseases of cereals (Yarham, 1988). Ploughing under of barley straw infected with *Pyrenophora teres* (net blotch) delayed the onset of the disease in the winter crop sown into that soil some 3 weeks later (Jordan and Allen, 1984). Where reduced cultivations, (i.e. disturbance of the soil just sufficient to produce a seedbed) were used, however, the severity of trash-borne cereal diseases increased (Yarham and Norton, 1981). Similarly, minimal tillage or no-till soyabean production resulted in higher weed populations which may have represented inoculum reservoirs for the carry-over of certain fungi, for example *Cercospora kikuchii* (purple seed stain) (McLean and Roy, 1988).

Spacing of Crops

Increased spacing between plants in a row (Agarwal *et al.*, 1975), or between rows of plants (Bisht *et al.*, 1982) or between commercial crops susceptible to the same pathogen will reduce disease spread from a seedborne source. This strategy can be very effective where pathogens are splash-borne (e.g. bacteria, pycnidial fungi) and are spread over relatively short distances. For example, spacing between crops was considered necessary in Australia to prevent splash spread of the halo blight bacterium (*Pseudomonas syringae* pv. *phaseolicola*) from the newly introduced tropical pasture legume siratro bean (*Macroptilium atropurpureum*) to commercial bean (*Phaseolus vulgaris*) seed crops produced in proximity to one another (Johnson, 1970). No distance is suggested by Johnson (1970), but Taylor (1972) minimized the spread of halo blight in the UK by placing bean plots at distances of 46–64 m apart.

Separation of crops also may reduce the spread of certain seedborne aphid-transmitted non-persistent viruses (e.g. lettuce mosaic and alfalfa mosaic viruses) which are retained for brief periods by the vector and are distributed over limited distances (see Chapter 5, p. 72,83). With other viruses, depending on the type of vector, different strategies may be required including the use of insecticides, oil sprays, repellants, alarm pheromones and anti-feedants (Tomlinson, 1987).

Separation of crops is less useful where the inoculum is air-borne and disseminated over considerable distances (see Chapter 5). Combined strategies are probably necessary in such circumstances because the variation in range and concentration of air-borne inoculum make crop spacings which will exclude organisms very difficult to establish with confidence. For example, distances of more than 50 m (Kublan, 1952), not less than 100 m (Oort, 1940) and probably not less than 150 m (Strass, 1964) are spacings between crops suggested for the limitation of transfer of wheat and barley loose smut spores.

However, such distances do not completely exclude inoculum and that which is transferred may still constitute a threat to the developing crop. Hewett (1968) suggested that even greater isolation distances, similar to those used to protect brassicas and beet from cross-pollination, i.e. 1000 m for basic seed production, might be necessary to protect susceptible barley cultivars from infection bridging by loose smut spores. This was part of a tripartite strategy which also included seed treatment and the setting of tolerance limits for seedborne inoculum (Hewett, 1968). The problem of cross-infection in the production of smut-free barley seed was resolved in Canada by seed treatment and the compulsory isolation, backed by special legislation, of some 300 farms in a seed control area from other field sources of loose smut (Clark, 1952).

Air-borne conidia of *Alternaria brassicicola* infective at 1000 m downwind from the source, created similar problems for the safe siting of brassica (*B. oleracea*) crops for seed production in an area of continuous cropping in the UK (Humpherson-Jones and Maude, 1982). However, the siting of crops at least 150 m upwind of an infected maturing crop gave considerable reduction of seedborne infection in the new crops. In this example the treatment of all basic seed, the siting of new crops for seed production upwind of maturing seed crops and fungicide spraying of seed pods was the combined strategy evolved to control the problem (Humpherson-Jones and Maude, 1982).

An alternative form of spacing is the grouping together of crops with similar problems into isolated blocks to dilute the effect of air-borne inoculum. This is a form of spatial rotation which may be suitable for growers with large areas of land at their disposal. But this method would be difficult to apply to crops commanding large acreages such as cereals and oilseed rape (Maude, 1988a).

Other Cultural Control Measures

Transmission of infection may sometimes be avoided by varying certain planting parameters. Early sowing of winter wheat produces plants which are past the susceptible stage before bunt (*Tilletia laevis, T. tritici*) teliospores can germinate and infect (McGee, 1981). Shallow planting of wheat into wet soil reduces the transmission of bunt (*T. laevis*) and flag smut (*Urocystis agropyri*) in countries such as Egypt where irrigation is practised immediately after sowing (Jones and Seif El-Nasr, 1940).

Use of Fire

If the land is not ploughed immediately after harvest then straw burning is a method of reducing inoculum of debris-borne cereal pathogens, such as

Pyrenophora teres (barley net blotch) (Jordan and Allen, 1984). Complete control is not achieved by this method because small unburnt pockets of straw remain on which inoculum is produced (Jordan and Allen, 1984). There is recent evidence (Jenkyn *et al.*, 1994) based on crop survey data which suggests that, apart from leaf blotch of barley (*Rhynchosporium secalis*), the main trash-borne necrotrophs of wheat and barley maybe only slightly reduced by burning.

However, burning of straw and stubble has been used to good effect to reduce inoculum levels of the monocyclic pathogens *Gloeotinia granigena* (Hardison, 1963) and *Claviceps* spp. (Wells *et al.*, 1958; Hardison, 1976). The use of fire and of mechanized burning i.e. 'thermal sanitization' and its control of plant pathogens in general and its control of some seedborne pathogens has been reviewed by Hardison (1976).

Elimination of Weed and Crop Plant Hosts

The contribution of alternate hosts, volunteer plants and plant debris to the initiation of disease outbreaks in otherwise healthy crops has been reviewed in Chapter 5. Herbicides will effectively eliminate biotrophic pathogens by killing their host plants but have little effect on necrotrophs which survive on dead plant tissue. Thus herbicides can be most useful in the elimination of the weed hosts of viruses present in crops or as volunteers on headlands where these have been shown to be of consequence in the infection of crop plants (see Chapter 5, p. 88). Transfer of cucumber mosaic virus (CMV), seedborne in chickweed, to lettuce can be prevented by killing the weed host with applications of selective herbicides (Tomlinson, 1987). However, some herbicides (paraquat and glyphosate) were not effective against necrotrophs and failed to prevent sporulation of *Septoria nodorum* (the asexual pycnidial state of *Leptosphaeria nodorum*) on wheat straw (Harris, 1979) and gave inconsistent results against *Rhynchosporium secalis* on barley stubble (Stedman, 1982). Glyphosate did give some suppression of *Pyrenophora teres* on barley stubble (Yarham, 1988).

There is a need to justify the application of control measures. The disease risk from necrotrophic fungi on grasses which also occur on cereals is not well established. The biotroph *Claviceps purpurea* on black grass is a possible source of infection to wheat in the UK, but there appears to be little threat to the main cereal crops from inoculum of this fungus on grass species (see Chapter 5, p. 85). Because of the considerable area of cereals, inoculum flow would appear to be more likely in the opposite direction. Where, however, grasses are the main crop species then practices such as the mowing of headland grasses and the topping of pastures before the conidial 'honeydew' stage is reached will reduce ergot (O'Rourke, 1976). Equally, where the volunteer plants are of a main cereal type, for example headland barley plants, infected with loose smut these are a threat to nearby main crops and should be buried or otherwise destroyed.

Many of the seedborne bacteria have pathovars with a high host specificity and some have cultivar-specific races limiting the number of natural hosts which can function as alternate sources of the crop disease. However, where the crop and alternative crop or weed host are of the same plant family then infection bridging can occur and destruction or avoidance of affected hosts is necessary to break the cycle (see Chapter 5). Where possible, the physical removal of affected plants, the use of herbicides where appropriate, rotational practices, ploughing under of volunteer plants and affected stubble, etc. are sanitary measures which will help to reduce these sources of infection.

Elimination of Other Sources of Disease

The accumulation of crop refuse into tips or dumps outside packing sheds are possible sources of pathogens for the secondary infection of nearby crops. Dumps or cull piles of onions infected with *Botrytis allii* (neck rot) exemplify such sources (see Chapter 5, pp. 83–84). Burning of refuse is one method of eliminating this source of the fungus but bulb onion material is difficult to dispose of in this way. In the UK, the levelling of such dumps and covering them with soil is suggested to prevent fungal sporulation and spore dispersal (Maude, 1981). This area of disease control, which here relates to seed-associated pathogens, receives a more general treatment from Walker (1969).

9 Detection of Seedborne Organisms

Introduction

The foregoing chapters have described the infection of seeds, the transmission of disease by infected seeds, and the control of disease by the reduction of seedborne inoculum in certification or quarantine programmes or by seed treatment. As a result, some direct and indirect methods for the detection of seedborne inoculum have been described already (see Chapter 6). Such methods were initially important for research purposes and these have been modified or were developed separately for use as standardized techniques for the detection of specific pathogens. They have been used to implement tolerance limits in certification and quarantine programmes and to make seed treatment decisions. In addition, standard seed detection techniques can also be used to determine seed quality and to investigate poor germination and emergence rates (Ball and Reeves, 1991; Anon., 1993b; Reeves, 1995).

General accounts of the methodology of detection are given in textbooks by Neergaard (1977) and Agarwal and Sinclair (1987). There have been specific reviews for seedborne viruses (Fulton, 1964; Bennett, 1969; Bos, 1977; Mandahar, 1981; Mink, 1993) and of the newer methods for the detection of seedborne pathogens (Kulik, 1984) and viruses in seeds (Lange, 1986). Technological advances in methods for the detection of seedborne bacteria were collated at the First International Workshop on Seed Bacteriology in France (Schaad, 1982) and more recently produced in manual form by Saettler *et al.* (1989). Authors have identified the need for speed and accuracy in testing for seedborne organisms (Neergaard, 1977; Irwin, 1987; Ball and Reeves, 1991, 1992; Ball, 1992) and both requirements are treated in reviews of established and developing techniques for the rapid detection of plant pathogens and seedborne organisms (Duncan and Torrance, 1992; Rasmussen and Reeves, 1992; Reeves, 1995). The principles supporting the general and specific diagnostic techniques used in plant pathology have been described by Fox (1993). This account reviews principles and progress in the development of diagnostic methods for the detection and identification of seedborne fungi, bacteria and viruses.

Seed Health Testing

Direct observations on seed samples and seed incubation tests (blotter and agar) are important basic laboratory methods for the detection of parasites in general seed pathological research and in seed health testing. Doyer (1938) laid the foundations of seed health testing technology which has developed and is now practised by official seed testing stations in different countries and which is subject to rules laid down by the International Seed Testing Association (Anon., 1993b). The methods used by Doyer (1938) were observational, by which seed samples were divided into two, namely those with and those without recognizable 'carriers' (i.e. propagules) of disease. The carriers of disease could be observed macroscopically or microscopically. Macroscopic carriers included those fungal structures intermingled with the seeds, (e.g. sclerotia and bunt kernels), seeds with distinct lesions (e.g. as of *Colletotrichum lindemuthianum*), damage caused by parasitic insects and the presence of storehouse pests (granary pests). Microscopic carriers include phases of fungus growth distinguishable on dry seeds with the aid of a binocular microscope (e.g. pycnidia of *Septoria petroselini*, etc.) and infection by fungus spores (e.g. *Tilletia tritici*). These observations resulted in the production of three test methods: (i) the examination of dry samples of seeds for the presence of pests, i.e. nematodes and fungal sclerotia; (ii) the examination of germinating seeds on a moist blotter, mainly for the presence of fruiting bodies of fungi (Fig. 29); and (iii) the examination of seed washings for the presence of spores typical of a range of smut fungi.

The seed health testing techniques which are currently practised have similar aims (Anon., 1993b). Health of seed refers primarily to the presence or absence of disease-causing organisms, i.e. fungi, bacteria and viruses, and animal pests, such as eelworms (nematodes) and insects, but physiological conditions such as trace element deficiencies also may be involved (Anon., 1993b).

Diagnostics in research and in routine seed health testing (for certification and quarantine purposes) have similar objectives but health tests have to be completed within a limited time span and standardization is necessary to achieve results which are consistent when applied in practice (see Chapter 6). Ball and Reeves (1991) list six main requirements for seed health testing.

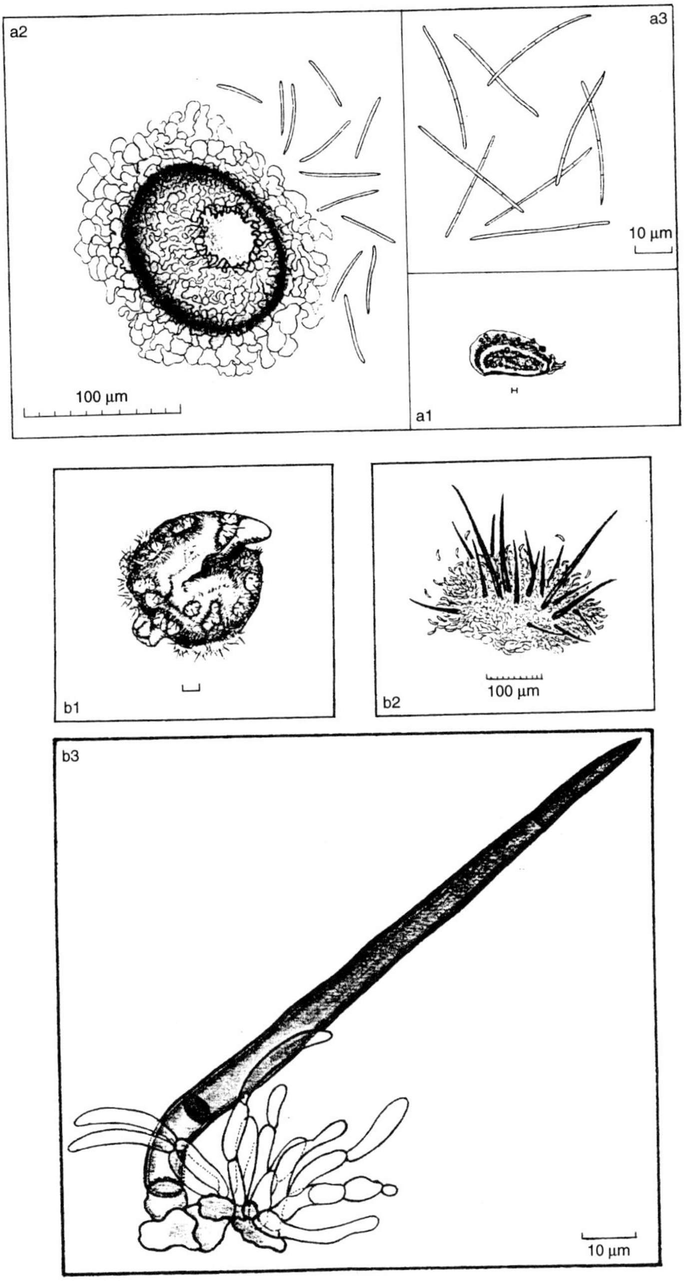

Fig. 29. Illustrations of seedborne fungal propagules recognizable by inspection. (a) *Septoria apiicola*: 1, pycnidia on celery seeds; 2, pycnidium and pycnidiospores; 3, pycnidiospores (enlarged). (b) *Colletotrichum spinacea*: 1, acervuli on spinach seed; 2, acevulus; 3, acevulus (enlarged) with spores. (Redrawn from Doyer, 1938.)

1. Specificity: the ability to distinguish the target pathogen from all organisms likely to occur on seeds from field or store, i.e. to avoid false positives (also see p. 202).
2. Sensitivity: the ability to detect organisms, which are potentially significant in field crops (see Chapter 6), at a low incidence in seed stocks.
3. Speed: in some cases small concessions to accuracy may be necessary to ensure rapid results, but such results should be followed by more definite testing.
4. Simplicity: the methodology should minimize the number of stages to reduce room for error and to enable tests to be performed by not necessarily highly qualified staff.
5. Cost effectiveness: test costs should form part of acceptable production margins for each crop.
6. Reliability: test methods must be sufficiently robust so that results are repeatable within and between samples of the same stock regardless of who performs the test (within the bounds of statistical probability and sample variation).

Current Techniques for the Detection of Fungal Seedborne Pathogens

Direct inspection

Seed samples may be examined dry for the presence of ergots, other sclerotia and smut balls without or with a stereomicroscope. The sample may be immersed in water or another liquid to make fungal fruiting bodies, for example pycnidia, or symptoms more visible or to encourage the liberation of spores. After immersion, the seeds are examined by means of a stereomicroscope (Fig. 29a,b). Seeds may be immersed in water containing a wetting agent or in alcohol and shaken to remove fungal spores, hyphae, etc. attached to or carried with the seeds. Excess liquid is then removed and the extract is examined at a higher magnification using a compound microscope.

Although seed inspection methods are useful for determining the incidence of easily visible and relatively superficial fungi or fungal bodies, they give no indication of their viability.

Further tests are necessary to establish pathogen viability and ultimately pathogenicity, for example *Septoria apiicola* where seedborne pycnidia may contain only non-viable spores; special methods are needed to obtain spores from celery seeds and to test them for viability and pathogenicity (Maude, 1963, 1964; Hewett, 1968).

Incubation methods

Agar testing

The agar test gives an indication of the viable inoculum present in an infected seed sample. This is done by placing seeds onto sterile potato dextrose or malt agar to encourage the growth of seedborne necrotrophs. Plates are incubated at 20°C in the dark for 7 days when the characteristic growth form (mycelium) of the fungus can be identified by eye or using a low or high power microscope. Near-ultraviolet light (NUV) may be used during the last 2 or 3 days of incubation to encourage the development of fruiting bodies (Limonard, 1968). Seeds are spaced on the Petri dishes to avoid cross-contamination. Where there is a considerable outgrowth of fungal saprophytes from the seeds concealing the pathogen or where it is desirable to identify internal infection, seeds may be pre-treated (surface sterilized in 1% free chlorine for 10 min) to free them of superficial microorganisms and then plated onto agar (see Chapter 4).

There are many variations of the agar test. Acidic agars (e.g. prune lactose yeast) may be used to reduce bacterial contaminants (Maude, 1963); agars may be made semi-selective by the addition of specific chemicals (Kritzman and Netzer, 1978) and/or antibiotics and/or fungicides (Kritzman and Netzer, 1978; Byford and Gambogi, 1985; Anon., 1993b).

There are standard seed health tests on agar for the detection of *Ascochyta pisi* on pea, *Botrytis cinerea* on flax and *Septoria nodorum* on wheat (Anon., 1993b). Agar tests are most effective when used for the detection of high-incidence pathogens such as *A. pisi* and *Botrytis allii* (onion neck rot) which occur in seed samples at levels greater than 1% and which can be detected by testing 200–400 seeds (see Chapter 6). Where relationships have been established between the incidence of these fungi in laboratory tests, disease transmission in the field and final infection levels in the harvested crop, the agar test can be used to impose tolerance limits for infection in samples of seed (see Chapters 4 and 6).

Blotter testing

The blotter test gives an indication of the infection of the seed, as shown by the presence of mycelium and fruiting bodies, and, in some tests, infection of the germinated seedlings as demonstrated by symptoms on the young plants. Standard detection methods vary depending on which fungus is being tested (Anon., 1993b). For example, in phaseolus beans, 400 seeds per sample are placed between water-soaked sheets of paper towelling and incubated for 7 days at 20°C, dark spots on the cotyledons are symptomatic of *Colletotrichum lindemuthianum* infection. In other tests the germination of seeds is deliberately suppressed to allow seedborne infection to develop. Thus brassica seeds (1000 per sample, see Table 13) are placed on blotters irrigated with the herbicide 2,4-dichlorophenoxyacetic acid (2,4-D) solution and incubated for 11 days at 20°C in alternating cycles of light and dark to allow the pycnidia of seedborne *Phoma lingam* to develop. Likewise, carrot seeds (400 per sample) are tested for *Alternaria dauci* and *A. radicina* on triple-layer blotters soaked in sterile distilled water by incubating them for 3 days at 20°C, then at −20°C overnight followed by 7 days at 20°C in alternating cycles of light and dark for production of conidia.

Deep freezing and 2,4-D solutions disrupt seed tissues and thereby increase the ease with which seedborne fungi grow from the seeds (Limonard, 1966; Jorgensen, 1977). The germination of seeds is inhibited by 2,4-D solutions, allowing greater numbers of seeds to be tested (Maguire *et al.*, 1978) (see Chapter 6).

Blotter tests are also used for the detection of *Bipolario oryzae*, *Pyricularia oryzae* and *Alternaria padwickii* on rice. The identity of the fruiting bodies of the pathogens (i.e. conidia, pycnidia, etc.) which develop is confirmed by microscope examination. Sterile media including sand, artificial composts, etc. can be used for the detection of certain pathogens. Results are based on the presence of symptoms typical of the organisms.

Staining methods

Staining methods are used for seedborne biotrophs which cannot be grown out on a nutrient (agar) or basal (blotter) substrate. The main standard method used in seed health testing is that of staining barley embryos of susceptible cultivars for the presence of loose smut (*Ustilago segetum* var. *tritici*) mycelium (see Chapter 6). The embryos of barley seeds (2000–4000 seeds per sample) are extracted over 24 h at 20°C in a 5% solution of sodium hydroxide. After clearing in lactophenol, the embryos are examined under magnification for the presence of the golden brown mycelium of the fungus in the scutellum of seeds (Anon., 1993b). In certain laboratories lactic acid is preferred to lactophenol as a clearing agent (J.C. Reeves, pers. comm.). Seedborne inoculum (mycelium and/or oospores) of downy mildew (*Peronosclerospora sorghi*) in maize is detected under the microscope following a process of maceration of the seeds in sodium hydroxide containing trypan blue stain (Rao *et al.*, 1984).

Effectiveness of laboratory detection methods for fungi

The methods used for the detection of fungi are relatively easy and do not require sophisticated and expensive equipment (Irwin, 1987). Blotter and agar tests still are important basic laboratory methods the strength and the weakness of which resides in their simplicity. Subjectivity can be a problem; although staff can be quickly taught to recognize the main causal organisms, individuals see and record organisms differently. The tests themselves require more critically defined objectives. In past years, test method objectives have varied, some assessed the total inoculum of seeds, others the viable inoculum and yet others the infective (disease-causing) inoculum. Infective inoculum relates to the transmission of the organism and most closely predicts field outbreaks of disease (see Chapter 4) and is the parameter which preferably should be assessed in laboratory tests for fungi (de Tempe, 1963). However, this may not always be possible. Formerly, individual laboratories developed seed testing methods especially suited to their own requirements, facilities and personnel (Noble, 1951). For example, three separate laboratories produced an agar plate test for the detection of *Aureobasidium lini* (syn. *Polyspora lini*, stem break and browning) of flax seeds. Known respectively as the Ottawa method (Groves and Skolko, 1944), the Ulster method (Muskett and Malone, 1941) and the New Zealand method (Newhook, 1947), they gave very different results when applied to seeds from the same samples mainly because of intrinsic differences in the methods themselves and in their application to seeds (Wallen and Skolko, 1951). Variation in technique is acceptable but in such cases the objectives must be clearly defined (Wallen and Skolko, 1951). There is also a national and international need for standardization in the technology of seed health testing.

Initiation of comparative testing and related workshops within the framework of the International Seed Testing Association in 1958, and their continuance on a 3-yearly basis, has resulted in limited improvements to empirically derived tests such as the blotter and agar fungal detection tests (Neergaard, 1986). Further improvements are unlikely because of inherent limitations within the methodology of such relatively simple tests. Limonard (1968), in an extensive comparative examination of blotter, agar, sand and soil incubation methods for detection of fungi, found that there were many factors affecting the methods used, and in the microfloral content of the seeds themselves, that caused an imbalance between fungi and bacteria resulting in inconsistencies in the reproducibility of tests. He considered that some redress was possible by greater use of pre-treatments, such as surface sterilization with a chlorine generator and use of selective media. This should introduce greater standardization to laboratory results and thereby establish a more accurate relationship between these tests and transmission of disease.

Detection of Seedborne Bacteria and Viruses

Many seedborne bacteria and viruses are infective at extremely low incidences and as such present problems of seed sampling and pathogen detection (e.g. the bacterium *Pseudomonas syringae* var. *phaseolicola* of phaseolus bean and lettuce mosaic virus (LMV) (see Table 11 and Chapter 6). Therefore, traditional methods for the detection of seedborne bacteria and viruses have involved sowing seeds in quantity in the glasshouse or field and then recording the development of symptoms on seedlings after emergence. Isolation and identification then followed using laboratory or host inoculation techniques (Phatak, 1974; Schaad, 1989a; Ball and Reeves, 1992).

Such methods are time consuming and demanding on space and Neergaard (1986) identified a need for more routine laboratory practices for the detection of seedborne bacteria and viruses. Much has been done to develop the appropriate technology. Over 50% of papers in the section on 'Methodology and testing techniques' at the International Symposium on Seed Pathology in Copenhagen in 1982 reported on novel methods for the detection of bacteria and viruses (Anselme, 1983). Progress in the detection of seedborne bacteria has involved two main steps: (i) improvement in extraction methods from seeds; and (ii) improvements in detection and identification technology. Even greater precision and speed of testing for seedborne bacteria and viruses has been obtained by the introduction of serological and nucleic acid techniques, first developed as a result of medical research but with which the plant pathology sector now has acquired its own momentum as the range of practical applications continues to expand (Smith, 1992).

Growing-on tests for bacteria and viruses

Suspect seed samples are sown in glasshouses or fields in isolation of like plant species to prevent the transfer of infection. They give an indication of the total potential transmission, i.e. from external and internal inoculum sources, of the seed stock under the environmental conditions of the test. Fluctuations in environmental conditions and in viability of seed stocks may influence test results. Growing-on tests are easy to perform and have been used in the past for the detection of many seedborne bacteria (Schaad, 1989a), for example, *P. syringae* pv. *phaseolicola* by Grogan and Kimble (1967) and *P. syringae* pv. *pisi* by Watson and Dye (1971). However, these tests are time consuming and can occupy considerable space, especially where large numbers of seeds require testing for the presence of low-incidence pathogens. The detection of LMV in commercial seed stocks in the USA was based on growing large numbers of lettuce seedlings in aphid-proof glasshouses and inspecting (indexing) them for the presence of virus symptoms (Grogan, 1983) (see Chapter 6). Growing-on tests and host inoculation tests (the *Chenopodium quinoa* test) (Zink *et al.*, 1956; Marrou and Messiaen, 1967; Rohloff and Marrou, 1980) were replaced by the enzyme-linked immunosorbent assay (ELISA) in several countries from 1985 (Dinant and Lot, 1992).

Laboratory tests for the detection of bacteria

Basic laboratory methods for the detection of seedborne bacteria involve: (i) their extraction from seeds; (ii) isolation into culture; and (iii) identification by a range of different methods.

Extraction

Extraction procedures attempt to optimize the recovery of the target bacterium. This is dependent upon the optimum release of pathogenic bacteria from a seed sample; this release may be affected by other factors, including the saprophytic microflora of seeds and inhibitory compounds in seeds. Roth (1989) reviews the factors which influence the procedure. Pre-washing or surface sterilization have been used to reduce the surface microflora of seeds and increase the efficiency of extraction of target organisms and thereby the sensitivity of the test. However, by use of this procedure superficial inoculum, which may form a component of disease transmission (Taylor *et al.*, 1979a; Weller and Saettler, 1980) (see Chapters 2 and 4), will have been removed. Total inoculum (comprised of superficial and deep-seated bacteria) may be obtained by grinding or milling untreated seeds into a flour before extraction (Taylor *et al.*, 1979a).

Bacteria are extracted from seed flour or seeds in a liquid medium; sterile tap water, distilled water, various buffers and enrichments have been tested (Roth, 1989). Generally, a sterile buffered saline is most commonly used which affords some protection to the bacterial cells. With most methods, the volume and agitation of the liquid medium, and the duration and temperature of soaking, are critical in optimizing recovery of the target bacterium. With many extraction procedures, sterile plastic bags are used to contain the seeds; the amount of saline which is added will depend on a number of factors, i.e. seed type, soak temperature, inoculum concentration, etc., and this needs to be determined in a series of preliminary tests.

Rehydration of seeds followed by extraction may be damaging to bacterial cells and there is a potential for false negative results if the bacteria die during extraction (Roth, 1989).

Selective enrichment of the extraction medium (Trujillo and Saettler, 1979) may be used where inoculum density is low or where the bacterium is slow growing but the benefits of this treatment can be reduced by a concomitant increase in contaminating bacteria (Roth, 1989).

Short-duration incubation of extraction media (*c.* 4 h) and low temperatures (4–6°C), by reducing contaminating organisms, may increase the incidence of target bacteria (Roth, 1989).

Isolation

Bacteria are isolated by plating small volumes of the extraction medium onto general or semi-selective agar media (Schaad, 1989a).

Identification

Standard tests

Once isolated into culture, plant pathogenic bacteria are identified by means of classic, morphological, physiological and nutritional tests (Stead, 1992). Such tests are described in books on diagnosis by Fahey and Persley (1983), Lelliot and Stead (1987) and Schaad (1988).

Agar plating

Identification of some seedborne bacteria is possible following isolation onto general plating media. For example, *Pseudomonas syringae* pv. *phaseolicola* was successfully isolated from a water suspension of phaseolus bean flour plated onto King's medium B (KB) (King *et al.*, 1954) on which colonies of the pathogen produced a characteristic fluorescent pigment (Taylor, 1970c). Another example is the identification of *Xanthomonas campestris* pv. *campestris* by the production of typical starch-hydrolysing colonies on nutrient-starch cycloheximide agar (Schaad and Donaldson, 1980).

These methods are most effective when pathogenic bacteria are present in large numbers in extracts and bacterial saprophytes are at a low incidence. When the reverse is true then semi-selective media can be used to advantage. Semi-selective media are those which have been amended by the addition of chemical agents to reduce the growth of saprophytes without adversely affecting the growth of the pathogen. Schaad (1989a) lists a number of these media including KBC for the selective isolation of *P. syringae* pv. *syringae* from contaminated bean seed. KBC is a modification of KB to which boric acid, cycloheximide and cephalexin are added; more than 90% of seed-associated saprophytes were inhibited on KBC compared with KB (Mohan and Schaad, 1987). Semi-selective media also can be used in seed health testing to determine the incidence of particular seedborne bacteria by the direct plating of seeds onto agar followed by incubation and examination for colonies characteristic of the organism on that medium surrounding individual seeds (Schaad, 1989a,b). However, because the disease threshold incidence of seedborne bacterial pathogens is frequently very low, the method is not in general use.

Further identification of the bacteria which produce characteristic colonies which develop on general and specific agar media may be obtained using morphological and biochemical tests (Stead, 1992) or by use of immunodetection and nucleic acid methods (see pp. 196–211). All these tests give presumptive diagnoses which may require confirmation by means of a host pathogenicity test. In certain cases, the seed extract medium may be used for this purpose and can be inoculated into the host plants or seeds to test the pathogenicity of the organisms (Schaad, 1989a). Otherwise pure cultures of bacteria are inoculated into plants by a number of different methods including injection, spray, following leaf abrasion, vacuum infiltration, etc. (Ball and

Reeves, 1992). Finer definition of pathovar and race identity is achieved by inoculation of the appropriate host species and a differential series of cultivars of the host species (Taylor *et al.*, 1989).

Bacteriophage tests

Bacteriophages are viruses which attack and kill bacteria. The relationship between the phage and its host bacterium may be specific and this can be used to detect and identify the bacteria. Several methods are employed (Sheppard, 1983). Suspensions of selected specific phages are spotted onto plates which have been seeded with pure cultures of bacterial isolates requiring identification; where bacteria are killed (lysed) by phages, clear zones develop (Fig. 30) which are confirmatory of the identity of the bacterium. Lysis zones are recorded after the incubation of plates for 24 h at 25°C and the method provides a rapid means of the identification of bacteria cultured from seed extracts provided highly specific phages are available (Taylor, 1970c).

An alternative use of phages is the rapid phage-plaque count method. An increase in phage concentration (titre) after 8–12 h in mixtures of selected phage culture with macerates of surface-sterilized seeds and soaked seeds is taken as evidence of the presence of a particular bacterium in the seed sample (Katznelson, 1950). The technique can detect as few as 3 bacteria ml^{-1} of ground bean tissue (Katznelson and Sutton, 1951).

Phage tests have been used to identify a number of seedborne bacteria (Schaad, 1989a) but a major limitation is that bacteriophages are rarely truly species specific, also bacteria may be resistant to phages and false negatives frequently cause failure with the method (Sheppard *et al.*, 1989). For example, the failure of bacteriophages to detect a specific brown pigment-producing

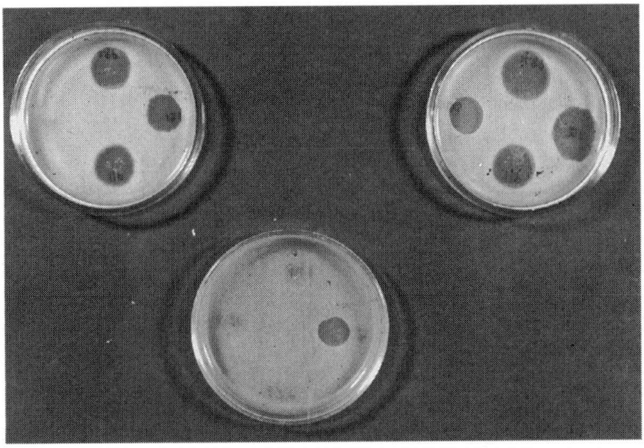

Fig. 30. Bacteriophage tests: lysis zones after 24 h of incubation (from Taylor, 1970a).

strain (i.e. fuscous blight) of bacterial blight (*X. campestris* pv. *phaseoli*) in bean seeds resulted in severe outbreaks of disease in seed fields in Canada in 1962 (Wallen and Sutton, 1965). However, Irwin (1987) suggests that where other highly specific detection methods, for example serological tests, are not available the use of pathovar-indicative bacteriophages (Cupples, 1984) might have application for the detection of certain seedborne bacteria.

Immunodiagnostic and Nucleic Acid Laboratory Methods for the Detection of Bacteria, Fungi and Viruses

Serology

Serological tests for plant pathogenic bacteria have been known since 1918 (Schaad, 1979).

The methods are based on the immunological principle that foreign molecules (i.e. immunizing agents or antigens) injected into the bloodstream of mammals stimulate the immune system of those mammals to produce specific antibodies which will recognize and bind to the antigens (Schaad, 1979; Fox, 1993). Such antibodies recognize many chemical sites, referred to as epitopes, on target antigens; these are polyclonal antibodies. Methods for the production of antibodies in quantity as immune serum or antiserum are described by Fox (1993). The types of bacterial antigens, for example live whole cells, dead cells, sonicated cells, etc., are summarized by Sheppard (1983). Immunoassays utilizing antisera produced against purified pathogens or extracts of pathogens have been effective in the detection of viruses (Torrance, 1992) but have had a more limited value for the detection of bacteria and particularly fungi. Bacteria and fungi are more complex organisms and contain many non-specific antibodies which may cause cross-reactions with related and unrelated species concealing the effects of specific antibodies (Dewey, 1992; Miller *et al.*, 1992).

The development of monoclonal antibodies (MAbs) has greatly increased specificity. Monoclonal antibodies recognize one chemical site (epitope) on target antigens and in that respect are homogenous and free from the variability common to polyclonal antisera (Fox, 1993). They are prepared by cell culture techniques and produced by a hybridoma (a single lymphocyte cell hybridized with a tumorigenic cell) specific for a single epitope (Stead, 1992). Thus they can be selected to act at generic, specific, pathovar or strain levels. Franken and Van Vuurde (1990) have reviewed the most commonly used serological methods for the detection and identification of seedborne bacteria including agglutination, double diffusion, ELISA and immunofluorescence microscopy techniques. Lange (1986) examined serological and nucleic acid techniques for the detection of seedborne viruses.

Agglutination, precipitin and immunodiffusion methods

Some of the earlier serological detection methods using polyclonal antibodies remain relatively effective. They include agglutination tests (tube and slide: Figs 31 and 32) where the antigen/antibody reaction results in an agglutination or clumping of particulate antigen (Figs 31 and 32); precipitin tests (tube and slide) where the antigen is precipitated out of solution by the antibody, and

Fig. 31. Tube agglutination test, bacterial suspension (left), bacterial agglutination after the addition of antiserum (right) (from Taylor, 1970a).

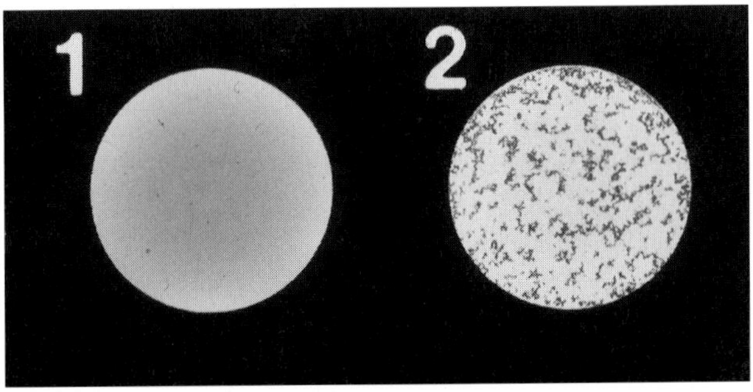

Fig. 32. Slide agglutination test. Agglutination with *Staphylococcus aureus* conjugated antiserum 1, positive granular agglutination; 2, negative granular agglutination. (From Lyons and Taylor, 1990.)

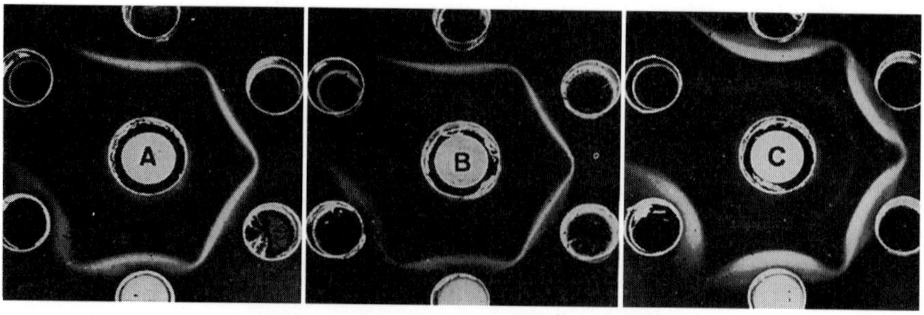

Fig. 33. Ouchterlony agar double-diffusion test. Precipitation lines indicate positive antiserum antigen reactions for *Pseudomonas syringae* pv. *phaseolicola*. (A) Race 1 antiserum; (B,C) race 2 antisera. The peripheral wells, beginning at the top and moving clockwise, contain antigen for race 2, race 1, race 2, race 1, race 2 and *P. fluorescens*. (From Guthrie, 1968.)

immunodiffusion (Ouchterlony double-diffusion method) where the antiserum diffuses from a central well through the agar to precipitate, or not, against individual antigens diffusing from the surrounding wells (Fig. 33). If similar precipitation lines formed then the antigens were considered to be identical (Fig. 33).

Accepting their limitations, such techniques have been used to detect *P. syringae* pv. *phaseolicola* of phaseolus bean seeds in the USA, UK, France and the Netherlands (Guthrie *et al.*, 1965; Taylor, 1970c; Trigalet and Bidaud, 1978; Van Vuurde and Van Den Bovenkamp, 1981), *X. campestris* pv. *phaseoli* in the USA (Saettler, 1971; Trujillo and Saettler, 1979) and *P. syringae* pv. *pisi* in compound peas in the UK (Ball and Reeves, 1992) (see Chapter 6).

Enzyme-linked immunosorbent assay

The ELISA technique, modified for the detection of plant viruses by Clark and Adams (1977), has had considerable impact on the diagnosis of virus and bacterial diseases of vegetatively produced crops with a lesser but nonetheless important role in the diagnosis of seedborne diseases (Lange, 1986). The development and principles of immunosorbent assays have been reviewed by Clark (1981).

There are many different forms of ELISA (Ball and Reeves, 1992). Three of the main types are the direct (double antibody sandwich (DAS-ELISA)) method, the indirect method and the competitive assay (Lange, 1986). The direct method is most often used in seed health assays (Lange, 1986). Using the direct method, seed or seedling (Van Vuurde and Maat, 1983) extract (i.e. the antigen) is selectively trapped and immobilized by solid-phase-specific antibody in microtitre wells. Enzyme-labelled antibody is then reacted with the immobilized antigen and, after removing unreacted enzyme-labelled antibody, the retained enzyme is assayed by adding a suitable substrate (Clark, 1981). Qualitative assays may be made by eye and visualization may be improved by the addition of substrates that give coloured hydrolysates. Quantitative assays are made using colorimetric or spectrophotometric equipment (Clark, 1981). Descriptions and schematic diagrams (Fig. 34) of the processes involved in direct and indirect ELISA methods are given by Fox (1993).

Polyclonal antibodies are effective when testing for viruses where extreme specificity is not needed and the detection of strains is seldom required. However, they fail to achieve the specificity necessary for the detection of fungi and bacteria and also may have cross-reactivity for non-target organisms (Irwin, 1987). A major effect of the introduction of single site-specific monoclonal antibodies has been to improve the reproducibility of ELISA in the detection of pathogens of seeds and vegetatively propagated material (Fox, 1993). When using monoclonal antibodies in direct ELISA tests, polyclonals are commonly applied first as trapping antibodies.

Seed testing using ELISA techniques is based on the use of polystyrene microtitre plates, each containing up to 96 wells, which are especially suitable for tests involving the indexing of numerous samples and sub-samples of seeds.

Lange (1986) describes modifications to the original ELISA method of Clark and Adams (1977) necessary to make it effective for the detection of seedborne viruses. The introduction of the method for the indexing of seedborne viruses in certification schemes was a significant advance. This was the outcome of research, such as that which demonstrated that the ELISA test was more sensitive (Jafarpour *et al.*, 1979) and could be performed more quickly and cheaply than plant indicator and growing-on tests for the detection of LMV in seed samples (Jafarpour *et al.*, 1979; Falk and Purcifull, 1983; Van Vuurde and Maat, 1983). Maury-Chovelon (1984) showed that it was possible, using standard DAS-ELISA, to detect one LMV infected seed in 7000 healthy ones. The ELISA test may be applied to detect to a tolerance level of zero infected seeds out of 30,000 (Dinant and Lot, 1992), the tolerance limit which is imposed to control LMV in the USA (see Chapter 6). High correlations were achieved between plant indicator, growing-on tests and ELISA tests (Falk and Purcifull, 1983; Van Vuurde and Maat, 1983) and over 95% positives were obtained in photometric tests (Van Vuurde and Maat, 1983). Similarly high correlations between ELISA testing and growing-on tests for seedborne peanut mottle virus (PeMoV) were obtained by Bharathan *et al.* (1984) and the method was also used to detect squash mosaic virus (SMV) in parts of squash seeds and in seed lots of cucurbits (Nolan and Campbell, 1984). ELISA is the preferred method of indexing for 25 of the 35 seed-transmitted viruses of legumes listed in the technical guidelines for the safe movement of legume germplasm internationally (Frison *et al.*, 1990).

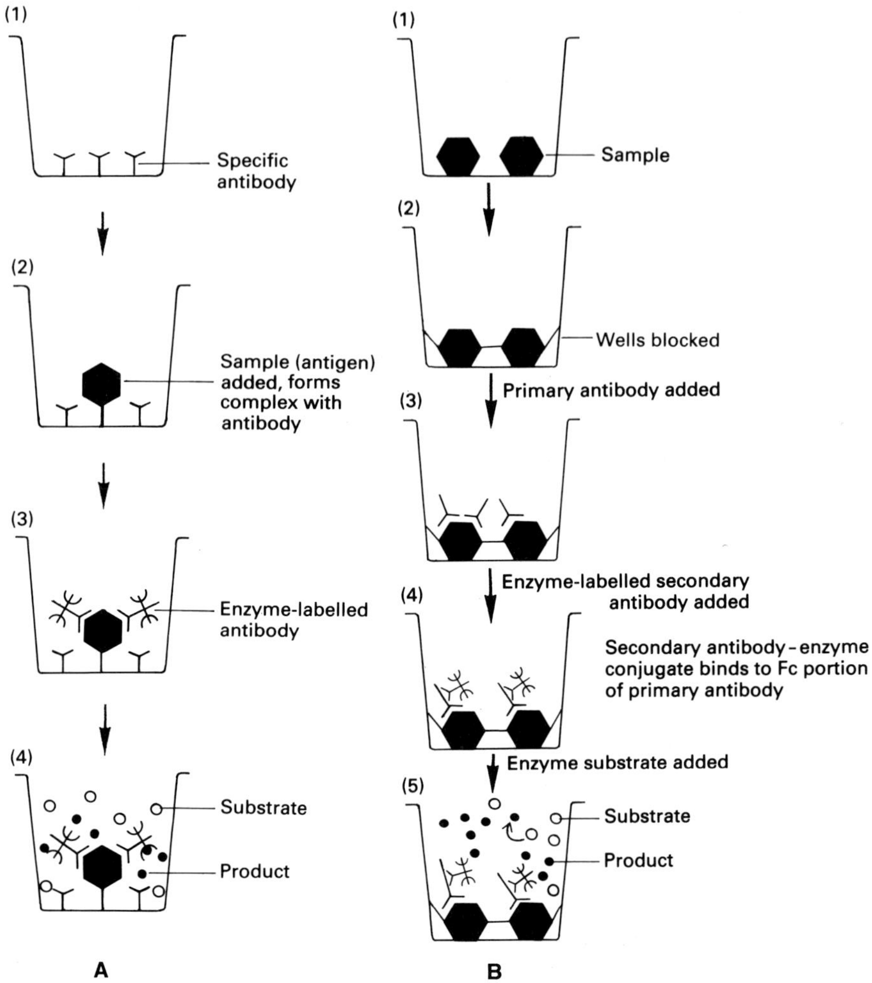

Fig. 34. Enzyme-linked immunosorbent assay (ELISA) procedures. (a) Direct sandwich ELISA. A specific antibody is absorbed to the well (1), to which a test sample, including the pathogen, is added (2), followed by enzyme-labelled specific antibody (3) and then enzyme substrate, leading to a coloured product in proportion to concentration of the pathogen (4). (b) Indirect ELISA using antigen-coated wells. A test sample, including the pathogen, is used to coat the well (1) and the remaining sites are blocked (2); specific primary antibody (3), enzyme-labelled secondary antibody (4) and then enzyme substrate are added, leading to a coloured product in proportion to the concentration of the pathogen (5). (Redrawn from Fox, 1993.)

Highly specific monoclonal antibodies have been developed to *P. syringae* pv. *pisi* (bacterial blight of pea) and have been tested using both indirect and competitive ELISA assays to distinguish between strains of the bacterium (Candlish *et al.*, 1988). As a result, a commercial kit to detect seedborne infection has been developed and marketed (J.D. Taylor, pers. comm., 1994) to give seed producers an approximate indication of the disease status of their seed (Ball and Reeves, 1991).

There are monoclonal antibody techniques now which can be used to detect seedborne fungi of spruce seeds (Mitchell, 1988) and certain seedborne fungi (Dewey, 1992) and bacteria (Alvarez *et al.*, 1993) of rice seeds. Recently MAbs have been raised with limited success in detecting seedborne *Pyrenophora graminea* (leaf stripe of barley) (Burns *et al.*, 1994).

Other immunodiagnostic methods

There are modifications of the ELISA test and there are other immunodiagnostic methods, some of which have been developed for seedborne pathogens in response to a perceived need for an increase in sensitivity of detection, for example of LMV in lettuce seed samples. Simultaneous addition of seed extract and enzyme-conjugated antibodies was reported to enhance test sensitivity in this respect (Van Vuurde and Maat, 1985). Immunoassays such as the biotin-avidin ELISA and the enzyme-linked fluorescent assay (ELFA) were equally effective and superior to standard ELISA in the detection of seedborne LMV (Dolores-Talens *et al.*, 1989).

Radioimmunosorbent assay (RIA) is based on DAS-ELISA but uses radioactive gamma globulin instead of a gamma globulin conjugate and can detect down to three infected seeds in 30,000 (Ghabrial and Shepherd, 1982). The method employs sophisticated equipment and a potentially dangerous isotope, ^{125}I (Dinant and Lot, 1992). A variation of this method based on the solid-phase radioimmune assay (SPRIA) and is as sensitive a detection method as RIA (Kulik, 1984).

Immunosorbent electron microscopy (ISEM) methods involve the use of antibodies bonded onto carbon-coated grids which are then floated onto crude seed extracts. Antigen (virus) antibody bonding occurs and, after staining, the adsorbed virus can be seen and photographed using transmission electron microscopy. Using this technique it was possible to detect one part of LMV-infected seed extract in 100 parts of healthy lettuce seed extract, and barley stripe mosaic virus (BSMV), tobacco ringspot virus (TRSV) and soyabean mosaic virus (SoyMV) were all detected at a ten times lower incidence (Brlansky and Derrick, 1979). ISEM is a rapid, reliable method for testing seed for virus infection. It is well suited for screening programmes in which several viruses are included, but it is unlikely to be used where numerous samples require testing for the same virus as each grid has to be handled separately (Lange, 1986).

Indirect (Malin *et al.*, 1983) and direct (Franken and Van Vuurde, 1990) immunofluorescence staining involves microscope recognition under ultraviolet light of target antigen after staining with homologous antibodies which have been conjugated with a fluorescent dye (e.g. fluorescein isothiocyanate (FITC)). Using the direct method, seed leachates from beans are mixed with agar in Petri dishes and, when colonies are formed 48 h later, the agar is incubated with the FITC-conjugated antiserum. After several washing steps the agar can be viewed by low power ultraviolet microscopy for the presence of fluorescing colonies which identify the organism (Franken and Van Vuurde, 1990). The amount of fluorescence is proportional to the concentration of bacteria in the seed preparation.

Xanthomonas campestris pv. *phaseoli* cells were identified in bean seed samples containing one infected seed in 10,000 (0.01%) using indirect immunofluorescence (Malin *et al*., 1983) and good detection results are claimed for seedborne *Pseudomonas syringae* pv. *phaseolicola* (halo blight of bean) and *Clavibacter michiganensis* subsp. *michiganensis* (bacterial canker of tomato) using direct immunofluorescence (Franken and Van Vuurde, 1990).

Dot-ELISA (dot blot, dot immunobinding or spot immunodetection) is a further modification of the ELISA technique (Lange, 1986; Stead, 1992). This is an indirect procedure where the target antigen is bound to a membrane (nylon or nitrocellulose), for example, by touching the cut surface of a presoaked virus-infected seed against it. After blocking, in a mixture of skimmed milk and detergent, the membrane is treated with antiserum specific for the virus, washed and then treated with enzyme-coagulated immunoglobulin (Lange, 1986). A positive result appears as a coloured dot on a white background. Dot-ELISA was found to be slightly more sensitive for the detection of pea seedborne mosaic virus (PSbMV) than ELISA (Lange, 1986). However, in 1986 (Lange, 1986) dot-ELISA, though simple and apparently effective, remained to be proved in seed health testing. This technique is not the same but is based on broadly similar principles to nucleic acid dot blotting (see pp. 204–205).

There is a critical appraisal of some serological methods including ELISA, Ouchterlony double diffusion, immunofluorescence microscopy and slide agglutination for the detection of seedborne bacteria (Franken and Van Vuurde, 1990) (Table 24). In this comparison analytical sensitivity and specificity, respectively, describe the minimum amount of antigen detectable and the degree of cross-reactivity in a test (Sheppard *et al*., 1986). Standardization (Table 24) of working procedures is important (Trigalet *et al*., 1978; Schaad, 1979). Some methods are more amenable to standardization than others. The ELISA technique can be easily automated and as a result standardized and is also suited for screening large numbers of seeds (Sheppard, 1983; Sheppard *et al*., 1986). In this respect ELISA may have a performance advantage over the other methods for the detection of seedborne bacteria (Table 24).

Table 24. Some important properties of serological techniques commonly used in the detection of seedborne bacteria (Franken and Van Vuurde, 1990).

Properties	ELISA	Double diffusion	Immunofluorescence microscopy	Slide agglutination
Analytical specificity*	2–3†	1	2	3
Analytical sensitivity*	3‡	4	2	3
Potential standardization	1†	1	2	1–2
For testing large numbers of seeds	1†	2–3	2	2

* Terminology of Sheppard *et al.* (1986) is used (see p. 202).
†1, very high; 2, high; 3, moderate; 4, low.
‡1, very high = 10^2–10^3 cells ml^{-1}; 2, high = 10^3–10^5 cells ml^{-1}; 3, moderate = 10^5–10^7 cells ml^{-1}; 4, low = > 10^7 cells ml^{-1}.

Nucleic acid methods

The exploitation of nucleic acids in practical methods for the detection and/or identification of plant pathogens is in its infancy (Oliver, 1993). This is especially true in relation to the detection of seedborne organisms, but some progress has been made. Lange (1986) examined nucleic acid techniques for the detection of seedborne viruses, and their use for the detection of plant pathogenic bacteria (Rasmussen and Reeves, 1992) and seedborne diseases (Reeves, 1995) has been reviewed.

Use of probes

Ribonucleic acid (RNA) and deoxyribonucleic acid (DNA) contain the genetic information from which the cell compounds of all living organisms are derived. The identity of any organism results from the expression of its nucleic acids into protein and RNAs and the detection of a nucleic acid sequence is a positive identification (Oliver, 1993). Thus all viable propagules, for example virus particles, bacterial cells, fungal spores and mycelium, contain the entire nucleic complement for that organism. DNA occurs as a double helix, the two strands of which are held together by hydrogen bonding between base pairs; adenine (A) binds only to thymine (T) and guanosine (G) only to cytosine (C) (Oliver, 1993) (Fig. 35).

A specific sequence of DNA from a plant pathogen can be selected which only couples (i.e. hybridizes) with the identical sequence of DNA in that target

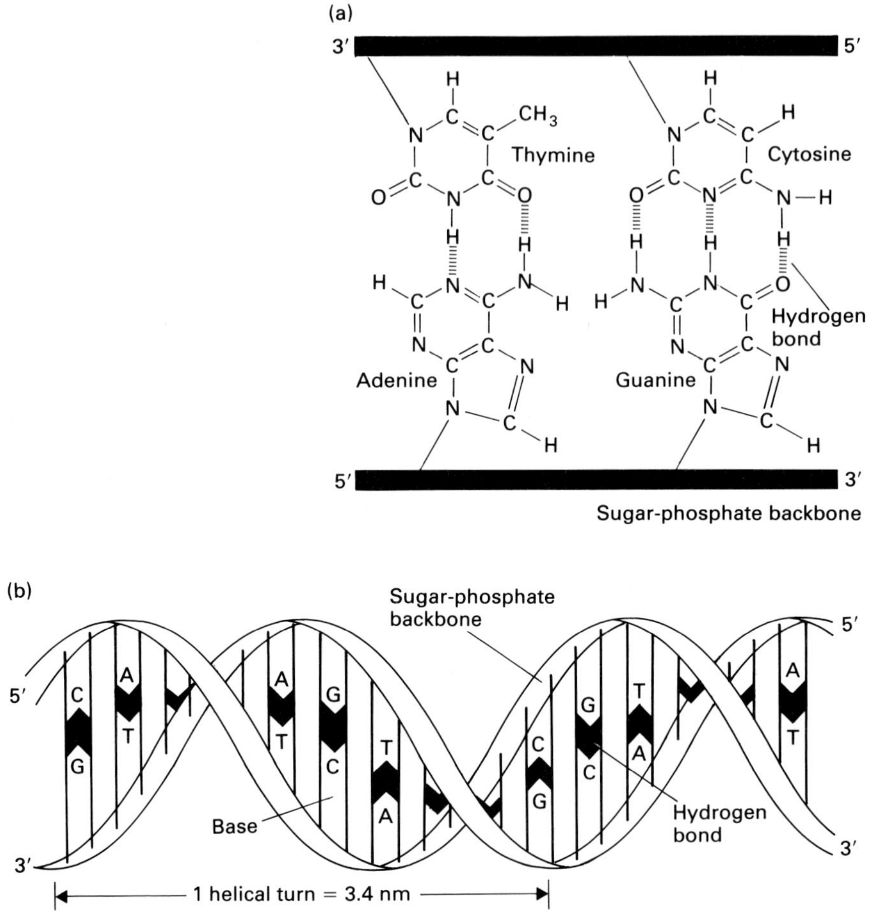

Fig. 35. Structure of DNA: base pairs (a) and the double helix (b) (redrawn from Oliver, 1993).

organism. This is a probe, and probes are labelled using radioisotopes or biochemically so that hybridization of the probe and target DNA can be detected visually (Ball and Reeves, 1991). Selection of probes with the necessary degree of specificity is critical for the identification of pathogens (Oliver, 1993). Nucleic acid methods may identify DNA from both living and dead organisms and plant inoculations may be necessary to verify the pathogenicity of microorganisms.

Target DNA is dotted (dot blots) onto and bound to nylon or nitrocellulose filters. These filters are incubated in the presence of probe DNA. The filters are then autoradiographed and if there is a positive reaction, i.e. the probe and target DNA hybridize, the label appears as a spot on photographic film. The dot blot method is quick, simple and a number of samples can be processed at the same time. Where the target organism can be cultured, other forms of hybridization, including colony blotting, may be used (Rasmussen and Reeves, 1992; Oliver, 1993).

Detection of bacteria

Probes developed in the 1980s were used for the detection and identification of bacteria in medical and microbiological areas of research (Reeves, 1995). Their use for the identification of plant pathogenic bacteria from plant extracts followed (Rasmussen and Reeves, 1992) but there has been less progress in their development for the detection of seedborne pathogens in seed testing programmes (Reeves, 1995). Reeves (1995) suggests that there are two main ways in which DNA probes could be used in a seed test for the detection of bacteria. Seed extracts could be probed and dot blotted directly or pure colonies from isolation plates could probed and dot blotted or colony blots could be made. The first method involves the use of the probe for detection in which case sensitivity and accuracy are required. For dot blotting of colonies and for colony blots, some extraction and purification of the organism has been done and the probe is being used essentially for identification; sensitivity is not important because the organism is present in quantity (Reeves, 1995).

Both approaches were applied by Schaad *et al.* (1989), who demonstrated that a DNA probe carrying a fragment of a gene involved in phaseolotoxin production by *Pseudomonas syringae* pv. *phaseolicola* hybridized reliably for pathogen detection and the identification of colonies of the bacterium from seed soak liquid assay plates or for assay or probing of maceration fluids from diseased lesions. However, the probe failed to detect the bacterium in bulked cell suspensions from seeds naturally or artificially contaminated with *P. syringae* pv. *phaseolicola*, but gave positive results when the suspensions were concentrated. Thus, in testing for seedborne bacteria a DNA probe may have a practical use after the bacteria have been extracted from the seeds and isolated onto agar. It could then be used for the identification of pure cultures by colony hybridization or by dot blotting cells lifted from from single colonies (Reeves, 1995).

A problem of direct probing using dot blotting detection methods is that the sensitivity of detection may be adversely affected by contaminating DNA from other microorganisms which are present on seeds and from seed DNA.

Detection of fungi

Nucleic acid techniques have been applied to fungal pathogens to investigate genetic diversity within fungal species and for identification and taxonomic purposes (Coddington *et al.*, 1987; Kistler *et al.* 1987; Manicom *et al.*, 1990; Flood *et al.*, 1992). However, it is not apparent how these methods could be adapted for use in the detection of fungi on or in seeds, particularly where large-scale seed testing is required (Reeves, 1995). Most work involving the use of DNA probes has been for the detection of fungal pathogens in infected plant material or in soil, particularly for those fungi which are difficult to detect or identify using traditional methods (Reeves, 1995). However, some progress can be claimed for the detection of seedborne fungi. Yao *et al.* (1990) successfully used a probe comprising the total genomic DNA of *Peronosclerospora sorghi* (downy mildew of sorghum) to detect this seedborne fungus by hybridization with target DNA extracted by grinding up sorghum seeds. No DNA hybridization occurred in tests of a number of other fungal sorghum seed contaminants and pathogens. They suggested that, because DNA extraction and hybridization procedures were simple, single seed or bulked samples could be used to detect even low frequencies of contaminated seeds.

Detection of viruses

Nucleic acid probe methods are in routine use for the detection of many plant viruses and viroids (Salazar and Querci, 1992) but as yet have not been used for this purpose in seed pathology (Lange, 1986; Reeves, 1995). Probe technology is widely used in the routine testing of potato samples for the detection of tuber viruses (see Reeves, 1995). The methods involved are similar to those which could be used for the testing of samples of true seeds.

Conclusions

Nucleic acid probes have been applied widely and quickly in research programmes in plant pathology but they have not been developed as yet for use in routine seed testing. The potential for the use of probes in seed testing is considerable. Some associated problems have been resolved but others await solution. One of the major technological disadvantages, i.e. the radioisotope labelling of probes (Lange, 1986; Schaad *et al.*, 1989; Ball and Reeves, 1992), has been overcome by the introduction of non-radioactive labelling methods (Reeves, 1995). However, before probes can be applied, organisms have to be extracted from seeds, and in many cases they have to be purified. Probes are extremely fast and accurate means of identification using pure colonies of pathogens but, where seed extracts are probed directly, contaminating DNA may cause problems in the accurate detection of organisms.

Polymerase chain reaction technology

The polymerase chain reaction (PCR) is a biochemical method of amplification that multiplies the concentration of the target sequence of DNA by up to a millionfold (Saiki *et al*., 1988; Oliver, 1993), thereby transforming previously undetectable amounts into detectable quantities.

It is possible to use part or all of a target sequence of DNA as primers to increase the concentration or amplify part or all of the sequence. Amplification involves the two primers which flank the DNA segment to be amplified and exposed to repeated cycles of heat denaturation of the DNA, annealing (joining) the primers to their complementary sequences, and extension of the annealed primers with the thermostable *Taq* (from the bacterium *Thermus aquaticus*) DNA polymerase enzyme (Fig. 36). The primers hybridize to opposite strands of the target sequence and are so orientated that DNA synthesis by the polymerase proceeds across the region between the primers, effectively doubling the amount of that DNA segment. Since the extension products are also complementary to, and capable of, binding primers each successive cycle essentially doubles the amount of DNA synthesized in the previous cycle (Saiki *et al*., 1988). The method creates such large increases in concentration that it makes detection of a DNA sequence by gel electrophoresis (Old and Primrose, 1985) or by a specific probe more effective. Reactions take place in small tubes placed into a block in a programmable thermal cycling machine; typically about 30 cycles are used, each taking about 2–5 min (Ball and Reeves, 1991; Fox, 1993).

Potentially, the method could be used on washings from seed samples to identify seed pathogens (Vivian, 1992) without the need for extraction and isolation of the organism (Ball and Reeves, 1991). At present, however, although the method is effective with purified DNA there may be problems with seed samples having large amounts of extraneous DNA because this could result in considerable non-specific hybridization with primer DNA (Vivian, 1992). However, non-target DNA did not affect the sensitivity of PCR detection of *Pseudomonas syringae* pv. *phaseolicola* DNA from phaseolus bean samples (Prosen *et al*., 1993).

Therefore, the choice of primers is important and it is suggested that primers with longer sequences of bases (up to about 20) are less likely to be present at random, even in the large genomes of plants (Fox, 1993), and will achieve specific hybridizations.

Detection of bacteria

PCR was used by Rasmussen and Wulff in 1990 to detect *P. syringae* pv. *pisi* at a level of 10^2 cells ml^{-1} of ground pea extract. In this test, however, compounds in the extract inhibited the *Taq* polymerase enzyme used in the PCR reaction (Rasmussen and Wulff, 1991) and a lengthier nucleic acid extraction process was necessary to achieve a reliable result. Enzyme inhibition did not occur if seeds were not ground but were soaked for periods of less than 6 hours (Rasmussen and Wulff, 1991; Reeves *et al*., 1994). False positives which occurred may have been due to contamination problems and

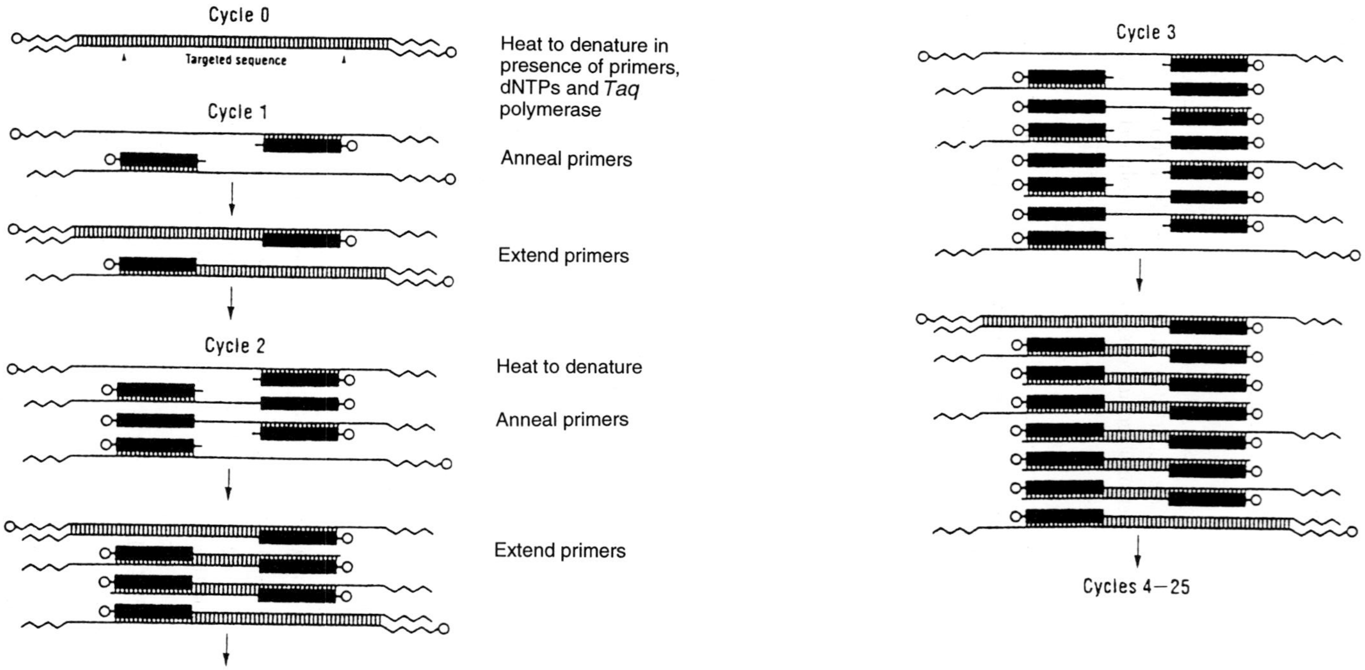

Fig. 36. The polymerase chain reaction based on Ball and Reeves (1991). Figure provided by Dr J.C. Reeves.

a final pathogenicity test may be necessary to eliminate these (Reeves *et al.*, 1994). The probe was effective in identifying *P. syringae* pv. *pisi* from other bacterial species present on pea seeds and a seed test is being developed based on colony hybridization (Reeves *et al.*, 1994).

The PCR technique also has been applied to the detection of the halo blight bacterium *P. syringae* pv. *phaseolicola* in phaseolus bean seeds (Prosen *et al.*, 1991; Tourte and Manceau, 1991). More recently (Prosen *et al.*, 1993) a segment of the phaseolotoxin gene complex was amplified using PCR and was used to detect the pathogen in water extracts from soaked bean seeds. The detection threshold was 10–20 colony forming units (cfu) ml^{-1}. Presumptive epiphytes present at concentrations of $\geq 10^4$ cfu ml^{-1} did not interfere with detection even when the pathogen was present at concentrations of ≤ 30 cfu ml^{-1} (Prosen *et al.*, 1993). The method detected the pathogen in a commercial seed lot that failed to yield the bacterium by conventional techniques (Prosen *et al.*, 1993).

Detection of fungi

Although PCR techniques have been used for the characterization and identification of fungi, little attention has been given to the detection of seedborne pathogens (Reeves, 1995).

PCR has been used to refine existing DNA probe techniques, such as as the probe developed to detect *Phoma tracheiphila* in lemon trees (Rollo *et al.*, 1987). PCR too has been refined, for example the use of nested PCR, where a second round of amplification used primers internal to those of the first round (Schesser *et al.*, 1991). This was done to overcome non-specific amplification products from other fungi and from host tissues in the detection of *Gaeumannomyces graminis* in wheat plants (Schesser *et al.*, 1991).

A further PCR method involves the use of random amplified polymorphic DNA or RAPD (Williams *et al.*, 1990). These are very short primers, not having identified DNA sequences, and usually with 8–12 bases (Caetano Anolles *et al.*, 1992), which will hybridize to a large number of arbitrary sites in the genome (Fox, 1993). When the products from such a reaction are run on an electrophoresis gel, banding patterns may emerge which can be used to distinguish between species or varieties (Ball and Reeves, 1991). RAPD primers have been applied successfully to fungi as a taxonomic instrument to distinguish between species of *Pyrenophora* causing leaf diseases of barley (Reeves and Ball, 1991), and between species of *Mycosphaerella* causing Sigatoka disease of banana (Johanson *et al.*, 1994). They have been used to separate pathotypes of *Leptosphaeria maculans* (Goodwin and Annis, 1991).

RAPD cannot be used directly to detect seedborne organisms because of non-specific hybridization of the primer which may occur with extraneous DNA from seed extracts (Reeves, 1995). However, RAPD can be used as a rapid preliminary sieve to identify DNA fragments which could be tested for their specificity as probes and if suitable could be sequenced to produce longer primers for use with conventional PCR techniques (Reeves, 1995). This approach was used in the development of a PCR method for the detection of the bacterial pathogen of maize, *Erwinia stewartii* (Blakemore *et al.*, 1992; Blakemore and Reeves, 1993), and is being tested for the detection of *Phomopsis* spp. in soyabean seeds (Jaccoud Filho and Reeves, 1993).

Detection of viruses

The majority of viruses have RNA genomes which initially made them unsuitable for PCR (Fox, 1993). However, this has been overcome following the development of a reverse transcription/PCR assay for the identification of plum pox potyvirus (Levy and Hadidi, 1991). A reverse transcriptase enzyme can be used immediately before PCR and this has been done successfully in the detection of pea seedborne mosaic virus (PSbMV) (Kohen *et al.*, 1992) and cucumber mosaic virus (CMV) in lupin seeds (Wylie *et al.*, 1993).

Multiplex PCR and the ligase chain reaction (LCR)

Other chain reaction nucleic acid methods which may have a role in seed diagnostics in the future include multiplex PCR and the LCR (Reeves, 1995). Multiplex PCR allows the simultaneous testing for more than one organism by the use of mixtures of primers which are specific for the individual organisms. The LCR, like PCR, is a DNA amplification system which is highly specific and very sensitive (Lee, 1993) since even a single mismatch between bases will prevent the reaction. The method has been used in human medicine but although it is well suited to automation it has not, as yet, been used in plant pathology (Reeves, 1995). Its extreme specificity, however, may prevent it working effectively as a diagnostic tool for plant pathogens.

Conclusions

The time taken to complete a diagnostic test can be critical for the release of seed. It is necessary to determine the number and frequency of tests that can be completed within a limited time span, a constraint introduced in certain circumstances by the need to sow the seed bulk within a matter of weeks (see Chapter 7). The development of rapid and more advanced methods for the detection of seedborne organisms will increase the frequency of some diagnostic tests and in that respect will benefit commercial seed testing. Nucleic acid-based methods tend to be relatively expensive to apply and the advantages of these more rapid methods have to be considered against the lesser cost, but greater inconvenience to the grower, of the longer period required to achieve identification of a pathogen using isolation-based methods.

It is important that, as new and more advanced detection methods are introduced, their performance should be subjected to critical appraisal to establish the limits of their analytical and diagnostic sensitivity and specificity (Sheppard *et al.*, 1986; Sheppard, 1993), ideally to demonstrate that they are at least as effective as the techniques which they seek to replace. At present, a major disadvantage of nucleic acid methods in seed health testing is that of quantification (Reeves, 1995). It is not possible to determine the incidence of infected seeds in a seed sample directly using these methods and to develop nucleic acid techniques for practical use in diagnostics it may be necessary to link them with a seed assay such as the most probable number method (see Chapter 6) which would produce a quantitative result (Reeves, 1995). This technology remains to be put in place.

References

Aebi, H. and Rapin, J. (1954) Lutte contre la cercosporiose de la betterave sucrière. *Revue romande d'Agriculture, de Viticulture et d'Arboriculture, Lausanne* 10, 45–47.

Agarwal, V.K. (1983) Quality seed production at Pantnagar, India. *Seed Science and Technology* 11, 1071–1078.

Agarwal, V.K. and Sinclair, J.B. (1987) *Principles of Seed Pathology 1 and 2.* CRC Press, Boca Raton, Florida.

Agarwal, V.K., Singh, O.V., Thapliyal, P.N. and Malhotra, B.K. (1974) Control of purple stain disease of soybean. *Indian Journal of Mycology and Plant Pathology* 4, 1–4.

Agarwal, V.K., Singh, O.V. and Modgal, S.C. (1975) Influences of different doses of nitrogen and spacing on the seedborne infections of rice. *Indian Phytopathology* 28, 38–40.

Agrawal, S.C. and Nema, S. (1989) Effect of carbendazim on *Macrophomina* leaf blight of black gram and green gram. *Indian Journal of Plant Protection* 17, 147–149.

Ahmed, A.H.M. and Tribe, H.T. (1977) Biological control of onion white rot *(Sclerotium cepivorum)* by *Coniothyrium minitans. Plant Pathology* 26, 75–78.

Ahmed, Q.A. and Ahmed, N. (1983) Production of high quality seed for important crops in Bangladesh. *Seed Science and Technology* 11, 1079–1086.

Akhtar, M.A. and Khokhar, L.K. (1987) Effect of agrochemicals for control of complete bunt of wheat. *Pakistan Journal of Agricultural Research* 8, 119–120.

Alabouvette, C. and Brunin, B. (1970) Recherches sur la maladie du colza due à *Leptosphaeria maculans* (Desm.): Ces et de Not. 1. Rôle des restes de culture dans la conversation et le dissémination du parasite. *Annales de Phytopatologie* 2, 463–475.

Alagarsamy, G. and Jeyarajan, R. (1989) Chemical control of seedling disease of cotton. *Indian Journal of Plant Protection* 17, 255–257.

Alcock, N.L. (1931) Notes on common diseases sometimes seed borne. *Transactions of the Royal Society of Edinburgh* 30, 332–337.

Alexander, M. (1985) Enhancing nitrogen fixation by use of pesticides: a review. *Advances in Agronomy* 38, 267–282.

Allcock, H.J. and Jones, J.R. (1950) *The Nomogram.* Pitman, London, UK.

Allen, D.J. (1983) *The Pathology of Tropical Food Legumes.* John Wiley & Sons, New York.

Allen, J.D. and Smith, H.C. (1961) Dry-rot (*Leptosphaeria maculans*) of brassicas: seed transmission and treatment. *New Zealand Journal of Agricultural Research* 4, 676–685.

Aluko, M.O. (1983) Seed health testing in Nigeria. *Seed Science and Technology* 11, 1239–1248.

Alvarez, A.M., Bendict, A.A. and Gnanamanickam, S.S. (1993) Identification of seed-borne bacterial pathogens of rice with monoclonal antibodies. In: *Proceedings of the First ISTA Plant Disease Committee Symposium on Seed Health Testing.* Agriculture Canada Central Seed Laboratory, Ottawa, Canada, pp. 3–8.

Amin, K.S. (1981) Pea stem rot and its chemical control. *Indian Phytopathology* 34, 224–225.

Ammon, H.U. (1963) Uber einige Arten aus den Gattungen *Pyrenophora* Fries und *Cochliobolus* Drechsler mit *Helminthosporium* als Nebenfructform. *Phytopathologische Zeitschrift* 47, 244–300.

Anahosur, K.H. and Lakshman, M. (1989) Evaluation of metalaxyl-ZM (280 FW) for the control of sorghum downy mildew. *Karnataka Journal of Agricultural Sciences* 2, 51–53.

Anderson, J.D. (1973) Dichloromethane and lettuce seed germination. *Science* 179, 94–95.

Anonymous (1944) Plant diseases. Notes contributed by the Biological Branch. *Agricultural Gazette of New South Wales* 55, 493–498.

Anonymous (1950) Definitions of some terms used in plant pathology. *Transactions of the British Mycological Society* 33, 154–160.

Anonymous (1954) *Rotations.* Bulletin 85, Ministry of Agriculture and Fisheries. Her Majesty's Stationery Office, London, UK.

Anonymous (1965a) Pea, leaf and pod spot. *Proceedings of the International Seed Testing Association* 30, 1079–1080.

Anonymous (1965b) Cabbage etc., black leg, dry rot, canker. *Proceedings of the International Seed Testing Association* 30, 1109–1110.

Anonymous (1976) *The Fodder Plant Seeds Regulations 1976.* Her Majesty's Stationery Office, London, UK.

Anonymous (1978) *Dry Bulb Onions, Horticultural Enterprises Booklet 1,* 4th edn. Ministry of Agriculture, Fisheries and Food, London, UK.

Anonymous (1980) *Plant Health – The Import and Export (Plant Health) (Great Britain) Order 1980, No.420.* Her Majesty's Stationery Office, London, UK.

Anonymous (1985a) *The Cereal Seed Regulations 1985.* Statutory Instruments No. 976. Her Majesty's Stationery Office, London, UK.

Anonymous (1985b) *The Vegetable Seed Regulations 1985* Statutory Instruments No. 979. Her Majesty's Stationery Office, London, UK.

Anonymous (1985c) International Rules for Seed Testing 1985. *Seed Science and Technology* 13 (2), 520 pp.

Anonymous (1986) *Guidelines for the Importation of Fodder Pea Seeds.* Ministry of Agriculture, Fisheries and Food, Plant Health Division, London, UK.

Anonymous (1988) *Mind Your Peas*. Ministry of Agriculture, Fisheries and Food, Alnwick, UK.

Anonymous (1989) *1992: Health Controls and the Internal Market*. Select Committee on the European Communities, 9th Report. Her Majesty's Stationery Office, London, UK, 121 pp.

Anonymous (1990) *The Fodder Plant Seeds (Amendment) Regulations 1990, No.1352* Her Majesty's Stationery Office, London, UK.

Anonymous (1991) Barley leaf stripe – code of practice introduced. *National Farmers Union of Scotland – Press Release No.91/8, C/15/1*. National Farmers Union, Edinburgh, UK.

Anonymous (1993a) *The Vegetable Seed Regulations 1993*. Statutory Instruments No. 2008. Her Majesty's Stationery Office, London, UK.

Anonymous (1993b) International rules for seed testing: rules 1993. *Seed Science and Technology* 21 (Suppl.).

Anselme, C. (1983) Section 1. Methology and testing techniques. International Symposium on Seed Pathology, Copenhagen 11–16 October 1982. *Seed Science and Technology* 11, 477–691.

Anselme, C., Hewett, P.D. and Champion, R. (1970) L'analyse sanitaire des semences de pois pour la recherche de l'*Ascochyta pisi* Lib. Méthodes d'analyse et limite de tolérance. *International Congress of Plant Protection* 7, 89–91.

Ark, P.A. (1958) Longevity of *Xanthomonas malvacearum* in dried cotton plants. *Plant Disease Reporter* 11, 1293.

Ark, P.A. and Thompson, J.P. (1958) Prevention of antibiotic injury with Na-K-chlorophyllin. *Plant Disease Reporter* 42, 1203–1205.

Arndt, C.H. (1946) Effect of storage conditions on survival of *Colletotrichum gossypii*. *Phytopathology* 36, 24–29.

Arndt, C.H. (1953) Survival of *Colletotrichum gossypii* on cotton seeds in storage. *Phytopathology* 43, 220.

Asher, M.J.C. and Dewar, A.M. (1994) Control of pests and diseases in sugar beet by seed treatments. In: Martin, T.J. (ed.) *Seed Treatment: Progress and Prospects*. BCPC Monograph No. 57. BCPC Publications, Farnham, UK, pp. 151–158.

Ashida, J. (1965) Adaptation of fungi to metal toxicants. *Annual Review of Phytopathology* 3, 153–174.

Aujla, S.S., Sharma, I. and Bir, B.S. (1986) Effect of various fungicides on teliospore germination of *Neovossia indica*. *Journal of Research, Punjab Agricultural University* 23, 442–443.

Aujla, S.S., Kaur, S. and Sharma, I. (1989) Chemical seed treatment for control of Karnal bunt of wheat. *Plant Protection Bulletin (Faridabad)* 41, 20–21.

Aulakh, K.S., Mathur, S.B. and Neergaard, P. (1974) Comparison of seedborne infection of *Drechslera oryzae* as recorded on blotter and in soil. *Seed Science and Technology* 2, 385–391.

Aveling, T.A.S. and Robbertse, P.J. (1990) Evaluation of antibiotics against *Xanthomonas campestris* causing black rot of *Brassica*. *Phytophylactica* 22, 229–231.

Awoderu, V.A., Larinde, M.A. and Botchey, S. (1983) Production of high quality in the West Africa Rice Development Association. *Seed Science and Technology* 11, 1093–1101.

Ayers, W.A. and Adams, P.B. (1981) Mycoparasitism and its application to biological control of plant diseases. In: Papavizas, G.C. (ed.) *Biological Control in Crop Production.* Allanheld, Osmun, Totowa, New Jersey, pp. 91–103.

Aylor, D.E. (1990) The role of intermittent wind in the dispersal of fungal pathogens. *Annual Review of Phytopathology* 28, 73–92.

Bachy, A. and Fehling, C. (1957) La fusariose du palmier a huile en Cote d'Ivoire. *Journal d'Agriculture Tropicale et de Botanique Appliquee* 4, 228–240.

Bacon, J.R. and Clayton, P.B. (1986) Protection for seeds: a new film coating technique. *Span* 29, 54–56.

Bacon, J.R., Brocklehurst, P.A., Gould, A., Mahon, R., Martin, N.C.J. and Wraith, M.J. (1988) Evaluation of a small scale fluidised bed seed treatment apparatus. In: Martin, T.J. (ed.) *Application to Seeds and Soil.* BCPC Monograph No. 39. BCPC Publications, Thornton Heath, UK, pp. 237–241.

Bagegni, A.M., Sleper, D.A., Kerr, H.D. and Morris, J.S. (1990) Viability of *Acremonium coenophialum* in tall fescue seed after ionising radiation treatments. *Crop Science* 30, 1272–1275.

Bailiss, K.W. and Offei, S.K. (1990) Alfalfa mosaic virus in lucerne seed during seed maturation and storage, and in seedlings. *Plant Pathology* 39, 539–547.

Baker, K.F. (1948) Fusarium wilt of garden stock (*Matthiola incana*). *Phytopathology* 38, 399–403.

Baker, K.F. (1950) Bacterial fasciation disease of ornamental plants in California. *Plant Disease Reporter* 34, 121–126.

Baker, K.F. (1956) Development and production of pathogen-free seed of three ornamental plants. *Plant Disease Reporter* Supplement 238, 68–71.

Baker, K.F. (1962a) Thermotherapy of planting material. *Phytopathology* 52, 1244–1255.

Baker, K.F. (1962b) Principles of heat treatment of soil and planting material. *Journal of the Institute of Agricultural Science* 28, 118–126.

Baker, K.F. (1969) Aerated steam treatment of seed for disease control. *Horticultural Research* 9, 59–73.

Baker, K.F. (1972) Seed pathology. In: Kozlowski, T.T. (ed.) *Seed Biology*, Vol. 2. Academic Press, New York, pp. 317–416.

Baker, K.F. (1980) Pathology of flower seeds. *Seed Science and Technology* 8, 575–589.

Baker, K.F. (1981) Biological control. In: Mace, M.E., Bell, A.A. and Beckman, C.H. (eds) *Fungal Wilt Diseases of Plants.* Academic Press, New York, pp. 523–560.

Baker, K.F. and Cook, R.J. (1974) *Biological Control of Plant Pathogens.* W.H. Freeman, San Francisco.

Baker, K.F. and Davis, L.H. (1950a) Some diseases of ornamental plants in California caused by species of *Alternaria* and *Stemphylium. Plant Disease Reporter* 34, 402–413.

Baker, K.F. and Davis, L.H. (1950b) Heterosporium disease of nasturtium and its control. *Phytopathology* 40, 553–566.

Baker, K.F. and Snyder, W.C. (1950) Plant diseases. Restrictive effects of California climate on some vegetable, flower, grain diseases. *California Agriculture* 8, 15–16.

Ball, S. (1992) New technology in disease and varietal recognition. *The Agronomist* 1, 10–11.

Ball, S.F.L. and Reeves, J.C. (1991) The application of new techniques in the rapid testing for seed-borne pathogens. *Plant Varieties and Seeds* 4, 169–176.

Ball, S. and Reeves, J. (1992) Application of rapid techniques to seed health testing – prospects and potential. In: Duncan J.M. and Torrance, L. (eds) *Techniques for the Rapid Detection of Plant Pathogens.* Blackwell Scientific Publications, Oxford, pp. 193–207.

Bant, J.H. and Storey, I.F. (1952) Hot water treatment of celery seed in Lancashire. *Plant Pathology* 1, 81–83.

Bant, J.H., Beaumont, A. and Storey, I.F. (1950) Hot water treatment of broccoli seed. *N. A. A. S. Quarterly Review* 9, 43–46.

Bartlett, D.H. (1994) Review of current and future seed treatment usage in oilseed rape. In: Martin, T.J. (ed.) *Seed Treatment: Progress and Prospects,* BCPC Monograph No. 57. BCPC Publications, Farnham, UK, pp. 159–168.

Bassey, E.O. and Gabrielson, R.L. (1983) The effects of humidity, seed infection level, temperature and nutrient stress on cabbage seedling disease caused by *Alternaria brassicicola. Seed Science and Technology* 11, 403–410.

Basu Chaudhary, K.C. and Mathur, S.B. (1979) Infection of sorghum seeds by *Colletotrichum graminicola* and transmission of the pathogen. *Seed Science and Technology* 7, 87–92.

Bateman, G.L., Ehle, H. and Wallace, H.A.H. (1986) Fungicidal treatment of cereal seeds. In: Jeffs, K.A. (ed.) *Seed Treatment.* BCPC Publications, Thornton Heath, UK, pp. 83–111.

Batts, C.C.V. (1955) Observations on the infection of wheat by loose smut (*Ustilago tritici* (Pers.) Rostr.). *Transactions of the British Mycological Society* 38, 465–475.

Batts, C.C.V. (1956) The control of loose smut in wheat and barley. *Annals of Applied Biology* 44, 437–452.

Batts, C.C.V. and Jeater, A. (1958a) The reaction of wheat varieties to loose smut as determined by embryo, seedling and adult plant tests. *Annals of Applied Biology* 46, 23–29.

Batts, C.C.V. and Jeater, A. (1958b) The development of loose smut in susceptible varieties of wheat, and some observations on field infection. *Transactions of the British Mycological Society* 41, 115–124.

Becker, G.J.F. (1956) Verslag over de werkzaamheden betreffende de afrijpingsziekten van granen in 1956. *Stichting Nederlands Graan-Centrum, Werkgroep Graanziekten – Commissie Afrijpingsziekten,* 14 pp.

Belletti, P. and Tamietti, G. (1982) Use of dry heat in the treatment of bean seeds infected by *Pseudomonas phaseolicola. Informatore Fitopatologico* 32, 59–61.

Bel'skii, A.I. and Mazulenko, N.N. (1984) Effects of presowing laser treatment of barley seeds on the incidence of fungal diseases of the plants. *Mikologiya i Fitopatologiya* 18, 312–316.

Bennett, C.W. (1969) Seed transmission of plant viruses. *Advances in Virus Research* 14, 221–261.

Bent, K.J. (1978) Chemical control of plant diseases: some relationships to pathogen ecology. In: Scott, P.R. and Bainbridge, A. (eds) *Plant Disease Epidemiology.* Blackwell Scientific Publications, Oxford, UK, pp. 177–186.

Bertus, A.L. (1972) The eradication of seed-borne *Septoria lactucae* Pass. from lettuce with aerated steam. *Journal of Horticultural Science* 47, 259–261.

Besri, M. (1983) Seed-borne diseases of wheat and barley in Morocco in relation to seed certification programmes. *Seed Science and Technology* 11, 1103–1113.

Bewley, J.D. and Black, M. (1994) *Seeds: Physiology of Development and Germination*, 2nd edn, Plenum Press, New York, 445 pp.

Beyries, A. (1969) Effectiveness of some systemic fungicides against *Corticium solani* by seed treatment of bean and radish. *Phytiatrie-Phytopharmacie* 18, 107–116.

Bharathan, N., Reddy, D.V.R., Rajeshwari, R., Murthy, V.K., Rao, V.R., and Lister, R.M. (1984) Screening peanut germ plasm lines by enzyme-linked immunosorbent assay for seed transmission of peanut mottle virus. *Plant Disease* 68, 757–758.

Biddle, A.J. (1981) Pea seed treatments to control *Ascochyta*. *Tests of Agrochemicals and Cultivars, Annals of Applied Biology* 97 (Suppl. 2), 34–35.

Biddle, A.J. (1994) Seed treatment usage on peas and beans in the UK. In: Martin, T.J. (ed.) *Seed Treatment: Progress and Prospects*. BCPC Monograph No. 57. BCPC Publications, Farnham, UK, pp. 143–149.

Bisht, V.S., Sinclair, J.B., Hummel, J.W. and McClary, R.D. (1982) Effect of tillage systems on yield components and diseases of soybeans. *Phytopathology* 72, 1134.

Blakemore, E.J.A. and Reeves, J.C. (1993) PCR used in the development of a new seed health test to identify *Erwinia stewartii*, a bacterial pathogen of maize. In: *Proceedings of the First ISTA Plant Disease Committee Symposium on Seed Health Testing*. Agriculture Canada Central Seed Laboratory Ottawa, Ontario, Canada, pp. 19–22.

Blakemore, E.J.A., Reeves, J.C. and Ball, S.F.L. (1992) Research note: polymerase chain reaction used in the development of a DNA probe to identify *Erwinia stewartii*, a bacterial pathogen of maize. *Seed Science and Technology* 20, 331–335.

Blith, W. (1652) *The English Improver Improved*. John Wright, London.

Blodgett, E.C. (1946) Observation on blasting of onion heads in Idaho. *Plant Disease Reporter* 30, 77–81.

Bochow, H. and Bottcher, H. (1978) Zur Bekampfung von *Botrytis alli* Munn durch Einsatz von Fungiziden. *Nachrichtenblatt fur den Pflanzenschutz in der DDR* 32, 135–137.

Bollen, G.J. and Fuchs, A. (1970) On the specificity of the *in vitro* and *in vivo* antifungal activity of benomyl. *Netherlands Journal of Plant Pathology* 76, 229–312.

Bolley, H.L. (1897) New work upon the smuts of wheat, oats and barley, a resume of treatment experiments for the last three years. *North Dakota Agricultural Experiment Station Bulletin* 27, 109–164.

Bolton, N.J.E. and Smith, J.M. (1988) Strategies to combat fungicide resistance in barley powdery mildew. *Proceedings of the Brighton Crop Protection Conference – Pests and Diseases* 1, 367–372.

Bondarzew, A.A. (1914) A new fungus disease of clover blossom. *Journal of Plant Diseases* (Russian) V111, 1.

Bonman, J.M. and Gabrielson, R.L. (1981) Localized infections of siliques and seed of cabbage by *Phoma lingam*. *Plant Disease* 65, 868–869.

Bos, L. (1977) Seed-borne viruses. In: Hewitt, W.B. and Chiarappa, L. (eds) *Plant Health and Quarantine in International Transfer of Genetic Resources*. CRC Press, Cleveland, Ohio, pp. 39–69.

Bos, L. (1981) Wild plants in the ecology of virus diseases. In: Maramorosch, K. and Harris, K.F. (eds) *Plant Diseases and Vectors: Ecology and Epidemiology*. Academic Press, New York, pp. 1–33.

Bottcher, H. (1988) Efficacy of fungicides against rotting in stored onion (*Allium cepa*). *Archiv für Phytopathologie und Pflanzenschutz* 24, 125–131.

Bowling, C.C. (1986) Treatment of rice seeds. Part 11 – In the USA and South America. In: Jeffs, K.A. (ed.) *Seed Treatment.* Publications, Thornton Heath, UK, pp. 155–162.

Bowman, J.G. (1988) The contribution and value of resistant cultivars to disease control in oilseed rape. In: Clifford B.C. and Lester, E. (eds) *Control of Plant Diseases.* Blackwell Scientific Publications, Oxford, pp. 93–102.

Bradford, K.J. (1986) Manipulation of seed water relations via osmotic priming to improve germination under stress conditions. *HortScience* 21, 1105–1112.

Braun, H. (1920) Presoak method of seed treatment: a means of preventing seed injury due to chemical disinfectants and of increasing germicidal efficiency. *Journal of Agricultural Research* 19, 363–392.

Brian, P.W. (1957) Effects of antibiotics on plants. *Annual Review of Plant Physiology* 8, 413–426.

Brian, P.W. and McGowan, J.C. (1945) Viridin: A highly fungistatic substance produced by *Trichoderma viride. Nature* 156, 144–145.

Briggs, D.E., Waring, R.H. and Hackett, A.M. (1974) The metabolism of carboxin in growing barley plants. *Pesticide Science* 5, 599–607.

Brinkerhoff, L.A. and Fink, G.B. (1964) Survival and infectivity of *Xanthomonas malvacearum* in cotton plant debris and soil. *Phytopathology* 54, 1198–1201.

Brinkerhoff, L.A. and Hunter, R.E. (1963) Internally infected seed as a source of inoculum for the primary cycle of bacterial blight of cotton. *Phytopathology* 53, 1397–1401.

Brlansky, R.H. and Derrick, K.S. (1979) Detection of seedborne plant viruses using serologically specific electron microscopy. *Phytopathology* 69, 96–100.

Broadbent, L. (1957) *Investigations of Virus Diseases of Brassica Crops.* Cambridge University Press, London, UK, 94 pp.

Broadbent, L. (1965) The epidemiology of tomato mosaic. X1. Seed transmission of TMV. *Annals of Applied Biology* 56, 177–205.

Broadbent, L. (1976) Epidemiology and control of tomato mosaic virus. *Annual Review of Phytopathology* 14, 75–96.

Broadbent, L., Tinsley, T.W., Buddin, W. and Roberts, E.T. (1951) The spread of lettuce mosaic in the field. *Annals of Applied Biology* 38, 689–706.

Broadbent, L., Read, W.H. and Last, F.T. (1965) The epidemiology of tomato mosaic virus X. Persistence of TMV-infected debris in soil, and the effects of soil partial sterilisation. *Annals of Applied Biology* 55, 471–483.

Brodal, G. (1993) Fungicide treatment of cereal seeds according to needs in the Nordic Countries. *Proceedings Crop Protection in Northern Britain,* 7–16.

Brokenshire, T. (1980) Control of pea downy mildew with seed treatment and sprays. *Tests of Agrochemicals and Cultivars, Annals of Applied Biology* 94 (Suppl. 1), 34–35.

Brooks, D.H. (1970) Powdery mildew of barley and its control. *Outlook on Agriculture* 6, 122–127.

Brooks, D.H. (1972) Observations on the effect of mildew, *Erysiphe graminis,* on the growth of spring and winter barley. *Annals of Applied Biology* 70, 149–156.

Brooks, D.H. and Buckley, N.G. (1977) Results in practice – 1. Cereals and grasses. In: Marsh, R.W. (ed.) *Systemic Fungicides,* 2nd edn. Longman, London, UK, pp. 213–219.

Budrina, A. (1935) Methods for the phytopathological examination of seeds. *Plant Protection (Zaschita Rastenii ot Vreditelei)* 6, 13–22.

Bujalski, W. and Nienow, A.W. (1991) The large scale priming of onion seeds: a comparison of different strategies for oxygenation. *Scientia Horticulturae* 46, 13–24.

Bujalski, W., Nienow, A.W. and Petch G.M. (1991) The bulk priming of leek seeds – the influence of oxygen enriched air. *Process Biochemistry* 26, 281–286.

Bujalski, W., Nienow, A.W., Petch G.M., Drew, R.L.K. and Maude, R.B. (1992) The process engineering of leek seeds: a feasibility study. *Seed Science and Technology* 20, 129–139.

Buller, A.H.R. (1950) *Researches on Fungi*, Vol. 7. University of Toronto Press, Toronto, Canada, pp. 415–428.

Burgess, L.W. and Griffen, D.M. (1968) The recovery of *Gibberella zeae* from wheat straws. *Australian Journal of Experimental Agriculture and Animal Husbandry* 8, 364–370.

Burkholder, W.H. (1948) Bacteria as plant pathogens. *Annual Review of Microbiology* 2, 389–412.

Burns, R., Vernon, M.L. and George, E.L. (1994) Monoclonal antibodies for the detection of *Pyrenophora graminea*. In: Schots, A., Dewey, F.M. and Oliver, R. (eds) *Modern Assays for Plant Pathogenic Fungi – Identification, Detection and Quantification.* CAB International, Wallingford, UK, pp. 199–203.

Butler, E.J. and Jones, S.G. (1949) *Plant Pathology.* Macmillan, London, UK.

Buttress, F.A. and Dennis, R.W.G. (1947) The early history of cereal seed treatment in England. *Agricultural History* 21, 93–103.

Byford, W.J. and Gambogi, P. (1985) Phoma and other fungi on beet seed. *Transactions of the British Mycological Society* 84, 21–28.

Byrde, R.J.W. and Archer, S.A. (1979) Diseases of plants. In: Hawker, L.E. and Linton, A.H. (eds) *Micro-organisms – Function, Form and Environment*, 2nd edn, Edward Arnold, London, UK, pp. 288–299.

Caetano Anolles, G., Bassam, B.J. and Gresshoff, P.M. (1992) Primer–template interactions during DNA amplification fingerprinting with single arbitary oligonucleotides. *Molecular and General Genetics* 235, 157–165.

Cafati, C.R. and Saettler, A.W. (1980) Role of nonhost species as alternate inoculum sources of *Xanthomonas phaseoli*. *Plant Disease* 64, 194–196.

Caldwell, J. (1934) The physiology of virus diseases in plants. V. The movement of the virus agent in tobacco and tomato. *Annals of Applied Biology* 21, 191–205.

Calvert, E.L. and Muskett, A.E. (1945) Blind-seed disease of rye-grass (*Phialea temulenta* Prill. & Delacr.). *Annals of Applied Biology* 32, 329–343.

Campbell, W.P. (1957) Studies on ergot infection in graminaceous hosts. *Canadian Journal of Botany* 35, 315–320.

Campbell, W.P. (1958) Infection of barley by *Claviceps purpurea*. *Canadian Journal of Botany* 36, 615–619.

Candlish, A.A.G., Taylor, J.D. and Cameron, J. (1988) Immunological methods as applied to bacterial pea blight. *Proceedings of the Brighton Crop Protection Conference – Pests and Diseases* 2, 787–794.

Carroll, T.W. (1983) Certification schemes against barley stripe mosaic. *Seed Science and Technology* 11, 1033–1042.

Carroll, T.W., Gossel, P.L. and Hockett, E.A. (1979) Inheritance of resistance to seed transmission of barley stripe mosaic virus in barley. *Phytopathology* 69, 431–433.

Carter, M.V. and Moller, W.J. (1961) Factors affecting the survival and dissemination of *Mycosphaerella pinodes* (Berk. & Blox.) Vestergr. in South Australian irrigated pea fields. *Australian Journal of Agricultural Research* 12, 878–888.

Castellani, E. (1988) *Pyricularia oryzae* (Briosi & Cavara). In: Smith, I.M., Dunez, J., Phillips, D.H., Lelliot, R.A. and Archer, S.A. (eds) *European Handbook of Plant Diseases*. Blackwell Scientific Publications, Oxford, UK, pp. 331–333.

Cercauskas, R.F., Dhingra, O.D., Sinclair, J.B. and Asmus, G. (1983) *Amanaranthus spinosa, Leonotis nepetaefolia* and *Leonurus sibiricus*: new hosts of *Phomopsis* spp. in Brazil. *Plant Disease* 67, 821–824.

Channon, A.G. (1969) Infection of flowers and seed of parsnip by *Itersonilia pastinacae*. *Annals of Applied Biology* 64, 281–288.

Channon, A.G. (1970) Tests of fungicides against *Alternaria brassicicola* and studies on some factors enhancing attack by the fungus. *Annals of Applied Biology* 65, 481–487.

Chapman, R.A. and Harris, C.R. (1990) Enhanced degradation of insecticides in soil. In: Racke, K.D. and Coats, J.R. (eds) *Enhanced Biodegradation of Pesticides in the Environment*. American Chemical Society Symposium Series No. 426. American Chemical Society, Washington DC, pp. 82–96.

Chatrath, M.S., Gupta, J.P. and Chiranjeevi, V. (1984) Uptake and distribution of carbendazim in urd bean (*Vigna mungo* (L.) Hepper). *Journal of Nuclear Agriculture and Biology* 13, 74–77.

Chauhan, M.S., Yadav, J.P.S. and Gangopadhyay, S. (1988) Chemical control of soilborne fungal pathogen complex of seedling cotton. *Tropical Pest Management* 34, 159–161.

Cheo, P.C. (1955) Effect of seed maturation on inhibition of southern bean mosaic virus in bean. *Phytopathology* 45, 17–21.

Chin, H.F. and Roberts, E.H. (1980) *Recalcitrant Crop Seeds*. Tropical Press, Kuala Lumpur, Malaysia.

Chirco, E.M. and Harman, G.E. (1979) The effect of *Alternaria brassicicola* infection on *Brassica* seed vigor and viability. *Journal of Seed Technology* 3, 12–22.

Christ, B.J. and Frank, J.A. (1989) Influence of foliar fungicides and seed treatments on powdery mildew, *Septoria*, and leaf rust epidemics on winter wheat. *Plant Disease* 73, 148–150.

Clark, F.S. (1952) The development of an isolated area for the production of smut-free barley seed. *Agricultural Institute Review* 7, 37–38.

Clark, M.F. (1981) Immunosorbent assays in plant pathology. *Annual Review of Phytopathology* 19, 83–106.

Clark, M.F. and Adams, A.N. (1977) Characteristics of the microplate method of enzyme-linked immunosorbent assay. *Journal of General Virology* 34, 475–483.

Clarke, B. (1988) Seed coating techniques. In: Martin, T.J. (ed.) *Application to Seeds and Soil*. BCPC Monograph No. 39. BCPC Publications, Thornton Heath, UK, pp. 205–211.

Clayton, C.N. (1948) Effects of several seed protectants on emergence and stand of okra. *Phytopathology* 38, 102–105.

Clayton, E.E. (1925) Second progress report on seed treatment for black leg (*Phoma lingam*) and black rot (*Pseudomonas campestris*) of cruciferous crops. *Phytopathology* 15, 49.

Cleene, M. de, Leyns F., Moster, M. van den, Swings, J. and Ley, J. de (1981) Reaction of grasses grown in Belgium to *Xanthomonas campestris* pv. *graminis*. *Parasitica* 37, 29–34.

Cochran, W.G. (1950) Estimation of bacterial densities by means of the 'most probable number'. *Biometrics* 6, 105–116.

Coddington, A., Matthews, P.M., Cullis, C. and Smith, K.H. (1987) Restriction digest patterns of total DNA from different races of *Fusarium oxysporum* f.sp. *pisi* – an improved method for race classsification. *Journal of Phytopathology* 118, 9–20.

Cohen, Y. and Sackston, W.E. (1974) Seed infection and latent infection of sunflowers by *Plasmopara halstedii*. *Canadian Journal of Botany* 52, 231–238.

Coléno, A., Trigalet, A. and Digat, B. (1976) Détection des lots des semences contaminés par une bacterie phytopathogène. *Annales de Phytopathologie* 8, 355–364.

Colhoun, J. (1970) Epidemiology of seed-borne *Fusarium* diseases of cereals. *Annales Academiae Scientiarum Fennicae A, IV. Biologica* 168, 31–36.

Colhoun, J. (1971) Cereal diseases. In: Western, J.H. (ed.) *Diseases of Crop Plants*. Macmillan, London, UK, pp. 181–225.

Colhoun, J. (1983) Measurement of inoculum per seed and its relationship to disease expression. *Seed Science and Technology* 11, 665–671.

Colhoun, J. and Park, D. (1964) *Fusarium* diseases of cereals 1. Infection of wheat plants, with particular reference to the effects of soil moisture and temperature on seedling infection. *Transactions of the British Mycological Society* 47, 559–572.

Colhoun, J., Taylor, G.S. and Tomlinson, R. (1968) *Fusarium* diseases of cereals 11. Infection of seedlings by *F. culmorum* and *F. avenaceum* in relation to environmental factors. *Transactions of the British Mycological Society* 51, 397–404.

Conner, R.L. and Kuzyk, A.D. (1990) Evaluation of seed-treatment fungicides for control of take-all in soft white spring wheat. *Canadian Journal of Plant Pathology* 12, 213–216.

Connick, W.J., Lewis, J.A. and Quimby, P.C. (1990) Formulation of biocontrol agents for use in plant pathology. In: Baker, R. and Dunn, P. (eds) *New Directions in Biological Control: Alternatives for Suppressing Agricultural Pests and Diseases*. Proceedings of a UCLA colloquium held at Frisco, Colorado, January 20–27, 1989. Alan R. Liss, New York, pp. 345–372.

Cook, A.A., Larson, R.H. and Walker, J.C. (1952) Relation of the black rot pathogen to cabbage seed. *Phytopathology* 42, 316–320.

Cook, J.C. (1968) *Fusarium* root and foot rot of cereals in the Pacific Northwest. *Phytopathology* 58, 127–131.

Coolbear, P., McGill, C.R. and Sakunnarak, N. (1991) Susceptibility of pea seeds to acetone toxicity: interactions with seed moisture content and ageing treatments. *Seed Science and Technology* 19, 519–526.

Cooper, J.I. and Harrison, B.D. (1973) The role of weed hosts and the distribution and activity of vector nematodes in the ecology of tobacco rattle virus. *Annals of Applied Biology* 73, 53–66.

Cooper, V.C. (1976) Studies on the seed transmission of cherry leaf roll virus. PhD thesis, University of Birmingham, Birmingham, UK.

Copeland, L.O., Adams, M.W. and Bell, D.C. (1975) An improved bean seed programme for maintaining disease-free seed of field beans. *Seed Science and Technology* 3, 719–724.

Corbett, J.R., Wright, K. and Baillie, A.C. (eds) (1984) *The Biochemical Mode of Action of Pesticides*, 2nd edn. Academic Press, New York.

Corbineau, F. and Come, D. (1988) Storage of recalcitrant seeds of four tropical species. *Seed Science and Technology* 16, 97–103.

Costa, A.S. and Duffus, J.E. (1958) Observations on lettuce mosaic in California. *Plant Disease Reporter* 42, 583–586.

Cromarty, A.S., Ellis, R.H. and Roberts, E.H. (1982) *The Design of Seed Storage Facilities for Genetic Conservation*. International Board for Plant Genetic Resources, Rome, Italy.

Crosier, W.F., Nash, G.T. and Crosier, D.C. (1970) Differential reaction of *Pythium* sp. and five isolates of *Rhizoctonia solani* to fungicides on pea seeds. *Plant Disease Reporter* 54, 349–352.

Cross, J.E., Kennedy, B.W., Lambert, J.W. and Cooper, R.L. (1966) Pathogenic races of the bacterial blight pathogen of soybeans, *Pseudomonas glycinea*. *Plant Disease Reporter* 50, 557–560.

Crowley, N.C. (1959) Studies on the time of embryo infection by seed-transmission. *Virology* 8, 116–123.

Croxall, H.E. and Ogilvie, L. (1942) Experiments with protectant seed dressings, 1940–42. In: *The Annual Report of the Agricultural and Horticultural Research Station*. AHRS, Bristol, pp. 65–76.

Cruger, G. (1979) *Phoma lingam* – seed transmission with cabbage. *Eucarpia 'Cruciferae 1979' Conference*. PUDOC, Wageningen, pp. 85–92.

Cruickshank, I.A.M. (1954) Thermo-chemical seed treatment. *Nature* 173, 217–218.

Crute, I.R. (1980) Evaluation of lettuce downy mildew with metalaxyl seed treatment. *Tests of Agrochemicals and Cultivars, Annals of Applied Biology* 94 (Suppl. 1), 28–29.

Crute, I.R. (1988) The contribution and value of resistant cultivars to disease control in vegetables. In: Clifford B.C. and Lester, E. (eds) *Control of Plant Diseases*. Blackwell Scientific Publications, Oxford, UK, pp. 77–91.

Cuero, R.G., Smith, J.E. and Lacey, J. (1986) The influence of gamma irradiation and sodium hypochlorite sterilisation on maize seed microflora and germination. *Food Microbiology* 3, 107–113.

Cunfer, B.M. (1983) Epidemiology and control of seed-borne *Septoria nodorum* on wheat. *Seed Science and Technology* 11, 707–718.

Cupples, D.A. (1984) The use of pathovar-indicative bacteriophages for rapidly detecting *Pseudomonas syringae* pv. *tomato* in tomato fruit and leaf lesions. *Phytopathology* 74, 891–894.

Damicone, J.P. and Cooley, D.R. (1981) Benomyl in acetone eradicates *Fusarium moniliforme* and *F. oxysporum* from asparagus seeds. *Plant Disease* 65, 892–893.

Darnell-Smith, G.P. (1915) The use of copper carbonate as a fungicide. *Agricultural Gazette of New South Wales* 26, 242–243.

Darnell-Smith, G.P. (1917) The prevention of bunt. Experiments with various fungicides. *Agricultural Gazette of New South Wales* 28, 185–189.

Davidse, L.C. (1973) Antimitotic activity of methyl benzimidazol-2yl carbamate. *Pesticide Biochemistry and Physiology* 3, 317–325.

Davidse, L.C. (1986) Benzimidazole fungicides: mechanism of action and biological impact. *Annual Review of Phytopathology* 24, 43–65.

Davidson, R.L. (ed.) (1980) *Handbook of Water Soluble Gums and Resins.* McGraw Hill, New York.

de Tempe, J. (1963) On methods of seed health testing – principles and practice. *Proceedings of the International Seed Testing Association* 28, 97–105.

de Tempe, J. (1968) The quantitative evaluation of seed-borne pathogenic infection. *Proceedings of the International Seed Testing Association* 33, 573–581.

Deacon, J.W. and Henry, C.M. (1978) Mycoparasitism by *Pythium oligandrum* and *P. acanthicum. Soil Biology and Biochemistry* 10, 409–415.

Dearman, J.A., Brocklehurst, P.A. and Drew, R.L.K. (1986) Effect of osmotic priming and ageing on onion seed germination. *Annals of Applied Biology* 108, 639–648.

Dekker, J. (1957) Internal disinfection of peas infected by *Ascochyta pisi* by means of the antibiotics rimocidin and pimaricin and some aspects of the parasitism of the fungus. *Tijdschrift over Plantenziekten* 63, 65–114.

Dekker, J. (1977) Resistance. In: Marsh, R.W. (ed.) *Systemic Fungicides,* 2nd. edn. Longman, London, UK, pp. 176–197.

Delanoe, D. (1972) Biologie et epidemiologie du mildiou du tournesol (*Plasmopara helianthi* Novot.). *CETIOM Informations Technique* 29, 1–49.

Derbyshire, D.M. (1961) A study of seed-borne infection of tomato by *Didymella lycopersici* Kleb. *Proceedings of the International Seed Testing Association* 26, 61–67.

Dewey, F.M. (1992) Detection of plant-invading fungi by monoclonal antibodies. In: Duncan J.M. and Torrance, L. (eds) *Techniques for the Rapid Detection of Plant Pathogens.* Blackwell Scientific Publications, Oxford, UK, pp. 47–63.

deZeeuw, D.J., Andersen, A.L. and Guyer, G.E. (1956) Comparison of fungicide and fungicide–insecticide seed treatment of peas and beans. *Phytopathology* 46, 10.

Dhingra, O.D. and da Silva, J.F. (1978) Effect of weed control on internally seedborne fungi in soybean seeds. *Plant Disease Reporter* 62, 513–516.

Dhingra, O.D. and Muchovej, J.J. (1982a) Infusion of fungicides into soybean seed with intact seed walls by organic solvents. *Seed Science and Technology* 10, 109–117.

Dhingra, O.D. and Muchovej, J.J. (1982b) Organic solvent seed treater. *Seed Science and Technology* 10, 105–108.

Dhitaphichit, P. and Jones, P. (1991) Virulent and fungicide tolerant races of loose smut (*Ustilago nuda* and *U. tritici*) in Ireland. *Plant Pathology* 40, 508–514.

Dickens, J.S.W. and Pemberton, A.W. (1983) Quarantine measures for seed in the United Kingdom. *Seed Science and Technology* 11, 1175–1180.

Dickens, J.S.W. and Sharp, M.K. (1970) Mercury-tolerant *Pyrenophora avenae* in seed oat samples from England and Wales. *Plant Pathology* 19, 93–94.

Dickinson, C.H. and Lucas, J.A. (1977) *Plant Pathology and Plant Pathogens. Basic Microbiology,* Vol. 6. Blackwell Scientific Publications, London, UK, 161 pp.

Dickinson, C.H. and Sheridan, J.J. (1968) Studies on the survival of *Mycosphaerella pinodes* and *Ascochyta pisi. Annals of Applied Biology* 62, 473–483.

Dillard, H.R. and Abawi, G.S. (1988) Occurrence of *Aphanomyces euteiches* f.sp. *phaseoli* on beans in New York State. *Plant Disease* 72, 912.

Dimond, A.E. and Horsfall, J.G. (1960) Prologue – inoculum and the diseased population. In: Horsfall, J.G. and Dimond, A.E. (eds) *Plant Pathology – an Advanced Treatise.* Academic Press, New York, pp. 1–22.

Dinant, S. and Lot, H. (1992) Lettuce mosaic virus: a review. *Plant Pathology* 41, 528–542.

Dixon. G.R. (1981) *Vegetable Crop Diseases.* Macmillan, London, UK, 404 pp.

Doling, D.A. (1965a) Some ecological factors influencing the proportion of loose smut (*Ustilago nuda* (Jens.) Rostr.) in wheat crops. *Annals of Applied Biology* 55, 303–306.

Doling, D.A. (1965b) Single-bath hot-water treatment for the control of loose smut (*Ustilago nuda*) in cereals. *Annals of Applied Biology* 55, 295–301.

Dolores-Talens, A.C., Hill, J.H. and Durand, D.P. (1989) Application of enzyme-linked fluorescent assay (ELFA) to detection of lettuce mosaic virus in lettuce seeds. *Journal of Phytopathology* 124, 149–154.

Doorn, A.M. van, Koert, J.L. and Kreyger, J. (1962) Investigations on the occurrence and control of neck rot (*Botrytis allii* Munn) in onions. *Verslagen van Landbouwkundige Onderzoekingen van het Rijkslandbouwproefstation* A 68 (7), 83 pp.

Dorogin, G.N. (1923) *Instructions for Testing of Seeds for Contamination with Fungus Pests* (pamphlet edited by Petrograd Plant Protection Station), 23 pp. (Abstract in *Review of Applied Mycology* 3, 221.)

Doyer, L.C. (1938) *Manual for the Determination of Seed-borne Diseases* (edited by The International Seed Testing Association). H. Veenman and Zonen, Wageningen, The Netherlands.

Duben, J. and Fehrmann, H. (1979) Occurrence and pathogenicity of *Fusarium* species on winter wheat in West Germany. 11. Comparison of pathogenicity on seedlings, haulm bases and ears. *Zeitschrift für Pflanzenkrankheiten und Pflanzenschutz* 86, 705–728.

Duffus, J.E. (1971) Role of weeds in the incidence of virus diseases. *Annual Review of Phytopathology* 9, 319–340.

Duncan, J.M. and Torrance, L. (eds) (1992) *Techniques for the Rapid Detection of Plant Pathogens.* Blackwells Scientific Publications, Oxford, UK.

Dunleavy, J.M. (1981) Soybean. In: Spencer, D.M. (ed.) *The Downy Mildews.* Academic Press, New York, pp. 515–529.

Durgapal, J.C. (1985) Self-sown plants from bacterial blight-infected rice seeds – a possible source of primary infection in North-West India. *Current Science, India* 54, 1283–1284.

Durrant, M.J., Dunning, R.A. and Byford, W.J. (1986) Treatment of sugar beet seeds. In: Jeffs, K.A. (ed.) *Seed Treatment.* BCPC Publications, Thornton Heath, UK, pp. 217–238.

Durrant, M.J., Payne, P.A., Prince, J.W.F. and Fletcher, R. (1988) Thiram steep seed treatment to control *Phoma betae* and improve the establishment of the sugar-beet plant stand. *Crop Protection* 7, 319–326.

Duthie, J.A. and Hall, R. (1987) Transmission of *Fusarium graminearum* from seeds to stems of winter wheat. *Plant Pathology* 36, 33–37.

Dykstra, T.P. (1961) Production of disease-free seed. *The Botanical Review* 27, 445–500.

Ebbels, D.L. (ed.) (1993) *Plant Health and the European Single Market.* BCPC Monograph No. 54. BCPC, Farnham, UK, 416 pp.

Eckert, J.W. and Sommer, N.F. (1967) Control of diseases of fruits and vegetables by postharvest treatment. *Annual Review of Phytopathology* 5, 391–432.

Edgington, L.V. (1981) Structural requirements of systemic fungicides. *Annual Review of Phytopathology* 19, 107–124.

Elbert, A., Overbeck, H., Iwaya, K. and Tsuboi, S. (1990) Imidacloprid, a novel systemic nitromethylene analogue insecticide for crop protection. *Proceedings of the Brighton Crop Protection Conference – Pests and Diseases* 1, 21–28.

Ellerbrock, L.A. and Lorbeer, J.W. (1977) Etiology and control of onion flower blight. *Phytopathology* 67, 155–159.

Ellis, M.A. and Paschal, E.H. (1979) Effect of fungicide seed treatment on internally seedborne fungi, germination and field emergence of pigeon pea (*Cajanus cajan*). *Seed Science and Technology* 7, 75–81.

Ellis, M.A., Foor, S.R. and Sinclair, J.B. (1976) Dichloromethane: non-aqueous vehicle for systemic fungicides in soybean seed. *Phytopathology* 66, 1249–1251.

Elsworth, J.E. (1988) Engineering responses to changing requirements in the United Kingdom. In: Martin, T.J. (ed.) *Application to Seeds and Soil.* BCPC Monograph No. 39. BCPC Publications, Thornton Heath, UK, pp. 145–154.

Elsworth, J.E. and Harris, D.A. (1973) The 'Rotostat' seed treater – a new application system. *Proceedings 7th British Insecticide and Fungicide Conference.* 3, 349–355.

Engelke, C. (1902) Neue beobachtungen uber die vegetations-formen des mutterpilzes *Claviceps purpurea* Tulasne. *Hedwigia* 41, 221–222.

Entwistle, A.R. (1986) Loss of control of *Allium* white rot by fungicides and its implications. *Aspects of Applied Biology – Crop Protection in Vegetables* 12, 201–209.

Fahey, P.C. and Persley, G.J. (1983) *Plant Bacterial Diseases: A Diagnostic Guide.* Academic Press, London, UK.

Falk, B.W. and Purcifull, D.E. (1983) Development and application of an enzyme-linked immunosorbent assay (ELISA) test to index lettuce seeds for lettuce mosaic virus in Florida. *Plant Disease* 67, 413–416.

Falloon, R.E. (1988) Effects of soil temperature on the efficacy of ergosterol biosynthesis inhibitor fungicides for control of *Ustilago bullata. Pesticide Science* 22, 169–178.

Farrant, J.M., Pammenter, N.W. and Berjak, P. (1988) Recalcitrance – a current assessment. *Seed Science and Technology* 16, 155–166.

Fatmi, M. and Schaad, N.W. (1991) Seed treatments for eradicating *Clavibacter michiganense* subsp. *michiganensis* from naturally infected tomato seeds. *Plant Disease* 75, 383–385.

Felsot, A.S. (1989) Enhanced biodegradation of insecticides in soil: implications for agroecosystems. *Annual Review of Entomology* 34, 453–476.

Fett, W.F. and Sequeira, L. (1981) Further characterisation of the physiologic races of *Pseudomonas glycinea*. *Canadian Journal of Botany* 59, 283–287.

Finney, J.R. and Hall, D.W. (1972) The effect of an autumn attack of mildew (*Erysiphe graminis*) on the growth and development of winter barley. *Plant Pathology* 21, 73–76.

Fisher, D.E. (1954) Internal infection of tomato seed by *Didymella lycopersici* Kleb. *Nature* 174, 656–657.

Fitt, B.D.L., McCartney, H.A. and and Walklate, P.J. (1989) The role of rain in dispersal of pathogen inoculum. *Annual Review of Phytopathology* 27, 241–270.

Fletcher, J.T. and Wolfe, M.S. (1981) Insensitivity of *Erysiphe graminis* f.sp. *hordei* to triadimefon, triadimenol and other fungicides. *Proceedings of the 1981 British Crop Protection Conference – Pests and Diseases* 2, 633–640.

Fletcher, J.T., George, A.J. and Green, D.E. (1969) Cucumber green mottle mosaic virus, its effect on yield and its control in the Lea Valley, England. *Plant Pathology* 18, 16–22.

Flood, J., Mepsted, R. and Cooper, R.M. (1990a) Contamination of oil palm pollen and seeds by *Fusarium* spp. *Mycological Research* 94, 708–709.

Flood, J., Mepsted, R. and Cooper, R.M. (1990b) Potential spread of *Fusarium* wilt of oil palm on contaminated seed and pollen. *3rd International Conference on Plant Protection in the Tropics* 102–107.

Flood, J., Whitehead, D.S. and Cooper, R.M. (1992) Vegetative compatibility and DNA polymorphisms in *Fusarium oxysporum* f.sp. *elaeidis* and their relationship to isolate virulence and origin. *Physiological and Molecular Plant Pathology* 41, 201–215.

Forster, R.L. and Schaad, N.W. (1988) Control of black chaff of wheat with seed treatment and a foundation seed health program. *Plant Disease* 72, 935–938.

Fourest, E., Rehms, L.D., Sands, D.C. and Bjarko, M. (1990) Eradication of *Xanthomonas campestris* pv. *translucens* from barley seed with dry heat treatments. *Plant Disease* 74, 816–818.

Fox, R.T.V. (1993) *Principles of Diagnostic Techniques in Plant Pathology*. CAB International, Wallingford, UK.

Frank, B. (1883) Ueber einige neue und weniger bekannte Pflanzenkrankheiten. 1. Die Fleckenkrankheit der Bohnen, veranlasst durch *Gloeosporium lindemuthianum* Sacc. et Magnus. *Landwirtschaftliche Jahrbucher* 12, 511.

Frank, J.A. and Ayres, J.E. (1986) Effect of triadimenol seed treatment on powdery mildew epidemics on winter wheat. *Phytopathology* 76, 254–257.

Franken, A.A.J.M. and Van Vuurde, J.W.L. (1990) Problems and new approaches in the use of serology for seedborne bacteria. *Seed Science and Technology* 18, 415–426.

Frison, E.A., Bos, L., Hamilton, R.I., Mathur, S.B. and Taylor, J.D. (eds) (1990) *FAO/IBPGR Technical Guidelines for the Safe Movement of Legume Germplasm*. Food and Agriculture Organization of the United Nations, Rome/International Board for Plant Genetic Resources, Rome, Italy, 88 pp.

Frosheiser, F.I. (1974) Alfalfa mosaic virus transmission to seed through alfalfa gametes and longevity in alfalfa seed. *Phytopathology* 64, 102–105.

Fryda, S.J. and Otta, J.D. (1978) Epiphytic movement and survival of *Pseudomonas syringae* on spring wheat. *Phytopathology* 68, 1064–1067.

Fujii, H. (1983) Pre-sowing treatment of rice seeds in Japan. *Seed Science and Technology* 11, 951–957.

Fujita, Y. (1990) Ecology and control of purple stain of soybean caused by *Cercospora kikuchii*. *Bulletin of the Tohoku National Agricultural Experiment Station* 81, 51–109.

Fulton, R.W. (1964) Transmission. In: Corbett, M.K. and Sisler, H.D. (eds) *Plant Virology*. University of Florida Press, Gainesville, Florida, pp. 39–67.

Gabrielson, R.L. (1962) Survival of the celery late blight organism. *Phytopathology* 52, 361.

Gabrielson, R.L. (1983) Black leg disease of crucifers caused by *Leptosphaeria maculans* (*Phoma lingam*) and its control. *Seed Science and Technology* 11, 749–780.

Gabrielson, R.L. (1988) Inoculum thresholds of seedborne pathogens – fungi. *Phytopathology* 78, 868–872.

Gabrielson, R.L., Mulanax, M.W., Matsuoka, K., Williams, P.H., Whiteaker, G.P. and Maguire, J.D. (1977) Fungicidal eradication of seedborne *Phoma lingam* of crucifers. *Plant Disease Reporter* 61, 118–121.

Gabrielson, R.L., Bonman, J.M., Maguire, J.D., Mulanax, M.W. and Whiteaker, G.P. (1978) Epidemiology and control of *Phoma lingam* in crucifer seed crops. In: *Third International Congress of Plant Pathology, Abstracts of Papers*, Abstract no. 124. Paul Parey, Berlin, Germany.

Gaikward, D.G. and Sokhi, S.S. (1987) Detection of seed rot, root rot and seedling infection in naturally infected cowpea seed in Senegal and their control. *Plant Disease Research* 2, 127–128.

Gambogi, P. (1983) Seed transmission of *Fusarium oxysporum*: epidemiology and control. *Seed Science and Technology* 11, 815–827.

Gardner, M.W. (1924) A native weed host for bacterial blight of bean. *Phytopathology* 14, 340.

Garrett, S.D. (1963) A comparison of cellulose-decomposing ability in five fungi causing cereal foot rots. *Transactions of the British Mycological Society* 46, 572–576.

Garrett, S.D. (1966) Cellulose-decomposing ability of some cereal foot-rot fungi in relation to their saprophytic survival. *Transactions of the British Mycological Society* 49, 57–68.

Gates, L.A. and Hull, R. (1954) Experiments on black leg disease of sugar beet seedlings. *Annals of Applied Biology* 41, 541–561.

Gaumann, E. (1950) *Principles of Plant Infection.* Crosby Lockwood & Son, London, UK.

Gaur, R.B., Ahmed, S.R. and Kataria, P.K. (1984) Controlling *Xanthomonas phaseoli* (Smith) Dowson of moong seed through chemotherapy and heat therapy. *Hindustan Antibiotics Bulletin* 26, 23–26.

Geng, S., Campbell, R.N., Carter, M. and Hills, F.J. (1983) Quality-control programs for seedborne pathogens. *Plant Disease* 67, 236–242.

George, R.A.T. (1980) *Vegetable Seed Technology: a Technical Guide to Vegetable Seed Production, Processing, Storage and Quality Control.* FAO, Rome, Italy.

Georgopoulos, S.G. and Zaracovitis, C. (1967) Tolerance of fungi to organic fungicides. *Annual Review of Phytopathology* 5, 109–130.

Ghabrial, S.A. and Shepherd, R.J. (1982) Radioimmunosorbent assay for detection of lettuce mosaic virus in lettuce seed. *Plant Disease* 66, 1037–1040.

GIFAP (1989) *Catalogue of Pesticide Formulation Types and International Coding System.* Technical monograph no. 2, GIFAP, Brussels.

Gladders, P. and Musa, T.M. (1980) Observations on the epidemiology of *Leptosphaeria maculans* stem canker in winter oilseed rape. *Plant Pathology* 29, 28–37.

Godwin, J.R., Shephard, M.C. and Noon, R.A. (1988) The development of Ferrax seed treatment for control of diseases of barley. In: Martin, T.J. (ed.) *Application to Seeds and Soil.* BCPC Monograph No. 39. BCPC Publications, Thornton Heath, UK, pp. 63–74.

Goncalves, E.J., Muchovej, J.J. and Muchovej, R.M.C. (1991) Effect of kind and method of fungicidal treatment of bean seed on infections by the VA-mycorrhizal fungus

Glomus macrocarpum and by the pathogenic fungus *Fusarium solani.* 1. Fungal and plant parameters. *Plant and Soil* 132, 41–46.

Goodwin, P.H. and Annis, S.L. (1991) Rapid identification of genetic variation and pathotype of *Leptosphaeria maculans* by random amplified polymorphic DNA assay. *Applied and Environmental Microbiology* 57, 2482–2486.

Gordon-Lennox, G., Walther, D. and Gindrat, D. (1987) Utilisation d' antagonistes pour l'enrobage des semences: efficacité et mode d'action contre les agents de la fonte des semis. *EPPO Bulletin* 17, 631–637.

Graham, D.C. and Harrison, M.D. (1975) Potential spread of *Erwinia* spp. in aerosols. *Phytopathology* 65, 739–741.

Graham, D.C., Quinn, C.E. and Bradley, L.F. (1977) Quantitative studies on the generation of aerosols of *Erwinia carotovora* var. *atroseptica* by simulated rain-drop impaction on blackleg-infected potato stems. *Journal of Applied Bacteriology* 43, 413–424.

Grainger, J. and Simpson, D.E. (1950) Electronic heating and control of seed-borne diseases. *Nature* 165, 532–533.

Gray, D. (1994) Large-scale priming techniques and their integration with crop protection treatments. In: Martin, T.J. (ed.) *Seed Treatment: Progress and Prospects.* BCPC Monograph No. 57. BCPC Publications, Farnham, UK, pp. 353–362.

Gray, D., Rowse, H.R., Finch-Savage, W.E., Bujalski, W. and Nienow, A.W. (1992) Priming of seeds – scaling up for commercial use. *4th International Workshop on Seeds, Basic and Applied Aspects of Seed Biology* 3.

Gray, R.A. (1955) Inhibition of root growth by streptomycin and reversal of the inhibition by manganese. *American Journal of Botany* 42, 327–331.

Green, M.C.E. and Gooch, N. (1992) Myclobutanil seed treatment for cereal seed borne disease control in the UK. *Brighton Crop Protection Conference – Pests and Diseases* 2, 627–632.

Gregory, P.H., Guthrie, E.J. and Bunce, M.E. (1959) Experiments on splash dispersal of fungus spores. *Journal of General Microbiology* 20, 328–354.

Grewal, J.S. (1982) Control of important seed-borne pathogens of chickpea. *Indian Journal of Genetics* 42, 393–398.

Gries, G.A. (1946) Physiology of *Fusarium* foot rot of squash. *Bulletin of the Connecticut Agricultural Experiment Station* 500, 20 pp.

Griffen, D.M. (1972) *Ecology of Soil Fungi.* Syracuse University Press, Syracuse, New York.

Griffiths, D.C. (1986) Insecticidal treatment of cereal seeds. In: Jeffs, K.A. (ed.) *Seed Treatment.* BCPC Publications, Thornton Heath, UK, pp. 113–137.

Griffiths, E. and Hann, C.A.O. (1976) Dispersal of *Septoria nodorum* spores and spread of glume blotch of wheat in the field. *Transactions of the British Mycological Society* 67, 413–418.

Grogan, R.G. (1980) Control of lettuce mosaic with virus free seed. *Plant Disease* 64, 446–449.

Grogan, R.G. (1983) Lettuce mosaic virus control by use of virus-indexed seed. *Seed Science and Technology* 11, 1043–1049.

Grogan, R.G., Welch, C.E. and Bardin, R. (1952) Common lettuce mosaic and its control by use of mosaic-free seed. *Phytopathology* 69, 573–578.

Grogan, R.G. and Kimble, K.A. (1967) The role of seed contamination in the transmission of *Pseudomonas phaseolicola* in *Phaseolus vulgaris*. *Phytopathology* 57, 28–31.

Grondeau, C., Samson, R. and Poutier, F. (1991) Spread of *Pseudomonas syringae* pv. *pisi*: can prevailing winds help bacteria's progression? *Proceedings of the 4th International Working group on* Pseudomonas syringae *Pathovars* 211–213.

Grondeau, C., Ladonne, F., Fourmond, A., Poutier, F. and Samson, R. (1992) Attempt to eradicate *Pseudomonas syringae* pv. *pisi* from pea seeds with heat treatments. *Seed Science and Technology* 20, 515–525.

Groth, D. (1981) Field inspections for export seed. *Iowa Seed Science* 3, 14.

Groves, J.W. and Skolko, A.J. (1944) Notes on seed borne fungi. 1. *Stemphylium*. *Canadian Journal of Research* 22, 190–199.

Gupta, P.C., Maheshwari, S.K. and Suhag, L.S. (1989) Efficacy of some seed dressing fungicides in controlling the wilt and root rot complex of pea. *Indian Journal of Mycology and Plant Pathology* 18, 176.

Guthrie, J.W. (1968) The serological relationships of races of *Pseudomonas phaseolicola*. *Phytopathology* 58, 716–717.

Guthrie, J.W. (1970) Factors influencing halo-blight transmission from externally contaminated *Phaseolus vulgaris* seed. *Phytopathology* 60, 371–372.

Guthrie, J.W., Huber, D.M. and Fenwick, H.S. (1965) Serological detection of halo-blight. *Plant Disease Reporter* 49, 297–299.

Gyurk, I. (1988) Recent advances in seed treatment technology in Hungary. In: Martin, T.J. (ed.) *Application to Seeds and Soil*. BCPC Monograph No. 39. BCPC Publications, Thornton Heath, UK, pp. 155–161.

Hadar, Y., Harman, G.E., Taylor, A.G. and Norton, J.M. (1983) Effects of pre-germination of pea and cucumber seeds and of seed treatment with *Enterobacter cloacae* on rots caused by *Pythium* spp. *Phytopathology* 73, 1322–1325.

Hadar, Y., Harman, G.E. and Taylor, A.G. (1984) Evaluation of *Trichoderma koningii* and *T. harzianum* from New York soils for biological control of seed rot caused by *Pythium* spp. *Phytopathology* 74, 106–110.

Halfon-Meiri, A. (1983) Seed transmission of safflower rust (*Puccinia carthami*) in Israel. *Seed Science and Technology* 11, 835–851.

Halfon-Meiri, A., Kulik, M.M. and Schoen, J.F. (1979) Studies on *Gibberella zeae* carried by wheat seeds produced in the Mid-atlantic region of the United States. *Seed Science and Technology* 7, 439–448.

Hall, D.W. (1971) Control of mildew in winter barley with ethirimol. *Proceedings of the 6th British Insecticide and Fungicide Conference* 1, 26–32.

Hall, H.S. and Pondell, R.E. (1980) The Wurster process. In: Kydoniens, A.F. (ed.) *Controlled Release Technologies, Theory and Application*. CRC Press, Boca Raton, Florida, pp. 134–154.

Hall, T.J. (1980) Resistance at the Tm-2 locus in the tomato to tomato mosaic virus. *Euphytica* 29, 189–197.

Hall, T.J. and Taylor, G.S. (1983) Aerated-steam treatment for control of *Alternaria tenuis* on lobelia seed. *Annals of Applied Biology* 103, 219–228.

Halliwell, R.S. and Langston, R. (1965) Effects of gamma irradiation on sympton expression of barley stripe mosaic virus disease and on two viruses *in vivo*. *Phytopathology* 55, 1039–1040.

Halloin, J.M. (1986) Treatment of cotton seeds. In: Jeffs, K.A. (ed.) *Seed Treatment.* BCPC Publications, Thornton Heath, UK, pp. 201–215.

Halmer, P. (1988) Technical and commercial aspects of seed pelleting and film-coating. In: Martin, T.J. (ed.) *Application to Seeds and Soil.* BCPC Monograph No. 39. BCPC Publications, Thornton Heath, UK, pp. 191–204.

Halmer, P. (1994) The development of quality seed treatments in commercial practice – objectives and achievements. In: Martin, T.J. (ed.) *Seed Treatment: Progress and Prospects.* BCPC Monograph No. 57. BCPC Publications, Farnham, UK, pp. 363–374.

Hamilton, R.I. (1983) Certification schemes against seed-borne viruses in leguminous hosts, present status and further areas for research and development. *Seed Science and Technology* 11, 1051–1062.

Hampton, R., Waterworth, H., Goodman, R.M. and Lee, R. (1982) Importance of seedborne viruses in crop germplasm. *Plant Disease* 66, 977–978.

Hankin, L. and Sands, D.C. (1977) Microwave treatment of tobacco seed to eliminate bacteria on the seed surface. *Phytopathology* 67, 794–795.

Hardison, J.R. (1948) Field control of blind seed disease of perennial ryegrass in Oregon. *Phytopathology* 38, 404–419.

Hardison, J.R. (1960) Disease control in forage seed production. *Advances in Agronomy* 12, 96–106.

Hardison, J.R. (1963) Control of *Gloeotinia temulenta* in seed fields of *Lolium perenne* by cultural methods. *Phytopathology* 53, 460–464.

Hardison, J.R. (1976) Fire and flame for plant disease control. *Annual Review of Phytopathology* 14, 355–379.

Hare, W.W. and Walker, J.C. (1944) *Ascochyta* diseases of canning pea. *Research Bulletin. Wisconsin Agricultural Experiment Station* 150, 1–31.

Harman, G.E. (1991) Seed treatments for biological control of plant disease. *Crop Protection* 10, 166–171.

Harman, G.E. and Lumsden, R.D. (1990) Biological disease control. In: Lynch, J.M. (ed.) *The Rhizosphere.* John Wiley & Sons, Chichester, UK, pp. 259–279.

Harman, G.E. and Nash, G. (1978) Soaking brassica seeds in fungicide solutions to eradicate seed-borne fungi: a comparison of aqueous and solvent infusion techniques. *Plant Disease Reporter* 62, 408–412.

Harman, G.E. and Nelson, E.B. (1994) Mechanisms of protection of seed and seedlings by biological seed treatments: implications for practical disease control. In: Martin, T.J. (ed.) *Seed Treatment: Progress and Prospects.* BCPC Monograph No. 57. BCPC Publications, Farnham, UK, pp. 283–292.

Harman, G.E. and Taylor, A.G. (1988) Improving seedling performance by integration of biological control agents at favorable pH levels with solid matrix priming. *Phytopathology* 78, 520–525.

Harman, G.E. and Taylor, A.G. (1990) Development of an effective biological seed treatment system. In: Hornby, D., Cook, R.J., Henis, Y., Ko, W.H., Rovira, A.D., Schippers, B. and Scott, P.R. (eds) *Biological Control of Soil-borne Plant Pathogens.* CAB International, Wallingford, UK, pp. 415–426.

Harman, G.E., Chet, I. and Baker, R. (1981) Factors affecting *Trichoderma hamatum* applied to seeds as a biocontrol agent. *Phytopathology* 71, 567–572.

Harman, G.E., Norton, J.M. and Stasz, T.E. (1987) Nyolate seed treatment of *Brassica* spp. to eradicate or reduce black rot caused by *Xanthomonas campestris* pv. *campestris*. *Plant Disease* 71, 27–30.

Harman, G.E., Taylor, A.G. and Stasz, T.E. (1989) Combining effective strains of *Trichoderma harzianum* and solid matrix priming to improve biological seed treatment. *Plant Disease* 73, 631–637.

Harper, F.R. (1968) Control of root rot in garden pea with a soil fungicide. *Plant Disease Reporter* 52, 565–568.

Harris, D. (1979) Survival of the glume blotch fungus on stubble and straw residues. In: Grossbard, E. (ed.) *Straw Decay and its Effect on Disposal and Utilisation*. Wiley & Sons, Chichester, UK, pp. 21–28.

Harrison, B.D. and Winslow, R.D. (1961) Laboratory and field studies on the relation of arabis mosaic virus to its nematode vector *Xiphinema diversicaudatum* (Micoletzky). *Annals of Applied Biology* 49, 621–633.

Harrison, J.G. (1978) Role of seed-borne infection in epidemiology of *Botrytis fabae* on field beans. *Transactions of the British Mycological Society* 70, 35–40.

Hartke, S. and Buchenauer, H. (1985) On the effectiveness of Hg-free seed dressings and their active ingredients against carbendazim sensitive and carbendazim resistant *Gerlachia nivalis* strains on winter wheat. *Zeitschrift für Pflazenkrankheiten und Pflanzenschutz* 92, 247–257.

Hartman, G.L., Manandhar, J.B. and Sinclair, J.B. (1986) Incidence of *Colletotrichum* spp. on soybean and weeds in Illinois and pathogenicity of *Colletotrichum truncatum*. *Plant Disease* 70, 780–782.

Haskell, R.J. (1917) The spray method of applying concentrated formaldehyde solution in the control of oat smut. *Phytopathology* 7, 381–383.

Hassall, K.A. (1990) *The Biochemistry and Uses of Pesticides*, 2nd edn. Weinheim, New York.

Hayden, N.J. and Maude, R.B. (1992) The role of seedborne *Aspergillus niger* in transmission of black mould of onion. *Plant Pathology* 41, 573–581.

Hayward, H.H. (1951) *The Structure of Economic Plants*. Macmillan, New York, 674 pp.

Heald, F.D. (1921) The relation of spore load to the percent of stinking smut appearing in the crop. *Phytopathology* 11, 269–278.

Heaney, S.P., Hutt, R.T. and Miles, V.G. (1986) Sensitivity to fungicides of barley powdery mildew populations in England and Scotland: status and implications for fungicide use. *Proceedings of the 1986 British Crop Protection Conference* 2, 793–800.

Henry, A.W. and Campbell, J.A. (1938) Inactivation of seedborne plant pathogens in the soil. *Canadian Journal of Research* 16, 331–338.

Hepperly, P.R., Kirkpatrick, B.L. and Sinclair, J.B. (1980) *Abutilon theophrasti*: wild host for three fungal parasites of soybean. *Phytopathology* 70, 307–310.

Herd, G.W. and Phillips, A.J.L. (1988) Control of seed-borne *Sclerotinia sclerotiorum* by fungicidal treatment of sunflower seed. *Plant Pathology* 37, 202–205.

Hewett, P.D. (1962) Methods for detecting viable seed-borne infection with celery leaf spot. *Journal of the National Institute of Agricultural Botany* 9, 174–178.

Hewett, P.D. (1966) *Ascochyta fabae* Speg. on tick bean seed. *Plant Pathology* 15, 161–163.

Hewett, P.D. (1968) The multiplication and control of two seed-borne diseases of cereals. *Journal of the National Institute of Agricultural Botany* 11, 47–53.
Hewett, P.D. (1970) A note on the extraction rate in the embryo method for loose smut of barley, *Ustilago nuda* (Jens.) Rostr. *Proceedings of the International Seed Testing Association* 35, 181–183.
Hewett, P.D. (1972) Resistance to barley loose smut (*Ustilago nuda*) in the variety Emir. *Transactions of the British Mycological Society* 59, 330–331.
Hewett, P.D. (1973) The field behaviour of seedborne *Ascochyta fabae* and disease control in field beans. *Annals of Applied Biology* 74, 287–295.
Hewett, P.D. (1978) Epidemiology of seed-borne disease. In: Scott, P.R. and Bainbridge, A. (eds) *Plant Disease Epidemiology.* Blackwell Scientific Publications, Oxford, UK, pp. 167–176.
Hewett, P.D. (1979a) Reaction of selected spring barley cultivars to inoculation with loose smut. *Plant Pathology* 28, 77–80.
Hewett, P.D. (1979b) Regulating seed-borne diseases by certification. In: Ebbels, D.L. and King, J.E. (eds) *Plant Health.* Blackwell Scientific Publications, Oxford, UK, pp. 163–173.
Hewett, P.D. (1980) Loose smut in winter barley: comparisons between embryo infection and the production of diseased ears in the field. *Journal of the National Institute of Agricultural Botany* 15, 231–235.
Hewett, P.D. (1981) Seed standards for disease in certification. *Journal of the National Institute of Agricultural Botany* 15, 373–384.
Hewett, P.D. (1983) Epidemiology – fundamental for disease control. *Seed Science and Technology* 11, 697–706.
Hewett, P.D. (1987) Pathogen viablity on seed in deep freeze storage. *Seed Science and Technology* 15, 73–77.
Heydecker, W. and Coolbear, P. (1977) Seed treatments for improved performance – survey and attempted prognosis. *Seed Science and Technology* 5, 353–425.
Heydecker, W., Higgins, J and Gulliver, R.L. (1973) Accelerated germination by osmotic seed treatment. *Nature* 246, 42–44.
Hide, G. (1988) *Thanatephorus cucumeris* (Frank) Donk. In: Smith, I.M., Dunez, J., Phillips, D.H., Lelliot, R.A. and Archer, S.A. (eds) *European Handbook of Plant Diseases.* Blackwell Scientific Publications, Oxford, UK, pp. 505–507.
Hillocks, R.J., Chinodya, R. and Gunner, R. (1988) Evaluation of seed dressing and in-furrow treatments with fungicides for the control of seedling disease in cotton caused by *Rhizoctonia solani. Crop Protection* 7, 309–313.
Hiltner, L. (1915) Uber die Erfolge der im Herbst 1914 in Bayern durchgefuhrten Beizung des Saatgutes von Winterroggen und Winterweizenmit sublimathaltigen Mitteln. *Prakt Blatter Pflanzenbau und Pflanzenschutz* 13, 65–90.
Hirst, J.M. (1953) Changes in atmospheric spore content: diurnal periodicity and the effects of weather. *Transactions of the British Mycological Society* 36, 375–393.
Hoffman, J.A. (1971) Control of common and dwarf bunt of wheat with systemic fungicides. *Plant Disease Reporter* 55, 1132–1135.
Hoffman, J.A. and Waldher, J.T. (1981) Chemical seed treatments for controlling seedborne and soilborne common bunt of wheat. *Plant Disease* 65, 256–259.
Hoggan, I.A. and Johnson, J. (1936) Behaviour of ordinary tomato mosaic virus in the soil. *Journal of Agricultural Research* 52, 271–294.

Holmes, S.J.I. and Colhoun, J. (1971) Infection of wheat seedlings by *Septoria nodorum* in relation to environmental factors. *Transactions of the British Mycological Society* 57, 493–500.

Hooda, I. and Grover, R.K. (1990) Environmental factors affecting control of *Macrophomina phaseolina* by fungicides on mung bean. *Plant Disease Research* 5, 25–33.

Horner, E.L. (1988) 'SHR': coating technology for the application of pesticides to seeds. In: Martin, T.J. (ed.) *Application to Seeds and Soil.* BCPC Monograph No. 39. BCPC Publications, Thornton Heath, UK, pp. 271–276.

Horsfall, J.G. (1932) Dusting tomato seeds with copper sulfate monohydrate for combating damping-off. *New York State Agricultural Experimental Station Technical Bulletin* 198, 34 pp.

Horsfall, J.G. (1945) *Fungicides and their Action.* Chronica Botanica Company, Waltham, Massachusetts.

Horsfall, J.G. and Cowling, E.B. (eds) (1978) *Plant Disease – an Advanced Treatise, Vol. II How Disease Develops in Populations.* Academic Press, London, UK.

Howell, C.R. (1991) Biological control of *Pythium* damping-off of cotton with seed-coating preparations of *Gliocladium virens*. *Phytopathology* 81, 738–741.

Howles, R. (1978) Inactivation of lettuce mosaic virus in lettuce seed by dry heat treatment. *Journal of the Australian Institute of Agricultural Science* 44, 131–132.

Huang, H-Q. and Nielsen, J. (1985) Seven races identified in 15 field collections of loose smut of wheat from the People's Republic of China. *Canadian Journal of Plant Pathology* 7, 165–167.

Huang, T.C. and Lee, H.L. (1988) Hot acidified zinc sulphate as seed soaking agent for the control of crucifer black rot. *Plant Protection Bulletin, Taiwan* 30, 245–258.

Humaydan, H.S., Harman, G.E., Nedrow, B.L. and Dinitto, L.V. (1980) Eradication of *Xanthomonas campestris*, the causal agent of black rot, from brassica seeds with antibiotics and sodium hypochlorite. *Phytopathology* 70, 127–131.

Humpherson-Jones, F.M. (1984) Seed-borne disease interactions between oilseed rape and other brassicas. *Proceedings of the British Crop Protection Conference 1984 – Pests and Diseases* 2, 799–806.

Humpherson-Jones, F.M. (1992) Epidemiology and control of dark leaf spot of brassicas. In: Chelkowski, J. and Visconti, A. (eds), Alternaria *Biology, Plant Diseases and Metabolites*, Elsevier, Amsterdam, pp. 267–288.

Humpherson-Jones, F.M. and Hocart, M.J. (1983) *Alternaria* disease of brassica seed crops. *Report of the National Vegetable Research Station for 1982*, 63.

Humpherson-Jones, F.M. and Maude, R.B. (1982) Studies on the epidemiology of *Alternaria brassicicola* in *Brassica oleracea* seed production crops. *Annals of Applied Biology* 100, 61–71.

Ilyas, M.B., Dhingra, O.D., Ellis, M.A. and Sinclair, J.B. (1975) Location of mycelium of *Diaporthe phaseolorum* var. *sojae* and *Cercospora kikuchii* in infected soybean seeds. *Plant Disease Reporter* 59, 17–19.

Ingold, C.T. (1953) *Dispersal in Fungi.* Oxford University Press, London, UK, 197 pp.

Irwin, J.A.G. (1987) Recent advances in the detection of seedborne pathogens. *Seed Science and Technology* 15, 755–763.

Ivens, G.W. (ed.) (1994) *The UK Pesticide Guide 1994.* CAB International, Wallingford/British Crop Protection Council, Farnham, UK.

Jaccoud Filho, D.S. and Reeves, J.C. (1993) Detection and identification of *Phomopsis* species in soya bean seeds using PCR. In: *Proceedings of the First ISTA Plant Disease Committee Symposium on Seed Health Testing*. Agriculture Canada Central Seed Laboratory Ottawa, Ontario, Canada, pp. 34–37.

Jacks, H. (1951) The efficiency of chemical treatments of vegetable seeds against seed-borne and soil-borne organisms. *Annals of Applied Biology* 38, 135–168.

Jacobsen, B.J. and Williams, P.H. (1971) Histology and control of *Brassica oleracea* seed infection by *Phoma lingam*. *Plant Disease Reporter* 55, 934–937.

Jafarpour, B., Shepherd, R.J. and Grogan, R.G. (1979) Serological detection of bean common mosaic and lettuce mosaic viruses in seed. *Phytopathology* 69, 1125–1129.

James, E. (1967) Preservation of seed stocks. *Advances in Agronomy* 19, 87–106.

Janyska, A. and Rod, J. (1980) The effect of the treatment of onion (*Allium cepa* L) on the yield and on the occurrence of neck rot. *Zahradnictvi* 7, 197–208.

Jarvis, W.R. (1977) Botryotinia *and* Botrytis *Species: Taxonomy, Physiology and Pathogenicity.* Monograph No. 15. Information Division, Canada Department of Agriculture, Ottawa, Canada, 195 pp.

Jarvis, W.R. (1980) Epidemiology. In: Coley-Smith, J.R., Verhoeff, K. and Jarvis, W.R. (eds) *The Biology of* Botrytis. Academic Press, London, UK, pp. 219–245

Jeffs, K.A. (1974) Tests for retention of powders. *Rothamsted Experimental Station Report for 1973*, 177.

Jeffs, K.A. (ed.) (1986a) *Seed Treatment.* BCPC Publications, Thornton Heath, UK.

Jeffs, K.A. (1986b) A brief history of seed treatment. In: Jeffs, K.A. (ed.) *Seed Treatment.* BCPC Publications, Thornton Heath, UK, pp. 1–15.

Jeffs, K.A. and Tuppen, R.J. (1986) Application of pesticides to seeds. In: Jeffs, K.A. (ed.) *Seed Treatment.* BCPC Publications, Thornton Heath, UK, pp. 17–45.

Jeffs, K.A., Lord, K.A. and Tuppen, R.J. (1968) Insecticides on single seeds treated with liquid dressings. *Journal of the Science of Food and Agriculture* 19, 195–198.

Jenkyn, J.F. (1974) Effects of mildew on the growth and yield of spring barley. *Annals of Applied Biology* 78, 281–288.

Jenkyn, J.F., Gutteridge, R.J. and Jalaluddin, M. (1994) Straw disposal and cereal diseases. In: Blakeman, J.P. and Williamson, B. (eds) *Ecology of Plant Pathogens.* CAB International, Wallingford, UK, pp. 285–300.

Jensen, J.L. (1888) The propagation and prevention of smut in oats and barley. *Journal of the Royal Agricultural Society* 24, 397–415.

Jhooty, J.S. and Behar, D.S. (1972) Control of powdery mildew of pea and pumpkin by seed treatment with benomyl. *Indian Journal of Agricultural Science* 42, 505–508.

Jindal, K.K., Thind, B.S. and Soni, P S. (1989) Physical and chemical agents for the control of *Xanthomonas campestris* pv. *vignicola* from cowpea seeds. *Seed Science and Technology* 17, 371–382.

Johanson, A., Crowhurst, R.N., Rikkerink, E.H.A., Fullerton, R.A. and Templeton, M.D. (1994) The use of species-specific DNA probes for the identification of *Mycosphaerella fijjiensis* and *M. musicola*, the causal agents of Sigatoka disease of banana. *Plant Pathology* 43, 701–707.

Johnson, J.C. (1970) Bean halo blight on *Phaseolus atropurpureus*. *Queensland Journal of Agricultural and Animal Sciences* 27, 129–135.

Johnson, T. (1925) Studies on the pathogenicity and physiology of *Helminthosporium gramineum* Rab. *Phytopathology* 15, 797–804.

Johnson, T. and Ogden, W.B. (1929) The overwintering of tobacco mosaic virus. *Bulletin Wisconsin Agricultural Experimental Station* 95.

Johnsson, L. (1991) Experiments with seed-borne and soil-borne dwarf bunt (*Tilletia controversa* Kuhn) in winter wheat in Sweden. *Zeitschrift für Pflanzenkrankeiten und Pflanzenschutz* 98, 162–167.

Johnston, W.H. and Metcalf, D.R. (1961) Note on Keystone, a loose smut resistant feed barley. *Canadian Journal of Plant Science* 41, 874–875.

Jones, D.R. (1987) Seedborne diseases and the international transfer of plant genetic resources: an Australian perspective. *Seed Science and Technology* 15, 765–776.

Jones, D.R., Slade, M.D. and Briks, K.A. (1989) Resistance to organomercury in *Pyrenophora graminea*. *Plant Pathology* 38, 509–513.

Jones, G.H. and Seif El-Nasr, A.El-G. (1940) The influence of sowing depth and moisture on smut diseases, and the prospects of a new method of control. *Annals of Applied Biology* 27, 35–57.

Jones, J.P. (1961) A weed host of *Xanthomonas phaseoli* var. *sojense*. *Phytopathology* 51, 206.

Jordan, V.W.L. (1981) Aetiology of barley net blotch caused by *Pyrenophora teres* and some effects on yield. *Plant Pathology* 30, 77–87.

Jordan, V.W.L. and Allen, E.C. (1984) Barley net blotch: influence of straw disposal and cultivation methods on inoculum potential, and on incidence and severity of autumn disease. *Plant Pathology* 33, 547–560.

Jorgensen, J. (1977) Incidence of infections of barley seed by *Pyrenophora graminea* and *P. teres* as revealed by the freezing blotter method and disease counts in the field. *Seed Science and Technology* 5, 105–110.

Josephson, E.S. and Peterson, M.S. (eds) (1983) *Preservation of Food by Ionising Radiation*, Vol. 111. CRC Press, Boca Raton, Florida.

Jung, J. (1956) Sind Narbe und Griffel Eintrittspforten fur Pilzinfektionen. *Phytopathologische Zeitschrift* 27, 405–426.

Kabeere, F. (1983) Difficulties encountered in the production of high quality seeds in Uganda. *Seed Science and Technology* 11, 1087–1092.

Kahn, R.P. (1979) A concept of pest risk analysis. *EPPO Bulletin* 9, 119–130.

Kahn, R.P. (1983) Safeguards and the international exchange of plant germplasm as seeds. *Seed Science and Technology* 11, 1159–1173.

Kaiser, W.J. (1983) Plant introduction and related seed pathology research in the United States. *Seed Science and Technology* 11, 1197–1212.

Kaiser, W.J. and Hannan, R.M. (1987) Seed-treatment fungicides for control of seed-borne *Ascochyta lentis* on lentil. *Plant Disease* 71, 58–62.

Kaiser, W.J. and Hannan, R.M. (1988) Seed transmission of *Ascochyta rabiei* in chickpea and its control by seed-treatment fungicides. *Seed Science and Technology* 16, 625–637.

Karpati, J.F. (1983) Plant quarantine on a global basis. *Seed Science and Technology* 11, 1145–1157.

Katan, J. (1981) Solar heating (solarisation) for control of soilborne pests. *Annual Review of Phytopathology* 19, 211–236.

Katan, J. and DeVay, J.E. (eds) (1991) *Soil Solarisation*. CRC Press, Boca Raton, Florida, 267 pp.

Kataria, H.R. and Grover, R.K. (1978) Comparison of fungicides for the control of *Rhizoctonia solani* causing damping-off of mung bean (*Phaseolus aureus*). *Annals of Applied Biology* 88, 257–263.

Kataria, H.R. and Verma, P.R. (1990) Efficacy of fungicidal seed treatments against pre-emergence damping-off and post-emergence seedling root rot of growth chamber grown canola caused by *Rhizoctonia solani* AG-2-1 and AG-4. *Canadian Journal of Plant Pathology* 12, 409–416.

Katznelson, H. (1950) The detection of internally-borne bacterial pathogens of beans by a rapid phage-plaque count. *Science* 112, 645–647.

Katznelson, H. and Sutton, M.D. (1951) A rapid phage plaque count method for the detection of bacteria as applied to the demonstration of internally-borne bacterial infections of seed. *Journal of Bacteriology* 61, 689–701.

Kay, S.J. and Stewart, A. (1994) Evaluation of fungal antagonists for control of onion white rot in soil box trials. *Plant Pathology* 43, 371–377.

Kendrick, E.L. and Purdy, L.H. (1959) Influence of environmental factors on the development of wheat bunt in the Pacific Northwest. *Phytopathology* 49, 433–434.

Keon, J.P.R., Broomfield, J.A. and Hargreaves, J.A. (1992) Molecular genetic analysis of carboxin resistance in *Ustilago maydis*. *Proceedings of the Brighton Crop Protection Conference – Pests and Diseases* 1, 221–226.

Keyworth, W.G. and Howell, J.S. (1961) Studies on silvering of red beet. *Annals of Applied Biology* 49, 173–194.

Khan, A.A. (1992) Preplant physiological seed conditioning. *Horticultural Reviews* 14, 131–181.

Khan, A.A., Kar-ling, T., Borkowska, B. and Powell, L.E. (1978) Osmotic conditioning of seeds; physiological and biological changes. *Acta Horticulturae* 83, 267–278.

Khan, A.A., Peck, N.H. and Samimy, C. (1980–81) Seed osmoconditioning: physiological and biochemical changes. *Israel Journal of Botany* 29, 133–144.

Khan, A.A., Abawi, G.S. and Maguire, J.D. (1992) Integrating matriconditioning and fungicidal treatment of table beet seed to improve stand establishment and yield. *Crop Science* 32, 231–237.

Kimble, K.A. Grogan, R.G., Greathead, A.S., Paulus, A.O. and House, J.K. (1975) Development, application, and comparison of methods for indexing lettuce seed for mosaic virus in California. *Plant Disease Reporter* 59, 461–464.

King, E.O., Ward, M.K. and Raney, D.E. (1954) Two simple media for the demonstration of pyocyanin and fluorescin. *Journal of Laboratory and Clinical Medicine* 44, 301–307.

Kishun, R. (1984) Seed treatment for the control of cabbage black rot. *Journal of Turkish Phytopathology* 13, 81–86.

Kistler, H.C., Bosland, P.W., Benny, U., Leong, S. and Williams, P.H. (1987) Relatedness of strains of *Fusarium oxysporum* from crucifers measured by examination of mitochondrial and ribosomal DNA. *Phytopathology* 77, 1289–1293.

Klemm, M. (1926) Zur phytopatholgischen Untersuchung von Samen. *Pflanzenbau* 2, 242–243.

Klisiewicz, J.M. and Pound, G.S. (1961) Studies on control of black rot of crucifers by treating seeds with antibiotics. *Phytopathology* 51, 495–500.

Kmetz, K., Ellet, C.W. and Schmitthener, A.F. (1979) Soybean seed decay: sources of inoculum and nature of infection. *Phytopathology* 69, 798–801.

Knox-Davies, P.S. (1979) Relationships between *Alternaria brassicicola* and *Brassica* seeds. *Transactions of the British Mycological Society* 73, 235–248.

Knypl, J.S. Janas, K.M. and Radziwonowska-Jozwiak, A. (1980) Is enhanced vigour in soybean (*Glycine max*) dependent on activation in protein turnover during controlled hydration of seeds? *Physiologie Vegetale* 18, 157–161.

Kobayashi, T. (1983) Quarantine for seeds in Japan. *Seed Science and Technology* 11, 1181–1187.

Kobayashi, T. (1990) Impact of seed-borne pathogens on quarantine in Japan. *Seed Science and Technology* 18, 427–433.

Koch, E. and Leadbeater, A.J. (1992) Phenylpyrroles – a new class of fungicides for seed treatment. *Brighton Crop Protection Conference – Pests and Diseases* 3, 1137–1146.

Kohen, P.D., Doughterty, W.G. and Hampton, R.O. (1992) Detection of pea seedborne potyvirus by sequence specific enzymatic amplification. *Journal of Virological Methods* 37, 253–258.

Kolev, N. (1984) Heat treatment of kidney bean seeds. *Rastitelna Zashchita* 32, 7–8.

Kommedahl, T. and Windels, C.E. (1986) Treatment of maize seeds. In: Jeffs, K.A. (ed.) *Seed Treatment.* BCPC Publications, Thornton Heath, UK, pp. 163–183.

Konzak, C.F. (1953) Inheritance of resistance in barley to physiologic races of *Ustilago nuda*. *Phytopathology* 43, 369–375.

Koshiek, H. and Jeffs, K.A. (1986) Assessment of application methods and prevention of loss. In: Jeffs, K.A. (ed.) *Seed Treatment.* BCPC Publications, Thornton Heath, UK, pp. 46–50.

Kotasthane, S.R., Gupta, O. and Khare, M.N. (1987) Influence of fungicidal seed treatment and soil amendment on the development of *Fusarium udum* propagules in soil and pigeon pea wilt. *Indian Phytopathology* 40, 197–200.

Kozlowski, T.T. (ed.) (1972) *Seed Biology*, Vol. 1. Academic Press, New York, 416 pp.

Kraft, J.M. (1982) Field and greenhouse studies on pea seed treatments. *Plant Disease* 66, 798–800.

Kranz, J. and Hau, B. (1980) Systems analysis in epidemiology. *Annual Review of Phytopathology* 18, 67–83.

Kreitlow, K.W., Lefebre, C.L., Presley, J.T. and Zaumeyer, W.J. (1961) Diseases that seed can spread. In: Stefferud, A. (ed.) *Seeds – the Yearbook of Agriculture*. United States Department of Agriculture, Washington DC, pp. 265–272.

Kritzman, G. and Netzer, D. (1978) A selective medium for the isolation and identification of *Botrytis* spp. from soil and onion seed. *Phytoparasitica* 6, 3–7.

Krout, W.S. (1921) Treatment of celery seed for the control of *Septoria* blight. *Journal of Agricultural Research* 21, 369–372.

Kublan, A. (1952) Barley and wheat loose smut and its control. *Deutsche Landwirtschaft* 3, 353–354.

Kuhn, J. (1873) Die Anwendung des Kupfervitrioles als Schutzmittel gegen den Steinbrand des Weizens. *Botanische Zeitung* 31, 502–505.

Kulik, M.M. (1983) The current scenario of the pod and stem blight stem canker-seed decay complex of soybean. *International Journal of Tropical Plant Diseases* 1, 1–11.

Kulik, M.M. (1984) New techniques for the detection of seed-borne pathogenic viruses, viroids, bacteria and fungi. *Seed Science and Technology* 12, 831–840.

Kumar, V. and Shetty, H.S. (1983) Seed-borne nature and transmission of *Botryodiplodia theobromae* in maize (*Zea maydis*). *Seed Science and Technology* 11, 781–789.

Lakshman, P. and Mohan, S. (1989) Effect of fungicides against sorghum downy mildew in Tamil Nadu. *Indian Journal of Mycology and Plant Pathology* 19, 111–113.

Lambat, A.K., Wadhi, S.R., and Sanwal, K.C. (1983) Seed health testing at the National Bureau of Genetic Resources, India. *Seed Science and Technology* 11, 1249–1257.

Lang, W. (1917) Zur Angsteckung der Gerste durch *Ustilago nuda*. *Bericht der Deutschen botanischen Gesellscaft* 35, 4–20.

Langcake, P., Kuhn, P.J. and Wade, M. (1983) The mode of action of systemic fungicides. In: Hutson, D.H. and Roberts, T.R. (eds) *Progress in Pesticide Biochemistry and Toxicology*, Vol. 3. John Wiley & Sons, New York, pp. 1–109.

Lange, L. (1986) The practical application of new developments in test procedures for the detection of viruses in seed. *Developments in Applied Biology* 1, 269–281.

Lanter, J.M., McGuire, J.M. and Goode, M.J. (1982) Persistence of tomato mosaic virus in tomato debris and soil under field conditions. *Plant Disease* 66, 552–555.

Large, E.C. (1950) *The Advance of the Fungi*, 3rd edn. Jonathan Cape, London, UK.

Latch, G.C.M. (1966) Fungous diseases of ryegrasses in New Zealand. II Foliage, root and seed diseases. *New Zealand Journal of Agricultural Research* 9, 809–819.

Latch, G.C.M. (1983) Incidence of endophytes in seedlines and their control with fungicides. *Proceedings of the New Zealand Grassland Association* 44, 251–253.

Lawton, M.B. and Burpee, L.L. (1990) Seed treatments for typhula blight and pink snow mold of winter wheat and relationships among disease intensity, crop recovery, and yield. *Canadian Journal of Plant Pathology* 12, 63–74.

Leadbitter, N.J., Nyfeler, R. and Elmshueser, H. (1994) The phenylpyrroles: the history of their development at Ciba. In: Martin, T.J. (ed.) *Seed Treatment: Progress and Prospects.* BCPC Monograph No. 57. BCPC Publications, Farnham, UK, pp. 129–134.

Leben, C. (1965) Influence of humidity on the migration of bacteria on cucumber seedlings. *Canadian Journal of Microbiology* 11, 671–676.

Leben, C. (1974) Survival of plant pathogenic bacteria. *Special Circular, Ohio Agricultural Research Development Center, Wooster* 100, 21 pp.

Leben, C. and Daft, G.C. (1966) Migration of bacteria on seedling plants. *Canadian Journal of Microbiology* 12, 1119–1123.

Leben, C. and Sleesman, J.P. (1981) Bacterial pathogens: reducing seed and *in vitro* survival by physical treatments. *Plant Disease* 65, 876–878.

Leblond, D. (1948) Relation entre le nombre de spores de maladies portées par les grain de semence et la manifestation de ces maladies dans le champ. *Report Quebec Society of Plant Pathology*, 1945–1947 141–145.

Lee, H. (1993) Infectious disease testing by ligase chain reaction. *Clinical Chemistry* 39, 729–730.

Lehman, S.G. (1952) Survival of the purple stain fungus in soybean seeds. *Phytopathology* 42, 285.

Lelliot, R.A. (1988) *Xanthomonas campestris* pv. *phaseoli* (Smith) Dye. In: Smith, I.M., Dunez, J., Phillips, D.H., Lelliot, R.A. and Archer, S.A. (eds), *European Handbook of Plant Diseases.* Blackwell Scientific Publications, Oxford, UK, pp. 165–166.

Lelliot, R.A. and Stead, D.E. (1987) *Methods for the Diagnosis of Bacterial Diseases.* Blackwell Scientific Publications, Oxford, UK.

Leroux, P. (1986) Caractéristiques des souches *d'Ustilago nuda*, agent du charbon nu de l'orge, résistantes à la carboxine. *Agronomie* 6, 225–226.

Leroux, P. and Berthier, G. (1988) Resistance to carboxin and fenfuram in *Ustilago nuda* (Jens.) Rostr., the causal agent of barley loose smut. *Crop Protection* 7, 16–19.

Leukel, R.W. (1936) The present status of seed treatment with special reference to cereals. *Botanical Review* 2, 498–527.

Leukel, R.W. (1948) Recent developments in seed treatment. *Botanical Review* 14, 235–265.

Leukel, R.W. (1953) Treating seeds to prevent disease. In: Stefferud, A. (ed.) *Plant Diseases – the Yearbook of Agriculture.* United States Department of Agriculture, Washington DC, pp. 134–145.

Levy, L. and Hadidi, A. (1991) Development of a reverse transcription/polymerase chain reaction assay for the identification of plum pox potyvirus from microgram quantities of total nucleic acids. *Phytopathology* 81, 1154.

Liang, J. (1983) Plant quarantine in the People's Republic of China. *Seed Science and Technology* 11, 1213–1216.

Lifshitz, R., Windham, M.T. and Baker, R. (1986) Mechanisms of biological control of preemergence damping-off of pea by seed treatment with *Trichoderma* spp. *Phytopathology* 76, 720––725.

Lin, C.Y. (1981) Studies on black rot of cruciferous crops caused by *Xanthomonas campestris* pv. *campestris* in Taiwan. *Plant Protection Bulletin, Taiwan* 23, 162–167.

Limonard, T. (1966) A modified blotter test for seed health. *Netherlands Journal of Plant Palhology* 72, 319–321.

Limonard, T. (1968) Ecological aspects of seed health testing. *Proceedings of the International Seed Testing Association* 33, 343–513.

Lipps, P.E. (1983) Survival of *Colletotrichum graminicola* in infested corn residues in Ohio. *Plant Disease* 67, 102–104.

Lipps, P.E. and Madden, L.V. (1988) Effect of triadimenol seed treatment and triadimefon foliar treatment on powdery mildew epidemics and grain yield of winter wheat cultivars. *Plant Disease* 72, 887–892.

Lobik, A.I. (1926) A method for the determination of the degree of contamination of grain with smut spores. *Défense des Plantes (Leningrad)* 2, 602–604.

Locke, J.C., Papavizas, G.C. and Lewis, J.A. (1979) Seed treatment with fungicides for the control of *Pythium* seed rot of peas. *Plant Disease Reporter* 63, 725–728.

Locke, T. (1986) Current incidence in the United Kingdom of fungicide resistance in pathogens of cereals. *Proceedings 1986 British Crop Protection Conference – Pests and Diseases* 2, 781–786.

Longden, P.C. (1975) Sugar beet seed pelleting. *ADAS Quarterly Review* 18, 73–80.

Lord, K.A., Jeffs, K.A. and Tuppen, R.J. (1971) Retention and distribution of dry powder and liquid formulations of insecticides and fungicides on commercially dressed cereal seed. *Pesticide Science* 2, 49–55.

Lukezic, F.L. (1974) Dissemination and survival of *Colletotrichum trifolii* under field conditions. *Phytopathology* 64, 57–59.

Lutchmeah, R.S. and Cooke, R.C. (1985) Pelleting of seed with the antagonist *Pythium oligandrum* for biological control of damping-off. *Plant Pathology* 34, 528–531.

Luthra, J.C. (1953) Solar energy treatment of wheat loose smut *Ustilago tritici* (Pers.) Rostr. *Indian Phytopathology* 6, 49–56.

Luthra, J.C. and Sattar, A. (1934) Some experiments on the control of loose smut, *Ustilago tritici* (Pers.) Jens. of wheat. *Indian Journal of Agricultural Science* 4, 177–199.

Lynch, J.M. (1992) Environmental implications of the release of biocontrol agents. In: Tjamos, E.C., Papavizas, G.C. and Cook, R.J. (eds) *Biological Control of Plant Diseases*. Plenum Press, New York.

Lyons, N.F. and Taylor, J.D. (1990) Serological detection and identification of bacteria from plants by the conjugated *Staphylococcus aureus* slide agglutination test. *Plant Pathology* 39, 584–590.

Mace, E.M., Bell, A.A. and Beckman, C.H. (eds) (1981) *Fungal Wilt Diseases of Plants*. Academic Press, New York.

Machacek, J.E. and Greaney, F.J. (1935) Studies on the control of root-rot diseases of cereals caused by *Fusarium culmorum* (W. G. SM.) Sacc. and *Helminthosporium sativum* P., K., and B. 111. Effect of seed treatment on the control of root rot and on the yield of wheat. *Scientific Agriculture* 15, 607–620.

Mackie, W.W. and Briggs, F.N. (1923) Fungicidal dusts for the control of bunt. *Bulletin of the California Agricultural Experiment Station* 364, 533–572.

Macneill, B.H. and Howard, H.A. (1959) A pea wilt new to Ontario. *Proceedings of the Canadian Phytopathological Society* 26, 13.

Maddaloni, J., Sala, M., Carletti, S. and Marquez, R. (1989) Effects of systemic fungicides on the viability of the endophyte fungus (*Acremonium coenophialum* Morgan-Jones and Gams) in seed of fescue (*Festuca arundinacea* Schreb.). *Informe Tecnico, Estacion Experimental Regional Agropecuaria Pergamino, Argentina* 225, 10 pp.

Maden, S. (1983) Transmissions of seedborne infections of *Ascochyta rabiei* (Pass.) Labr. to seedlings and its control. *Journal of Turkish Phytopathology* 12, 77–82.

MAFF (Ministry of Agriculture, Fisheries and Food) (1991) The seedmens' return for the year ending 31st May, 1991. *Statistical Returns for Agricultural Seeds*. Ministry of Agriculture, Fisheries and Food, Cambridge, UK, pp. 8–10.

Maguire, J.D., Gabrielson, R.L., Mulanax, M.W. and Russell, T.S. (1978) Factors affecting the sensitivity of 2,4-D assays of crucifer seed for *Phoma lingam*. *Seed Science and Technology* 6, 915–924.

Maklakova, G.F. (1964) Latent clubroot infection of crucifers. *Agrobiologiya* 6, 945–947.

Malalasekera, R.A.P. and Colhoun, J. (1969) *Fusarium* diseases of cereals V. A technique for the examination of wheat seed infected with *Fusarium culmorum*. *Transactions of the British Mycological Society* 52, 187–193.

Malik M.M.S. and Batts, C.C.V. (1960a) The infection of barley by loose smut (*Ustilago nuda* (Jens.) Rostr.) *Transactions of the British Mycological Society* 43, 117–125.

Malik, M.M.S. and Batts, C.C.V. (1960b) The development of loose smut (*Ustilago nuda*) in the barley plant, with observations on spore formation in nature and in culture. *Transactions of the British Mycological Society* 43, 126–131.

Malik, M.M.S. and Batts, C.C.V. (1960c) The determination of the reaction of barley varieties to loose smut. *Annals of Applied Biology* 48, 39–50.

Malin, E.M., Roth, D.A. and Belden, E.L. (1983) Indirect immunofluorescent staining for detection and identification of *Xanthomonas campestris* pv. *phaseoli* in naturally infected bean seed. *Plant Disease* 67, 645–647.

Malone, J.P. (1968) Mercury-resistant *Pyrenophora avenae* in Northern Ireland in seed oats. *Plant Pathology* 17, 41–45.

Malone, J.P. and Muskett, A.E. (1964) Seed-borne fungi. *Proceedings of the International Seed Testing Association* 29, 179–383.

Mandahar, C.L. (1981) Virus transmission through seed and pollen. In: Maramorosch, K. and Harris, K.F. (eds) *Plant Diseases and Vectors: Ecology and Epidemiology*. Academic Press, New York, pp. 241–292.

Manicom, B.Q., Bar-Joseph, M., Kotze, J.M. and Becker, M.M. (1990) A restriction length polymorphism probe relating vegetative compatibility groups and pathogenicity in *Fusarium oxysporum* f.sp. *dianthi*. *Phytopathology* 80, 336–339.

Mantle, P. (1988a) *Pyricularia oryzae* (Briosi & Cavara). In: Smith, I.M., Dunez, J., Phillips, D.H., Lelliot, R.A. and Archer, S.A. (eds) *European Handbook of Plant Diseases*. Blackwell Scientific Publications, Oxford, UK, pp. 331–333.

Mantle, P. (1988b) *Claviceps purpurea* (Fr.) Tul. In: Smith, I.M., Dunez, J., Phillips, D.H., Lelliot, R.A. and Archer, S.A. (eds) *European Handbook of Plant Diseases*. Blackwell Scientific Publications, Oxford, UK, pp. 274–276.

Marinescu, G. (1975) Recherches sur la désinfection des semences de tomates contamines par *Corynebacterium michiganense*. *Bulletin de l'Académie des Sciences Agricoles et Forestiéres* 5, 101–109.

Marrou, J. and Messiaen, C.M. (1967) The *Chenopodium quinoa* test: a critical method for detecting seed transmission of lettuce mosaic virus. *Proceedings of the International Seed Testing Association* 32, 49–57.

Marsh, R.W. (ed.) (1977) *Systemic Fungicides*, 2nd edn. Longman, London, UK.

Marshall, G.M. (1959) The incidence of certain seed-borne diseases in commercial seed samples. 1. Loose smut of barley *Ustilago nuda* (Jens.) Rostr. *Annals of Applied Biology* 47, 232–289.

Martin, H. and Woodcock, D. (1983) *The Scientific Principles of Crop Protection*, 7th edn. Edward Arnold, London, UK.

Martin, R.A. and Edgington, L.V. (1980) Effect of temperature on efficacy of triadimenol and fenapanil to control loose smut of barley. *Canadian Journal of Plant Pathology* 2, 201–204.

Martin, T.J. (ed.) (1988) *Application to Seeds and Soil*. BCPC Monograph No. 39. BCPC Publications, Thornton Heath, UK.

Martin, T.J. (ed.) (1994) *Seed Treatment Progress and Prospects*. BCPC Monograph No. 57. BCPC Publications, Farnham, UK.

Martin, T.J., Morris, D.B. and Chipper, M.E. (1981) Triadimenol seed treatment on spring barley; results of a 60 site evaluation in the United Kingdom, 1980. *Proceedings of the 1981 British Crop Protection Conference – Pests and Diseases* 1, 299–306.

Mathre, D.E., Johnston, R.H. and Grey, W.E. (1990) Effect of seed treatment in preventing transmission of dwarf bunt of winter wheat to new areas. *Pesticide Science* 30, 217–222.

Mathur, K., Siradhana, B.S. and Lodha, B.C. (1987) Studies on seedling blight of sorghum caused by *Gloeocercospora sorghi*. *Seed Science and Technology* 15, 851–858.

Mathys, G. and Baker, E.A. (1980) An appraisal of the effectiveness of quarantines. *Annual Review of Phytopathology* 18, 85–101.

Matthews, P., Dawson, J.R.O. and Hills, G.H. (1981) Pea seed-borne mosaic virus (PSbMV). *John Innes Seventy-First Annual Report for 1980*, 31–32.

Maude, R.B. (1960) Leaf and pod spot of peas. *Report of the National Vegetable Research Station for 1959*, 42.

Maude, R.B. (1962) Seed transmission of *Didymella* stem-rot of tomatoes. *Annals of Applied Biology* 50, 105–111.

Maude, R.B. (1963) Testing the viability of *Septoria* on celery seed. *Plant Pathology* 12, 15–17.

Maude, R.B. (1964) Studies on *Septoria* on celery seed. *Annals of Applied Biology* 54, 313–326.

Maude, R.B. (1966a) Pea seed infection by *Mycosphaerella pinodes* and *Ascochyta pisi* and its control by seed soaks in thiram and captan suspensions. *Annals of Applied Biology* 57, 193–200.

Maude, R.B. (1966b) Testing steam/air mixtures for control of *Ascochyta pisi* and *Mycosphaerella pinodes* on pea seed. *Plant Pathology* 15, 187–189.

Maude, R.B. (1966c) Studies on the etiology of black rot, *Stemphylium radicinum* (Meier, Drechsl. & Eddy) Neerg., and leaf blight, *Alternaria dauci* (Kuhn) Groves & Skolko, on carrot crops; and on fungicide control of their seed-borne infection phases. *Annals of Applied Biology* 57, 83–93.

Maude, R.B. (1967) Thiram soak for the control of fungal seedborne disease. *Proceedings of the 4th British Insecticide and Fungicide Conference* 1, 259–264.

Maude, R.B. (1973) Seed-borne diseases and their control. In: Heydecker, W. (ed.) *Seed Ecology*. Butterworths, London, UK, pp. 325–335.

Maude, R.B. (1977) Neck rot (*Botrytis allii*) in the autumn-sown bulb onion crop. *Report of the National Vegetable Research Station for 1976*, 94.

Maude, R.B. (1981) *Onion Neck Rot*. Leaflet 779. Ministry of Agriculture Fisheries and Food (Publications), Alnwick, UK, 5 pp.

Maude, R.B. (1983a) The correlation between seed-borne infection by *Botrytis allii* and neck rot development in store. *Seed Science and Technology* 11, 829–834.

Maude, R.B. (1983b) Eradicative seed treatments. *Seed Science and Technology* 11, 907–920.

Maude, R.B. (1986) Treatment of vegetable seeds. In: Jeffs, K.A. (ed.) *Seed Treatment*. BCPC Publications, Thornton Heath, UK, pp. 239–261.

Maude, R.B. (1988a) The value of seed health testing and rotational practices in the control of vegetable diseases. In: Clifford, B.C. and Lester, E. (eds) *Control of Plant Diseases*. Blackwell Scientific Publications, Oxford, UK, pp. 123–134.

Maude, R.B. (1988b) Seed and transplant treatments for the production of healthy vegetables. In: Rudd-Jones, D. and Langton, F.A. (eds) *Healthy Planting Material: Strategies and Technologies*. BCPC Monograph No. 33. BCPC Publications, Thornton Heath, UK, pp. 81–91.

Maude, R.B. (1990a) Storage diseases of onions. In: Rabinowich, H.D. and Brewster, J.L. (eds) *Onions and Allied Crops. Vol. 11. Agronomy, Biotic Interactions, Pathology, and Crop Protection*. CRC Press, Boca Raton, Florida, pp. 273–296.

Maude, R.B. (1990b) Application of biocontrol agents to seeds. In: Annual Report 1989. AFRC Institute of Horicultural Research, East Malling, Kent, UK, p. 14.

Maude, R.B. (1990c) Seed treatment. *Pesticide Outlook* 1, 16–22.

Maude, R.B. and Bambridge, J.M. (1991) Evaluation of seed treatments against *Alternaria dauci* (leaf blight) of naturally infected carrot seeds. *Tests of Agrochemicals and Cultivars, Annals of Applied Biology* 118 (Suppl. 12), 30–31.

Maude, R.B. and Humpherson-Jones, F.M. (1979) Canker of brassicas – studies on ascospore release. *Report of the National Vegetable Research Station for 1978*, 70–71.

Maude, R.B. and Humpherson-Jones, F.M. (1980a) Studies on the seed-borne phases of dark leaf spot (*Alternaria brassicicola*) and grey leaf spot (*Alternaria brassicae*) of brassicas. *Annals of Applied Biology* 95, 311–319.

Maude, R.B. and Humpherson-Jones, F.M. (1980b) The effect of iprodione on the seed-borne phase of *Alternaria brassicicola*. *Annals of Applied Biology* 95, 321–327.

Maude, R.B. and Kyle, A.M. (1970) Seed treatments with benomyl and other fungicides for the control of *Ascochyta pisi* on peas. *Annals of Applied Biology* 66, 37–41.

Maude, R.B. and Presly, A.H. (1977) Neck rot (*Botrytis allii*) of bulb onions. 1. Seed-borne infection and its relationship to the disease in the onion crop. *Annals of Applied Biology* 86, 163–180.

Maude, R.B. and Shuring, C.G. (1970) The persistence of *Septoria apiicola* on diseased celery debris in soil. *Plant Pathology* 19, 177–179.

Maude, R.B. and Suett, D.L. (1986) Application of fungicide to brassica seeds using a film-coating technique. *Proceedings 1986 British Crop Protection Conference – Pests and Diseases* 1, 237–242.

Maude, R.B. and Thompson, A.R. (1989) Pests and diseases of vegetables. In: Scopes, N. and Stables, L. (eds) *Pest and Disease Control Handbook*, 3rd edn., BCPC, Thornton Heath, UK, pp. 261–321.

Maude, R.B., Vizor, A.S. and Shuring, C.G. (1969) The control of fungal seed-borne diseases by means of a thiram seed soak. *Annals of Applied Biology* 64, 245–257.

Maude, R.B., Presly, A.H. and Dudley, C.L. (1973) Canker of brassicas. *Report of the National Vegetable Research Station for 1972*, 94.

Maude, R.B., Bambridge, J.M. and Presly, A.H. (1982) The persistence of *Botrytis allii* in field soil. *Plant Pathology* 31, 247–252.

Maude, R.B., Humpherson-Jones, F.M. and Shuring, C.G. (1984) Treatments to control *Phoma* and *Alternaria* infections of brassica seeds. *Plant Pathology* 33, 525–535.

Maude, R.B., Bambridge, J.M., Presly, A.H. and Phelps, K. (1985) Effects of field environment and cultural practices on incidence of *Botrytis allii* (neck rot) in stored bulb onions. *Quaderni della scuola di Specializzazione in Viticoltura ed Enologia* 9, 241–243.

Maude, R.B., Presly, A.H. and Lovett, J.F. (1986) Demonstration of the adherence of thiram to pea seeds using a rapid method of spectrophotometric analysis. *Seed Science and Technology* 14, 361–369.

Maude, R.B., Drew, R.L.K., Gray, D., Petch, G.M., Bujalski, W. and Nienow, A.W. (1992) Strategies for control of seed-borne *Alternaria dauci* (leaf blight) of carrots in priming and process engineering systems. *Plant Pathology* 41, 204–214.

Maury-Chovelon, V. (1984) Recherches sur la détection du virus de la mosaique de la laitue par le méthode immunoenzymatique: adaptation au contrôle de l'état sanitaire des graines. PhD thésis, Montpellier, 103 pp.

Mayer, A.M. and Poljakoff-Mayber, A. (1989) *The Germination of Seeds*. Pergamon Press, Oxford, UK, 270 pp.

McCallan, S.E.A. and Wellman, R.H. (1942) Fungicidal versus fungistatic. *Contributions from the Boyce Thompson Institute* 12, 451–463.

McCartney, H.A., Lacey, M.E. and Rawlinson, C.J. (1986). Dispersal of *Pyrenopeziza brassicae* spores from an oil-seed rape crop. *Journal of Agricultural Science, Cambridge* 107, 299–305.

McDonald, J.G. and Hamilton, R.I. (1972) Distribution of southern bean mosaic virus in the seed of *Phaseolus vulgaris*. *Phytopathology* 62, 387–389.

McGee, D.C. (1981) Seed pathology: its place in modern seed production. *Plant Disease* 65, 638–642.

McGee, D.C. (1986) Treatment of soybean seeds. In: Jeffs, K.A. (ed.) *Seed Treatment*. BCPC Publications, Thornton Heath, UK, pp. 185–200.

McGee, D.C. (1990) *Maize Diseases: a Reference Source for Seed Technologists*, 2nd edn. APS Press, St Paul, Minnesota.

McGee, D.C. (1992) *Soybean Diseases: a Reference Source for Seed Technologists*. APS Press, St Paul, Minnesota.

McGee, D.C. and Kellock, A.W. (1974) *Fusarium avenaceum*, a seed-borne pathogen of subterranean clover roots. *Australian Journal of Agricultural Research* 25, 549–557.

McGee, D.C. and Petrie, G.A. (1978) Variability of *Leptosphaeria maculans* in relation to blackleg of oilseed rape. *Phytopathology* 68, 625–630.

McIntyre, J.L., Sands, D.C. and Taylor, G.S. (1978) Overwintering, seed disinfestation and pathogenicity studies of the tobacco hollow stalk pathogen, *Erwinia caratovora* var. *caratovora*. *Phytopathology* 68, 435–440.

McKeen, C.D. (1950) Arasan as a seed and soil treatment for the control of damping-off in certain vegetables. *Scientific Agriculture* 30, 261–270.

McLean, K.S. and Roy, K.W. (1988) Purple seed stain of soybean caused by isolates of *Cercospora kikuchii* from weeds. *Canadian Journal of Plant Pathology* 10, 166–171.

McMeekin, D. (1973) Streptomycin inhibition of *Peronospora parasitica* and its host reversed by manganese and calcium. *Phytopathology* 63, 34–36.

McQuilken, M.P., Whipps, J.M. and Cooke, R.C. (1990) Control of damping-off in cress and sugar beet by commercial seed-coating with *Pythium oligandrum*. *Plant Pathology* 39, 452–462.

Medina, A.C. and Grogan, R.G. (1961) Seed transmission of bean mosaic viruses. *Phytopathology* 51, 452–456.

Megahed, El-S. and Moore, J.D. (1969) Inactivation of necrotic ringspot and prune dwarf viruses in seeds of some *Prunus* spp. *Phytopathology* 59, 1758–1760.

Menzies, J.G. and Jarvis, W.R. (1994) The infestation of tomato seed by *Fusarium oxysporum* f.sp. *radicis-lycopersici*. *Plant Pathology* 43, 378–386.

Mew, T.W., Gergon, E.B. and Merca, S.D. (1990) Impact of seedborne pathogens in rice germplasm exchange. *Seed Science and Technology* 18, 441–450.

Meyer, H. and Mayer, A.M. (1973) Dichloromethane and lettuce seed germination. *Science* 179, 96.

Milborrow, B.V. (1963) Penetration of seeds by acetone solutes. *Nature*, 199, 710.

Miles, G.F. (1946) 300 years of chemical seed treatment. *Agricultural Chemicals* 1, 22–25.

Millar, C.S. and Colhoun, J. (1969) *Fusarium* diseases of cereals VI. Epidemiology of *Fusarium nivale* on wheat. *Transactions of the British Mycological Society* 52, 195–204.

Millard, W.A. (1945) Canker and mosaic of broccoli. *Agriculture* 52, 39–42.

Miller, M.W. and de Whalley, C.V. (1981) The use of metalaxyl seed treatments to control downy mildew. *Proceedings 1981 British Crop Protection Conference – Insecticides and Fungicides* 1, 341–348.

Miller, P.W. and McWhorter, F.P. (1948) The use of vapour-heat as a practical means of disinfecting seeds. *Phytopathology* 38, 89–101.

Miller, S.A., Rittenburg, J.H., Petersen, F.P. and Grothaus, G.D. (1992) From research bench to the market place: development of commercial diagnostic kits. In: Duncan J.M. and Torrance, L. (eds) *Techniques for the Rapid Detection of Plant Pathogens.* Blackwell Scientific Publications, Oxford, UK, pp. 208–221.

Mink, G.I. (1993) Pollen- and seed-transmitted viruses and viroids. *Annual Review of Phytopathology* 31, 375–402.

Mitchell, L.A. (1988) A sensitive dot immunoassay employing monoclonal antibodies for detection of *Sirococcus strobilinus* in spruce seed. *Plant Disease* 72, 664–667.

Moffett, M.L. (1973) Seed transmission of *Pseudomonas phaseolicola* (halo blight) in *Macroptilium atropurpureum* cv. *siratro*. *Tropical Grasslands* 7, 195–199.

Mohajir, A-R., Arny, D.C. and Shands, H.L. (1952) Studies on the inheritance of loose smut resistance in spring barleys. *Phytopathology* 42, 367–373.

Mohammadi, O. (1992) Mycostop biofungicide – present status. In: Tjamos, E.C., Papavizas, G.C. and Cook, R.J. (eds) *Biological Control of Plant Diseases*. Plenum Press, New York, pp. 207–210.

Mohan, S.K. and Schaad, N.W. (1987) An improved agar plating assay for detecting *Pseudomonas syringae* pv. *syringae* and *P. s.* pv. *phaseolicola* in contaminated bean seed. *Phytopathology* 77, 1390–1395.

Moore, M.S. (1979) New systemic fungicides for the control of seed-borne and foliar diseases. *Proceedings of the New Zealand Weed and Pest Control Conference* 272–277.

Moore, R.T. (1972) Ustomycota, a new division of higher fungi. *Antonie van Leeuwenhoek Journal of Microbiology and Serology* 38, 567–584.

Moore, W.C. (1946) Seed-borne diseases. *Annals of Applied Biology* 33, 228–231.

Morris, D.B., Wainwright, A. and Meredith, R.H. (1994) New seed treatments based on bitertanol and tebuconazole for seed-borne disease control in wheat. In: Martin, T.J. (ed.) *Seed Treatment: Progress and Prospects*. BCPC Monograph No. 57. BCPC Publications, Farnham, UK, pp. 97–102.

Morschel, J.R.G. (1971) *Introduction to Plant Quarantine.* Government Publication Service, Canberra, Australia, 71 pp.

Morschel, J.R.G. (1983) Quarantine procedures for seed imported into Australia. *Seed Science and Technology* 11, 1231–1237.

Moseman, J.G. (1968) *Fungicidal Control of Smut Diseases of Cereals, CR 42-68*. USDA Agricultural Research Service, Beltsville, Maryland, 36 pp.

Mouya, I. (1992) Caracterisation de plusiers *Fusarium oxysporum* pathogenes de trois espèces de palmiers par la techniques RFLP. In: Renard, J.L. (ed.) *Mechanismes des Reactions de Defense du Palmier a Huile vis-a-vis du* Fusarium oxysporum *f.sp.* elaeidis. Document No. 2415. Institut de Recherches pour les Huiles et Oleagineaux, Montpellier.

Muchovej, J.J. and Dhingra, O.D. (1979) Benzene and ethanol for treatment of soybean seeds with systemic fungicides. *Seed Science and Technology* 7, 449–456.

Munn, M.T. (1917) Neck rot disease of onions. *Bulletin of the New York Agricultural Experiment Station* 437, 365–455.

Munn, M.T. (1919) The seed analysts responsibility with reference to seed-borne diseases. *Proceedings of the Official Seed Analysts* 11, 31–35.

Murant, A.F. (1981) Nepoviruses. In: Kurstak, E. (ed.) *Handbook of Plant Virus Infections and Comparative Diagnosis*. Elsevier/North Holland, London, UK, pp. 197–238.

Murant, A.F. (1983) Seed and pollen transmission of nematode-borne viruses. *Seed Science and Technology* 11, 973–987.

Murant, A.F. and Lister, R.M. (1967) Seed transmission in the ecology of nematode-transmitted viruses. *Annals of Applied Biology* 59, 63–76.

Muskett, A.E. (1937) A study of the epidemiology and control of *Helminthosporium* disease of oats. *Annals of Botany New Series* 1, 763–784.

Muskett, A.E. (1950) Seed-borne fungi and their significance. *Transactions of the British Mycological Society* 33, 1–12.

Muskett, A.E. and Malone, J. (1941) The Ulster method for the examination of flax seeds for the presence of seed-borne parasites. *Annals of Applied Biology* 28, 8–13.

Nakamura, S. (1986) Treatment of rice seeds. Part 1 – In Japan. In: Jeffs, K.A. (ed.) *Seed Treatment.* BCPC Publications, Thornton Heath, UK, pp. 139–154.

Narkiewicz-Jodko, M. (1975) Observations sur l'état des semences de tréfle violet pendent leur conservation. *Seed Science and Technology* 3, 731–735.

Nash, S.M. and Snyder, W.C. (1964) Dissemination of the root rot *Fusarium* with bean seed. *Phytopathology* 54, 880.

Naumann, K. and Karl, H. (1988) Possibilities of disinfecting bean seeds infected with *Pseudomonas syringae* pv. *phaseolicola*. *Nachrichtenblatt für den Pflanzenschutz in der DDR* 42, 204–208.

Navaratnam, S.J., Shuttleworth, D. and Walker, D. (1980) The effect of aerated steam on six seed-borne pathogens. *Australian Journal of Experimental Agriculture and Animal Husbandry* 20, 97–101.

Neergaard, P. (1977) *Seed Pathology*, Vols. 1 & 2. Macmillan, London, UK.

Neergaard, P. (1981) Risks for the EPPO region from seed-borne pathogens. *EPPO Bulletin* 11, 207–212.

Neergaard, P. (1986) Screening for plant health. *Annual Review of Phytopathology* 24, 1–16.

Neill, J.C. (1940) The endophyte of rye-grass (*Lolium perenne*). *New Zealand Journal of Science and Technology (Section A)* 21, 280–291.

Neill, J.C. and Hyde, E.O.C. (1939) Blindseed disease of ryegrass. *New Zealand Journal of Science and Technology (Section A)* 20, 281–301.

Nelson, P.E. (1981) Life cycle and epidemiology of *Fusarium oxysporum*. In: Mace, M.E., Bell, A.A. and Beckman, C.H. (eds) *Fungal Wilt Diseases of Plants.* Academic Press, New York, pp. 81–111.

Netzer, D. and Dishon, I. (1966) Occurrence of *Botrytis allii* in onions for seed production in Israel. *Plant Disease Reporter* 50, 21.

Netzer, D. and Kenneth, R.G. (1969) Persistence and transmission of *Alternaria dauci* (Kuhn) Groves & Skolko in the semi-arid conditions of Israel. *Annals of Applied Biology* 63, 289–294.

Nevill, D. and Burkard, N. (1988) The effect of polymer binder on the activity of insecticides applied to maize seeds. In: Martin, T.J. (ed.) *Application to Seeds and Soil.* BCPC Monograph No. 39. BCPC Publications, Thornton Heath, UK, pp. 221–227.

Neweigy, N.A., El-Din, I.F.G., Ziedan, M.I., Hanafy, E.A. and Neame, M.A.A. (1985) Effect of seed dressing of peas and soybean on their growth in soil infested with *S. rolfsii* or a mixture of three pathogenic fungi (*R. solani* + *F.solani* + *S. rolfsii*). *Annals of Agricultural Science, Moshtohor* 23, 1091–1104. Moshtohor, Tukh Kalubia, Egypt.

Newhook, F.J. (1947) The detection of browning (*Polyspora lini*) and wilt (*Fusarium lini*) in seed lines of linen flax. *New Zealand Journal of Science and Technology* 29, 44–48.

Nicholson, J.F., Dhingra, O.D. and Sinclair J.B. (1972) Internal seed-borne nature of *Sclerotinia sclerotiorum* and *Phomopsis* sp. and their effects on soybean quality. *Phytopathology* 62, 1261–1263.

Nielsen, J. (1983) Spring wheats immune or highly resistant to *Ustilago tritici*. *Plant Disease* 67, 860–863.

Nienow, A.W., Bujalski, W., Petch, G.M., Gray, D. and Drew, R.L.K. (1991) Bulk drying and priming of leeks: the effects of two polymers of polyethylene glycol and fluidised bed drying. *Seed Science and Technology* 19, 107–116.

Nienow, A.W., Bujalski, W., Maude, R.B. and Gray, D. (1994) The process engineering of vegetable seeds: priming, drying and coating. In: Hashim, M. A. (ed.) *Bioproducts Processing Technologies for the Tropics.* ICHEME Symposium Series no. 137, University of Malaysia, Kuala Lumpur, pp. 203–207.

Noble, M. (1948) *Sclerotinia spermophila* and other seed-borne fungi on clover. *Proceedings of the International Seed Testing Association* 14, 182–186.

Noble, M. (1951) Seed pathology. *Nature* 168, 1–8.

Noble, M. (1957) The transmission of plant pathogens by seed. In: Horton-Smith, C. (ed.) *Biological Aspects of Transmission of Disease.* Oliver & Boyd, Edinburgh, UK, pp. 81–85.

Noble, M. (1971) *Seed Pathology.* In: Western, J.H. (ed.) Diseases of Crop Plants. Macmillan, London, UK, pp. 21–36.

Noble, M. and Gray, E.G. (1945) Blind seed disease of ryegrass. *Scottish Journal of Agriculture* 25, 94–97.

Noble, M., de Tempe, J. and Neergaard, P. (1958) *An Annotated List of Seed-borne Diseases.* Commonwealth Mycological Institute, Kew, UK (in collaboration with the International Seed Testing Association), 159 pp.

Noble, M., MacGarvie, Q.D., Hams, A.R. and Leafe, E.L. (1966) Resistance to mercury of *Pyrenophora avenae* in Scottish seed oats. *Plant Pathology* 15, 23–28.

Nolan, P.A. and Campbell, R.N. (1984) Squash mosaic virus detection in individual seeds and seed lots of cucurbits by enzyme-linked immunosorbent assay. *Plant Disease* 68, 971–975.

Noon, R.A. (1986) Cereal seed treatment with fluatriafol-based mixtures. In: Rudd-Jones, D. and Langton, F.A. (eds) *Symposium on Healthy Planting Material.* BCPC Monograph No. 33. BCPC publications, Farnham, UK, pp. 161–168.

Noon, R.A. and Jackson, D. (1992) Alternatives to mercury for control of cereal seed-borne diseases. *Proceedings of the Brighton Crop Protection Conference – Pests and Diseases* 3, 1127–1136.

Noor-Hassan, R., Krstic, B. and Tosic, M. (1988) Contribution to the knowledge of tomato seed infection by mosaic virus and possibility of disinfection. *Zastita-Bilija* 39, 393–399.

Novotelnova, N.S. (1963) Character of parasitism of the causal agent of downy mildew of sunflower at seed infection. *Botanischeskii Zhurnal SSSR* 48, 845–860.

Ogilvie, L. (1969) *Diseases of Vegetables.* Bulletin No. 123, Ministry of Agriculture, Fisheries and Food. Her Majesty's Stationery Office, London, UK, 110 pp.

Old, K.M. (1968) Mercury-tolerant *Pyrenophora avenae* in seed oats. *Transactions of the British Mycological Society* 51, 525–534.

Old, R.W. and Primrose, S.B. (1985) *Principles of Gene Manipulation – an Introduction to Genetic Engineering; Studies in Microbiology,* Vol. 2, 3rd edn. Blackwell Scientific Publications, London, UK, 409 pp.

Olembo, S. (1983) Seed health testing at the plant quarantine station at Muguga, Kenya. *Seed Science and Technology* 11, 1217–1223.

Oliver, R. (1993) Nucleic acid-based methods for detection and identification. In: Fox, R.T.V. (ed.) *Principles of Diagnostic Techniques in Plant Pathology.* CAB International, Wallingford, Oxford, UK, pp. 153–169.

Olvang, H. (1984) Benomyl resistance in *Gerlachia nivalis.* 1. Survey of diseased plants in the field. *Zeitschrift für Pflanzenkrankheiten und Pflanzenschutz* 91, 294–300.

Oort, A.J.P. (1940) The dissemination of the spores of wheat loose smut (*Ustilago nuda*) through the air. *Tijdschrift over Pflantenziekten* 46, 1–18.

Ordish, G. (1976) *The Constant Pest.* Peter Davies; London, UK, 240 pp.

O'Rourke, C.J. (1976) *Diseases of Grasses and Forage Legumes in Ireland.* The Agricultural Institute, Dublin, Ireland, 115 pp.

Orth, H. (1939) Untersuchungen uber die biologie und Bekampfung des Erregers der Stengelfaule der Tomate (*Didymella lycopersici* (Kleb.)). *Zentralblatt für Bakteriologie* 100, 211–244.

Orton, C.R. (1931) Seed-borne parasites – a bibliography. *West Virginia Agricultural Experiment Station Bulletin* 245, 47 pp.

Osburn, R.M., Schroth, M.N., Hancock, J.G. and Hendson, M. (1989) Dynamics of sugar beet colonisation by *Pythium ultimum* and *Pseudomonas* species: effects on seed rot and damping-off. *Phytopathology* 79, 709–716.

Oyarzun, P., Gerlagh, M., Hoogland, A.E. and Vos, I. (1990) Seed treatment of peas with fosetyl-Al against *Aphanomyces euteiches. Netherlands Journal of Plant Pathology* 96, 301–311.

Papavizas, G.C. and Lewis, J.A. (1976) Acetone infusion of pyroxychlor into soybean seed for control of *Phytophthora megasperma* var. *sojae. Plant Disease Reporter* 60, 484–488.

Parida, T., Nayak, M. and Sridhar, R. (1990) Fate of carbendazim in rice tissues after seed or foliage treatments. *Pesticide Science* 30, 303–308.

Parke, J.L. (1990) Population dynamics of *Pseudomonas cepacia* in the pea spermosphere in relation to biocontrol of *Pythium. Phytopathology* 80, 1307–1311.

Patel, P.N. and Walker, J.C. (1965) Resistance in *Phaseolus* to halo blight. *Phytopathology* 55, 889–894.

Paulitz, T. (1992) Biological control of damping-off diseases with seed treatments. In: Tjamos, E.C., Papavizas, G.C. and Cook, R.J. (eds) *Biological Control of Plant Diseases.* Plenum Press, New York, pp. 145–156.

Paulus, A.O. and Nelson, J. (1977) Systemic fungicides for control of Phycomycetes on vegetable crops applied as seed treatments, granular or foliar sprays. *Proceedings of the 9th British Insecticide and Fungicide Conference* 3, 929–935.

Paveley, N.D. and Davies, J.M.L. (1994) Cereal seed treatment – risks, costs and benefits. In: Martin, T.J. (ed.) *Seed Treatment: Progress and Prospects.* BCPC Monograph No. 57. BCPC Publications, Farnham, UK, pp. 27–35.

Payne, P.A. and Williams, G.E. (1990) Hymexazol treatment of sugar-beet to control seedling disease caused by *Pythium* spp. and *Aphanomyces cochlioides. Crop Protection* 9, 371–377.

Peach, L., Maude, R.B. and Petch, G.M. (1994) Biocontrol of seedborne *Botrytis allii* using an antagonistic bacterium. In: Martin, T.J. (ed.) *Seed Treatment: Progress and Prospects.* BCPC Monograph No. 57. BCPC Publications, Farnham, UK, pp. 345–350.

Pedersen, P.N. (1960) Methods of testing the pseudoresistance of barley to infection by loose smut *Ustilago nuda* (Jens.) Rostr. *Acta Agriculture Scandinavica* 10, 312–332.

Pedersen, P.N. (1956) Infection of barley by loose smut. *Friesia* 5, 341–348.

Pederson, V.D. (1982) The influence of carboxin on seedling vigour and growth of barley plants from healthy and *Ustilago nuda* infected seed. *Phytopathology* 72, 971.

Pelet, F. (1971) Essais de traitments par la chaleur des graines de laitues en vue d'éliminer le virus de la mosaique. *Revue Horticole Suisse* 44, 172–177.

Pelham, J. (1972) Strain-genotype interaction of tobacco mosaic virus in tomato. *Annals of Applied Biology* 71, 219–228.

Perry, D.A. and Howell, P.J. (1965) Symptoms and nature of hollow heart in pea seed. *Plant Pathology* 14, 111–116

Persley, G.J. (1979) Studies on the survival and transmission of *Xanthomonas manihotis* on cassava seed. *Annals of Applied Biology* 93, 159–166.

Petch, G.M., Maude, R.B. and White, J.G. (1991a) Effect of film-coat layering of metalaxyl on the germination of carrot seeds, their emergence and the control of cavity spot. *Crop Protection* 10, 117–120.

Petch, G.M., Maude, R.B., Bujalski, W. and Nienow, A.W. (1991b) The effects of re-use of polyethylene glycol priming osmotica upon the development of microbial populations and germination of leeks and carrots. *Annals of Applied Biology* 119, 367–372.

Petrie, G.A. (1979) Blackleg of rape. *Canada Agriculture* 24, 22–25.

Phatak, H.C. (1974) Seed-borne plant viruses – identification and diagnosis in seed health testing. *Seed Science and Technology* 2, 3–155.

Phatak, H.C. (1980) The role of seed and pollen in the spread of plant pathogens particularly viruses. *Tropical Pest Management* 26, 278–285.

Phillips, A.J.L. (1992) A comparison of dust and acetone immersion applications of tolclofos-methyl to bean seeds for the control of *Rhizoctonia solani. Plant Pathology* 41, 35–40.

Piening, L. (1968) Development of barley net blotch from infected straw and seed. *Canadian Journal of Plant Science* 48, 623–625.

Popp, W. (1951) Infections in seeds and seedlings of wheat and barley in relation to the development of loose smut. *Phytopathology* 41, 261–275.

Popp, W. (1958) A new procedure for embryo examination for mycelium of smut fungi. *Phytopathology* 48, 261–275.

Porter, R.H. (1949) Recent developments in seed technology. *Botanical Review* 15, 221–344.

Prabhu, A.S. (1985) Evaluation of pyroquilon seed treatment for blast (Bl) control in upland rice. *International Rice Research Newsletter* 10, 13.

Prasad, K.V.V., Khare, M.N. and Jain, A.C. (1985) Site of infection and further development of *Colletotrichum graminicola* (Ces.) Wilson in naturally infected sorghum grains. *Seed Science and Technology* 13, 37–40.

Presly, A.H. (1977) *Botrytis* species on overwintered salad onions. PhD Thesis, University of Birmingham, Birmingham, UK, 169 pp.

Presly, A.H. and Maude, R.B. (1980) The effect of seed treatment with benzimidazole-based fungicides on infection of the foliage of overwintered salad onions by *Botrytis cinerea*. *Annals of Applied Biology* 94, 185–196.

Prew, R.D. (1981) Cropping systems in relation to soil-borne and trash-borne diseases of cereals. In: Jenkyn, J.F. and Plumb, R.T. (eds) *Strategies for the Control of Cereal Diseases*. Blackwell Scientific Publications, London, UK, pp. 149–156.

Priestley, R.H. and Bayles, R.A. (1988) The contribution and value of resistant cultivars to disease control in cereals. In: Clifford B.C. and Lester, E. (eds) *Control of Plant Diseases*. Blackwell Scientific Publications, Oxford, pp. 53–65.

Pritchard, D.W. (1974) Eradication of *Helminthosporium maydis* inside popcorn seed. *Phytopathology* 64, 757–758.

Prosen, D., Hatziloukas, E., Panopoulos, N.J. and Schaad, N.W. (1991) Direct detection of the halo blight pathogen *Pseudomonas syringae* pv. *phaseolicola* in bean seed by DNA amplification (abstract). *Phytopathology* 81, 1159.

Prosen, D., Hatziloukas, E., Schaad, N.W. and Panopoulos, N.J. (1993) Specific detection of *Pseudomonas syringae* pv. *phaseolicola* DNA in bean seed by polymerase chain reaction-based amplification of a phaseolotoxin gene region. *Phytopathology* 83, 965–970.

Pugh, G.J.F. (1973) Saprophytic fungi and seeds. In: Heydecker, W. (ed.) *Seed Ecology*. Butterworths, London, UK, pp. 337–345.

Purdy, L.H. and Kendrick, E.L. (1957) Influence of environmental factors on the development of wheat bunt in the Pacific Northwest. 1. Effect of soil moisture and soil temperature on spore germination. *Phytopathology* 47, 591–594.

Racke, K.D. and Coats, J.R. (eds) (1990) *Enhanced Biodegradation of Pesticides in the Environment*. American Chemical Society Symposium Series No. 426. American Chemical Society, Washington DC.

Randhawa, H.S. and Aulakh, K.S. (1984) Efficacy of hot water treatment to control seed-borne fungi of raya (*Brassica juncea* Coss). *Indian Journal of Plant Pathology* 2, 73–76.

Ralph, W. (1977a) Problems in testing and control of seed-borne bacterial pathogens: a critical evaluation. *Seed Science and Technology* 5, 735–752.

Ralph, W. (1977b) Steam–air treatment of bean seed infected with *Pseudomonas phaseolicola*. *Seed Science and Technology* 5, 559–565.

Ramakrishna, V. and Chatrath, M.S. (1987) Translocation and quantitative distribution of seed applied carbendazim in chickpea (*Cicer arietinum* L.) *Journal of Nuclear Agriculture and Biology* 16, 177–182.

Randhawa, P.S., Singh, N.J. and Schaad, N.W. (1987) Bacterial flora of cotton seeds and biocontrol of seedling blight caused by *Xanthomonas campestris* pv. *malvacearum*. *Seed Science and Technology* 15, 65–71.

Rao, B.M., Prakash, H.S., Shetty, S. and Safeeula, K.M. (1984) Techniques to detect seedborne inoculum of *Peronosclerospora sorghi* in maize. *Seed Science and Technology* 12, 593–599.

Rasmussen, O.F. and Reeves, J.C. (1992) DNA probes for the detection of plant pathogenic bacteria. *Journal of Biotechnology* 25, 203–220.

Rasmussen, O.F. and Wulff, B.S. (1990) Identification and use of DNA probes for plant pathogenic bacteria. *Proceedings 5th European Congress on Biotechnology*, 693–698.

Rasmussen, O.F. and Wulff, B.S. (1991) Detection of *Ps.* pv. *pisi* using PCR. *Proceedings 4th International Working Group on* Pseudomonas syringae *pathovars*, 369–376.

Rast, A.Th.B. and Stijger, C.C.M.M. (1987) Disinfection of pepper seed infected with different strains of capsicum mosaic virus by trisodium phosphate and dry heat treatment. *Plant Pathology* 36, 583–588.

Raut, J.G. (1983) Transmission of seed-borne *Macrophomina phaseolina* in sunflower. *Seed Science and Technology* 11, 807–814.

Reed, H.E. (1959) Studies on *Rhynchosporium secalis* (Oud.) Davis causing scald of barley. *Dissertation Abstracts* 20, 49.

Reeves, J.C. (1995) Nucleic acid techniques in testing for seedborne diseases. In: Skerritt, J.H. and Appels, R. (eds) *New Diagnostics in Crop Sciences.* CAB International, Wallingford, UK, pp. 127–149

Reeves, J.C. and Ball, S.F.L. (1991) Research note: preliminary results on the identification of *Pyrenophora* species using DNA polymorphisms amplified from arbitary primers. *Plant Varieties and Seeds* 4, 185–189.

Reeves, J.C., Rasmussen, O.F. and Simpkins, S.A. (1994) The use of a DNA probe and PCR for the detection of *Pseudomonas syringae* pv. *pisi* in pea seed. *Plant Pathogenic Bacteria, 8th International Conference*, 383–390.

Reid, R.D. (1970) Pelleting of seed: a review. *British Sugar Beet Review* 39, 79–80, 83.

Rennie, W. (1988) *Barley* (Hordeum vulgare) *Loose Smut* (Ustilago nuda), 3rd edn. Working Sheet No. 25. International Seed Testing Association, Zurich, Switzerland.

Rennie, W. (1990) *Pyrenophora teres* – net blotch. In: Richardson, M.J. (ed.) *An Annotated List of Seed-borne Diseases.* International Seed Testing Association, Zurich, Switzerland.

Rennie, W.J. (1993) Control of seed-borne pathogens in certification schemes. In: Ebbels, D. (ed.) *Plant Health and the European Single Market.* BCPC Monograph No. 54. BCPC Publications, Farnham, UK, pp. 61–68.

Rennie, W.J. and Cockerell, V. (1993) A review of seed borne pathogens in the post-mercury period. *Proceedings of Crop Protection in Northern Britain* 17–24.

Rennie, W.J. and Cockerell, V. (1994) Strategies for cereal seed treatment. In: Martin, T.J. (ed.) *Seed Treatment: Progress and Prospects.* BCPC Monograph No. 57. BCPC Publications, Farnham, UK, pp. 17–26.

Rennie, W.J., Richardson, M.J. and Noble, M. (1983) Seed-borne pathogens and the production of quality seed in Scotland. *Seed Science and Technology* 11, 1115–1127.

Rennie, W.J., Cockerell, V., Don, R. and Sommerville, J. (1990) Assessing the germination performance of *Monographella nivalis* infected winter wheat seed. *Proceedings of Crop Protection in Northern Britain*, 101–108.

Rhodes, D.J. and Powell, K.A. (1994) Biological seed treatments – the development process. In: Martin, T.J. (ed.) *Seed Treatment: Progress and Prospects*. BCPC Monograph No. 57. BCPC Publications, Farnham, UK, pp. 303–310.

Richardson, L.T. (1954) The persistence of thiram in soil and its relationship to the microbiological balance and damping-off control. *Canadian Journal of Botany* 32, 335–346.

Richardson, M.J. (1986) An assessment of the need for routine use of organomercurial seed treatment fungicides. *Field Crops Research* 13, 3–24.

Richardson, M.J. (1990) *An Annotated List of Seed-borne Diseases*, 4th edn. International Seed Testing Association, Zurich, Switzerland.

Riehm, E. (1913) Prufung einiger Mittel zur Bekampfung des Steinbrandes. *Mitteilungen der Kaiserlichen Bilogischen Anstalt fur Land-u Forstwirtschaft* 14, 8–9.

Roberts, D.M., Rollet, A.C. and Morris, D.B. (1988) The Baytan flowable delivery system – a closed system for the metering of liquid seed treatments. In: Martin, T.J. (ed.) *Application to Seeds and Soil*. BCPC Monograph No. 39. BPCP Publications, Thornton Heath, UK, pp. 163–169.

Roberts, D.P. and Lumsden, R.D. (1990) Effects of extracellular metabolites from *Gliocladium virens* on germination of sporangia and mycelial growth of *Pythium ultimum*. *Phytopathology* 80, 461–465.

Roberts, E.H. and Roberts, D.L. (1972). Moisture content of seeds. In: Roberts, E.H. (ed.) *Viability of Seeds*. Chapman and Hall, London, pp. 424–437.

Roberts, S.J. (1992) Effect of soil moisture on the transmission of pea bacterial blight (*Pseudomonas syringae* pv. *pisi*) from seed to seedling. *Plant Pathology* 41, 136–140.

Roberts, S.J. (1993) The threat of bacterial blight to UK pea crops. In: Ebbels, D. (ed.) *Plant Health and the European Single Market*. BCPC Monograph No. 54. BCPC Publishers, Farnham, UK, pp. 123–133.

Roberts, S.J., Reeves, J.C., Biddle, A.J., Taylor, J.D. and Higgins, P. (1991) Prevalence of pea bacterial blight in UK seed stocks, 1986–1990. *Aspects of Applied Biology* 27, 327–332.

Roberts, S.J., Phelps, K., Taylor, J.D. and Ridout, M. S. (1993) Design and interpretation of seed health assays. In: *Proceedings of the First ISTA Plant Disease Committee Symposium on Seed Health Testing*. Agriculture Canada Central Seed Laboratory Ottawa, Ontario, Canada, pp. 115–125.

Rohloff, I. (1963) Trials to inactivate lettuce mosaic virus in seed. *Gartenbauwissenschaft* 28, 19–28.

Rohloff, I. and Marrou, J. (1980) Lettuce mosaic virus in *Lactuca sativa*. In: *ISTA Handbook on Seed Health Testing*, 2nd edn. Working Sheet No. 9. International Seed Testing Association, Zurich, Switzerland.

Rollo, F., Amici, A., Foresi, F. and di Silvestro, I. (1987) Construction and characterisation of a cloned probe of *Phoma tracheiphila* in plant tissues. *Applied Microbiology and Biotechnology* 26, 352–357.

Roth, D.A. (1989) Review of extraction and isolation methods. In: Saettler, A.W., Schaad, N.W. and Roth, D.A. (eds) *Detection of Bacteria in Seed and Other Planting Material*. APS Press, St Paul, Minnesota, pp. 3–8.

Rowse, H.R. (1991) Methods of priming seeds. UK Patent No. 2192781.

Rowse, H.R. (1992) Methods of priming seeds. US Patent No. 5119589.

Rudolph, B.A. and Harrison, G.J. (1945) The invasion of the internal structure of cotton seed by certain Fusaria. *Phytopathology* 35, 542–548.

Russell, P.E. and Mussa, A.E.A. (1977) An evaluation of potential seed treatments to control *Fusarium solani* f.sp. *phaseoli*, the cause of foot and root rot of *Phaseolus vulgaris*. *Journal of Agricultural Science, Cambridge* 89, 235–238.

Russell, R.C. (1950) The whole embryo method of testing barley for loose smut as a routine test. *Scientific Agriculture* 30, 361–366.

Russell, R.C. (1961) The influence of ageing of seed on the development of loose smut in barley. *Canadian Journal of Botany* 39, 1741–1746.

Russell, R.C. and Popp, W. (1951) The embryo test as a method of forecasting loose-smut infection in barley. *Scientific Agriculture* 31, 559–565.

Ryder, E.J. (1964) Transmission of common lettuce mosaic virus through the gametes of the lettuce plant. *Plant Disease Reporter* 48, 522–524.

Ryder, E.J. (1970) Inheritence of resistance to common lettuce mosaic. *Journal of the American Society of Horticultural Science* 95, 378–379.

Sackston, W.E. (1981) Downy mildew of sunflower. In: Spencer, D.M. (ed.) *The Downy Mildews*. Academic Press, London, UK, pp. 545–575.

Sackston, W.E. (1983) Epidemiology and control of seed-borne *Verticillium* spp. causing vascular wilt. *Seed Science and Technology* 11, 731–747.

Saettler, A.W. (1971) Seedling injection as an aid to identifying bean blight bacteria. *Plant Disease Reporter* 55, 703–706.

Saettler, A.W., Schaad, N.W. and Roth, D.A. (1989) *Detection of Bacteria in Seed and Other Planting Material*. APS Press, St Paul, Minnesota, 122 pp.

Saifutdinova, M. (1985) Root rots of cucumber and tomato under cover. *Zashchita Rastenii* 1, 50.

Saiki, R.K., Gelfand, D.H., Stoffel, S., Scharf, S.J., Higuchi, R., Horn, G.T., Mullis, K.B. and Erlich, H.A. (1988) Primer-directed enzymatic amplification of DNA with a thermostable DNA polymerase. *Science* 239, 487–491.

Salazar, L.F. and Querci, M. (1992) Detection of viroids and viruses by nucleic acid probes. In: Duncan J.M. and Torrance, L. (eds) *Techniques for the Rapid Detection of Plant Pathogens*. Blackwell Scientific Publications, Oxford, UK, pp. 129–144.

Salter, W.J. and Smith, J.M. (1986) Peas – control of establishment pests and diseases using metalaxyl based seed coatings. *Aspects of Applied Biology* 12, 135–148.

Sampson, A.J., Cazenave, A., Laffranque, J-P., Glyn Jones, R., Kumazawa, S. and Chida, T. (1992) Metconazole, an advance in disease control in cereals and other crops. *Proceeding of the Brighton Crop Protection Conference – Pests and Diseases* 1, 419–426.

Satiya, D.V. and Hooda, I. (1987) A note on fungicidal control of damping-off of tomato and chilli caused by *Rhizoctonia solani* and *R. bataticola*. *Haryana Journal of Horticultural Sciences* 16, 294–297.

Saunders, W., Bedford, S.A. and MacKay, A. (1893) *Canada Experimental Farms Report*, pp. 41–42.

Sauve, E.M. and Shiel, R.S. (1980) Coating seeds with polvinyl resins. *Journal of Horticultural Science* 55, 371–373.

Schaad, N.W. (1979) Serological identification of plant pathogenic bacteria. *Annual Review of Phytopathology* 17, 123–147.

Schaad, N.W. (1982) *Report on First International Workshop on Seed Bacteriology (4–9 October, 1982) Angers, France.* International Seed Testing Secretariat, Zurich, Switzerland, 55 pp.

Schaad, N.W. (1988) *Laboratory Guide for the Identification of Plant Pathogenic Bacteria,* 2nd edn., APS Press, St Paul, Minnesota.

Schaad, N.W. (1989a) *Detection and identification of bacteria.* In: Saettler, A.W., Schaad, N.W. and Roth, D.A. (eds) *Detection of Bacteria in Seed and Other Planting Material.* APS Press, St Paul, Minnesota, pp. 9–16.

Schaad, N.W. (1989b) Detection of *Xanthomonas campestris* pv. *campestris* in crucifers. In: Saettler, A.W., Schaad, N.W. and Roth, D.A. (eds) *Detection of Bacteria in Seed and Other Planting Material.* APS Press, St Paul, Minnesota, pp. 68–75.

Schaad, N.W. and Donaldson, R.C. (1980) Comparison of two methods for detection of *Xanthomonas campestris* in infected crucifer seeds. *Seed Science and Technology* 8, 383–391.

Schaad, N.W. and White, W.C. (1974) Survival of *Xanthomonas campestris* in soil. *Phytopathology* 64, 1518–1520.

Schaad, N.W., Gabrielson, R.L. and Mulanax, M.W. (1980a) Hot acidified cupric acetate soaks for eradication of *Xanthomonas campestris* from crucifer seeds. *Applied and Environmental Microbiology* 39, 803–807.

Schaad, N.W., Sitterly, W.R. and Humaydan, H. (1980b) Relationship of incidence of seed-borne *Xanthomonas campestris* to black rot of crucifers in the field. *Plant Disease* 64, 91–92.

Schaad, N.W., Azad, H., Peet, R.C. and Panopoulos, N.J. (1989) Identification of *Pseudomonas syringae* pv. *phaseolicola* by a DNA hybridization probe. *Phytopathology* 79, 903–907.

Scheffer, R.Y. (1994) The seed industry's view on seed treatments. In: Martin, T.J. (ed.) *Seed Treatment: Progress and Prospects.* BCPC Monograph No. 57. BCPC Publications, Farnham, UK, pp. 311–314.

Schesser, K., Luder, A. and Henson, J.M. (1991) Use of polymerase chain reaction to detect take-all fungus, *Gaeumannomyces graminis,* in infected wheat plants. *Applied and Environmental Microbiology* 57, 553–556.

Schimmer, F.C. (1953) *Alternaria brassicicola* on summer cauliflower seed. *Plant Pathology* 2, 16–17.

Schmeer, H.E., Bluett, D.J., Meredith, R. and Heatherington, P.J. (1990) Field evaluation of imidacloprid as an insecticidal seed treatment in sugar beet and cereals with particlar reference to virus vector control. *Proceedings of the Brighton Crop Protection Conference – Pests and Diseases* 1, 29–36.

Schmeling, B. von and Kulka, M. (1966) Systemic fungicidal activity of 1,4-oxathiin derivatives. *Science* 152, 659–660.

Schnathorst, W.C. (1981) Life cycle and epidemilogy of *Verticillium.* In: Mace, M.E., Bell, A.A. and Beckman, C.H. (eds) *Fungal Wilt Diseases of Plants.* Academic Press, New York, pp. 81–111.

Schulteis, D.T. (1989) Nu-Flow AD: Apron/Chloroneb flowable fungicide for cottonseed treatment. In: *Proceedings, Beltwide Cotton Production Conference, Nashville, Tennessee, USA, 3–4 January 1989.* National Cotton Council of America, Memphis, Tennessee.

Schuster. M.L. and Coyne, D.P. (1974) Survival mechanisms of phytopathogenic bacteria. *Annual Review of Phytopathology* 12, 199–221.

Schuster, M.L. and Smith, C.C. (1983) Seed transmission and pathology of *Corynebacterium flaccumfaciens* in beans (*Phaseolus vulgaris*). *Seed Science and Technology* 11, 867–875.

Schwinn, F. (1994) Seed treatment – a panacea for plant protection? In: Martin, T.J. (ed.) *Seed Treatment: Progress and Prospects.* BCPC Monograph No. 57. BCPC Publications, Farnham, UK, pp. 3–14.

Schwinn, F., Nakamura, M. and Handschin, G. (1979) CGA 49104, a new systemic fungicide against rice blast. *Proceedings of the IX International Congress of Plant Protection,* Abstract No. 479.

Scott, J.M. (1989) Seed coatings and treatments and their effect on plant establishment. *Advances in Agronomy* 42, 44–77.

Scott, S.W. and Evans, D.R. (1984) Sclerotia of *Sclerotinia trifoliorum* in red clover seed. *Transactions of the British Mycological Society* 82, 567–569.

Seaman, W.L. and Wallen, V.R. (1967) Effect of exposure to radio frequency electric fields on seed-borne microorganisms. *Canadian Journal of Plant Science* 47, 39–49.

Sevilla, E.P. and Guerrero, F.C. (1983) Production of quality seed in the Philippines. *Seed Science and Technology* 11, 1139–1143.

Sharma, S.R. and Varma, A. (1975) Cure of seed transmitted cowpea banding mosaic disease. *Phytopathologische Zeitschrift* 83, 144–151.

Sharma, Y.R. and Chohan, J.S. (1971) Control by thermotherapy of seed-borne vegetable marrow mosaic virus. *Plant Protection Bulletin, FAO* 19, 86–88.

Shearer, B.L. and Wilcoxson, R.D. (1977) Pathogenicity and development of *Septoria avenae* f.sp. *triticea* on winter and spring rye and on spring barley and wheat. *Plant Disease Reporter* 61, 438–446.

Shearer, B.L. and Zadoks, J.C. (1972) Observations on the host range of an isolate of *Septoria nodorum* from wheat. *Netherlands Journal of Plant Pathology* 78, 153–159.

Shephard, M.C., Brent, K.J., Woolner, M. and Cole, A.M. (1975) Sensitivity to ethirimol of powdery mildew from UK barley crops. *Proceedings of the 8th British Crop Protection Conference* 1, 59–66.

Shepherd, R.J. (1972) Transmission of viruses through seed and pollen. In: Kado, C.I. and Agrawal, H.O. (eds) *Principles and Techniques in Plant Virology.* Van Nostrand Reinhold Company, New York, pp. 267–292.

Sheppard, J.W. (1983) Detection of seed-borne bacterial blights of bean. *Seed Science and Technology* 11, 561–567.

Sheppard, J.W. (1993) Diagnostic sensitivity, specificity and predictive values in evaluation of new test methods. In: *Proceedings of the First ISTA Plant Disease Committee Symposium on Seed Health Testing.* Agriculture Canada Central Seed Laboratory Ottawa, Ontario, Canada, pp. 132–142.

Sheppard, J.W., Wright, P.F. and DeSavigny, D.H. (1986) Methods for the evaluation of EIA tests for use in the detection of seed-borne diseases. *Seed Science and Technology* 14, 49–59.

Sheppard, J.W., Roth, D.A. and Saettler, A.W. (1989) Detection of *Xanthomonas* pv. *phaseoli* in bean. In: Saettler, A.W., Schaad, N.W. and Roth, D.A. (eds) *Detection of Bacteria in Seed.* APS Press, St Paul, Minnesota, pp. 17–29.

Sherf, A.F. and MacNab, A.A. (1986) *Vegetable Diseases and their Control.* John Wiley & Sons, New York.

Sheridan, J.E. (1966) Celery leaf spot: sources of inoculum. *Annals of Applied Biology* 57, 75–81.

Sheridan, J.E. (1971) The incidence and control of mercury-resistant strains of *Pyrenophora avenae* in British and New Zealand seed oats. *New Zealand Journal of Agricultural Research* 14, 469–480.

Sheridan, J.J. (1973) The survival of *Mycosphaerella pinodes* on pea haulm buried in soil. *Annals of Applied Biology* 75, 195–203.

Sheridan, J.E., Tickle, J.H. and Chin, Y.S. (1968) Resistance to mercury of *Pyrenophora avenae* (conidial state *Helminthosporium sativum*) in New Zealand seed oats. *New Zealand Journal of Agricultural Research* 11, 601–606.

Sheridan, J.E., Grbavic, N., and Ballard, L. (1983) Strategies for controlling net blotch in barley. *Proceedings of the Thirty-sixth New Zealand Weed and Pest Control Conference*, 242–246.

Shetty, H.S., Mathur, S.B. and Neergaard, P. (1980) *Sclerospora graminicola* in pearl millet seeds and its transmission. *Transactions of the British Mycological Society* 74, 127–134.

Shuckla, P., Singh, R.R. and Mishra, A.N. (1979) Efficacy of seed dressing fungicides for control of pea wilts. *Pesticides* 13, 40–41.

Siddiqui, M.R. (1983) Production of healthy seed in India. *Seed Science and Technology* 11, 1063–1070.

Siddiqui, M.R., Mathur, S.B. and Neergaard, P. (1983) Longevity and pathogenicity of *Colletotrichum* spp. in seed stored at 5°C. *Seed Science and Technology* 11, 353–361.

Siegel, M.R., Latch, G.C.M. and Johnson, M.C. (1987) Fungal endophytes of grasses. *Annual Review of Phytopathology* 25, 293–315.

Siegel, M.R., Varney, D.R., Johnson, M.C., Nesmith, W.C., Buckner, R.C., Bush, L.P., Burrus, P.B. and Hardison, J.R. (1984) A fungal endophyte of tall fescue: evaluation of control methods. *Phytopathology* 74, 937–941.

Silow, R.A. (1934) A systemic disease of red clover caused by *Botrytis anthophila* Bond. *Transactions of the British Mycological Society* 18, 239–248.

Simmonds, P.M. (1946) Detection of the loose smut fungi in embryos of barley and wheat. *Scientific Agriculture* 26, 51–58.

Sinclair, J.B. (1991) Latent infection of soybean plants and seeds by fungi. *Plant Disease* 75, 220–224.

Singh, D., Mathur, S.B. and Neergaard, P. (1980) Histological studies of *Alternaria sesamicola* penetration in sesame seed. *Seed Science and Technology* 8, 85–93.

Singh, K.G. (1983) Regional Asean collaboration in plant quarantine. *Seed Science and Technology* 11, 1189–1195.

Singh, S. (1958) Physiology and epidemiology of *Helminthosporium teres*. *Dissertation Abstracts* 18, 1951–1952.

Singh, T. and Sinclair, J.B. (1986) Further studies on the colonisation of soybean seeds by *Cercospora kikuchii* and *Phomopsis* sp. *Seed Science and Technology* 14, 71–77.

Singh, U.S. (1989) Studies on the systemicity of [14C] metalaxyl in cowpea (*Vigna unguiculata* (L.) Walp.). *Pesticide Science* 25, 145–154.

Singh, U.S., Kumar, J. and Tripathi, R.K. (1985) Uptake, translocation distribution and persistence of 14C-metalaxyl in pea (*Pisum sativum*). *Zeitschrift für Pflanzenkrankeiten und Pflanzenschutz* 92, 164–175.

Skoric, V. (1927) Bacterial blight of pea: overwintering, dissemination, and pathological histology. *Phytopathology* 17, 611–627.

Skoropad, W.P. (1959) Seed and seedling infection of barley by *Rhynchosporium secalis*. *Phytopathology* 49, 623–626.

Skvortzoff, S.S. (1937) A simple method for detecting hyphae of loose smut in wheat grains. *Plant Protection, Leningrad* 15, 90–91. Abstract in *Review of Applied Mycology* (1938), 17, 382.

Slawson, D. and Gillespie, J. (1994) Efficacy and physical/mechanical data requirements for approval of seed treatments in the UK. In: Martin, T.J. (ed.) *Seed Treatment: Progress and Prospects*. BCPC Monograph No. 57. BCPC Publications, Farnham, UK, pp. 441–448.

Smilanick, J.L., Hoffman, J.A., Secrest, L.R. and Wiese, K. (1988) Evaluation of chemical and physical treatments to prevent germination of *Tilletia indica* spores. *Plant Disease* 72, 46–51.

Smith, I.M. (1979) EPPO: the work of a regional plant protection organisation, with particular reference to phytosanitary regulations. In: Ebbels, D.L. and King, J.E. (eds) *Plant Health*. Blackwell Scientific Publications, Oxford, UK, pp. 13–22.

Smith, I.M. (1988) *Macrophomina phaseolina* (Tassi) Goidanich. In: Smith, I.M., Dunez, J., Phillips, D.H., Lelliot, R.A. and Archer, S.A. (eds) *European Handbook of Plant Diseases*. Blackwell Scientific Publications, Oxford, UK, pp. 404–405.

Smith, I.M. (1992) Introduction: practical implications of the development of new techniques. In: Duncan J.M. and Torrance, L. (eds) *Techniques for the Rapid Detection of Plant Pathogens*. Blackwell Scientific Publications, Oxford, UK, pp. 1–4.

Smith, I.M., Dunez, J., Phillips, D.H., Lelliot, R.A. and Archer, S.A. (eds) (1988) Basidiomycetes 1: Ustilaginales. In: *European Handbook of Plant Diseases*. Blackwell Scientific Publications, Oxford, UK, pp. 462–472.

Smith, P.R. (1966) Seed transmission of *Itersonilia pastinacae* in parsnip and its elimination by steam-air treatment. *Australian Journal of Experimental Agriculture and Animal Husbandry* 6, 441–444.

Smith, R.W. and Crossan, D.F. (1958) The taxonomy, etiology, and control of *Colletotrichum piperatum* (E.& E.) E. & H. and *Colletotrichum capsici* (Syd.) B.& B. *Plant Disease Reporter* 42, 1099–1103.

Smith, S.N. and Snyder, W.C. (1972) Germination of *Fusarium oxysporum* chlamydospores in soils favourable and unfavourable to wilt establishment. *Phytopathology* 62, 273–277.

Smith, V.L., Wilcox, W.F. and Harman, G.E. (1990) Potential for biological control of *Phytophthora* root and crown rots of apple by *Trichoderma* and *Gliocladium* spp. *Phytopathology* 80, 880–885.

Snel, M. and Edgington, L.V. (1970) Uptake, translocation and decomposition of systemic oxathiin fungicides in bean. *Phytopathology* 60, 1708–1716.

Snyder, W.C. (1932) Seed dissemination in *Fusarium* wilt of pea. *Phytopathology* 22, 253–257.

Snyder, W.C. and Baker, K.F. (1950) Occurrence of *Phoma lingam* in California as a subterranean pathogen of certain crucifers. *Plant Disease Reporter* 34, 21–22.

Snyder, W.C. and Nash Smith, S. (1981) Current status. In: Mace, M.E., Bell, A.A. and Beckman, C.H. (eds) *Fungal Wilt Diseases of Plants.* Academic Press, New York, pp. 81–111.

Snyder, W.C. and Wilhelm, S. (1962) Seed transmission of *Verticillium* wilt of spinach (abstract). *Phytopathology* 52, 365.

Soper, D. (1991) *Product Guide to Seed Treatments Available in the UK,* 2nd edn. BCPC Publications, Farnham, UK, 29 pp.

Soper, D. (1992) *1991 Product Guide to Seed Treatments Available in the UK – 1992 Up-date,* 2nd edn update. BCPC Publications, Farnham, UK.

Soteros, J.J. (1979a) Detection of *Alternaria radicina* and *A. dauci* from imported carrot seed in New Zealand. *New Zealand Journal of Agricultural Research* 22, 185–190.

Soteros, J.J. (1979b) Pathogenicity and control of *Alternaria radicina* and *A. dauci* in carrots. *New Zealand Journal of Agricultural Research* 22, 191–196.

Spek, J. van der (1972) Internal carriage of *Verticillium dahliae* by seeds and its consequences. *Mededelingen van de Faculteit Landbouwwetenschappen Rijksuniversiteit, Gent* 37, 567–573.

Spencer, D.M. (ed.) (1981) *The Downy Mildews.* Academic Press, London. 636 pp.

Splittstosser, W.E. and Mohamed-Yasseen, Y. (1991) Scanning electron microscopic studies on onion (*Allium cepa* L.) seeds and their relation to viability. *Proceedings of the Interamerican Society for Tropical Horticulture* 35, 127–132.

Srinivasan, M.C., Neergaard, P. and Mathur, S.B. (1973) A technique for detection of *Xanthomonas campestris* in routine seed health testing of crucifers. *Seed Science and Technology* 1, 853–859.

Stanghellini, M.E. and Hancock, J.G. (1971) The sporangium of *Pythium ultimum* as a survival structure in the soil. *Phytopathology* 61, 157–164.

Stead, D.E. (1992) Techniques for detecting and identifying plant pathogenic bacteria. In: Duncan J. M. and Torrance, L. (eds) *Techniques for the Rapid Detection of Plant Pathogens.* Blackwell Scientific Publications, Oxford, UK, pp. 76–111.

Stead, D.E. and Pemberton, A.W. (1987) Recent problems with *Pseudomonas syringae* pv. *phaseolicola* in the UK. *EPPO Bulletin* 17, 291–294.

Stedman O.J. (1982) The effect of three herbicides on the number of spores of *Rhynchosporium secalis* on barley stubble and volunteer plants. *Annals of Applied Biology* 100, 271–279.

Stienstra, T., Kommendahl, T., Stromberg, C.A., Matyac, C.A., Windels, C.E. and Morgan, F. (1985) Suppression of corn head smut with seed and soil treatments. *Plant Disease* 69, 301–302.

Stijger, C.C.M.M. and Rast, A.Th.B. (1988) Prospects of dry heat treatment at 70°C for disinfection of pepper seed infected with capsicum mosaic virus. *Mededelingen van de Faculteit Landbouwwetenschappen Rijksuniversiteit Gent* 53, 473–478.

Stirrup, H.H. and Ewan, J.W. (1931) *Investigations on Celery Diseases and their Control.* Bulletin No. 25. Ministry of Agriculture and Fisheries, London, UK, 34 pp.

Stott, I.P.H., Noon, R.A. and Heaney, S.P. (1990) Fluatriafol, ethirimol and thiabendazole seed treatment – an update on field performance and resistance monitoring. *Proceedings of the Brighton Crop Protection Conference – Pests and Diseases* 3, 1169–1174.

Strandberg, J.O. (1983) Infection and colonisation of inflorescences and mericarps of carrot by *Alternaria dauci. Plant Disease* 67, 1351–1353.

Strandberg, J.O. (1984) Efficacy of fungicides against persistence of *Alternaria dauci* on carrot seed. *Plant Disease* 68, 39–42.

Strass, F. (1964) A contribution to the knowledge on the occurrence of barley loose smut. *Bayerisches Landwirtschaftliches Jahrbuch* 41, 306–321.

Strider, D.L. (1979) Detection of *Xanthomonas nigromaculans* f.sp. *zinniae* in zinnia seed. *Plant Disease Reporter* 63, 869–873.

Stubbs, L.L. and O'Loughlin, G.T. (1962) Climatic elimination of mosaic spread in lettuce seed crops in the Swan Hill region of the Murray Valley. *Australian Journal of Experimental Agriculture and Animal Husbandry* 2, 16–19.

Suett, D.L. (1988) Application to seeds and soil: recent developments, future prospects and potential limitations. *Proceedings of the Brighton Crop Protection Conference – Pests and Diseases* 2, 823–832.

Suett, D.L. (1990) The threat of accelerated degradation of pesticides – myth or reality? *Proceedings of the Brighton Crop Protection Conference – Pests and Diseases* 3, 897–906.

Suett, D.L. (1991) Enhanced insecticide degradation in soil – implications for effective pest control. *Pesticide Outlook* 2, 7–12.

Suett, D.L. and Maude, R.B. (1988) Some factors the unifomity of film-coated seed treatments and their implications for biological performance. In: Martin, T.J. (ed.) *Application to Seeds and Soil.* BCPC Monograph No. 39. BCPC Publications, Thornton Heath, UK, pp. 25–32.

Suett, D.L., Whitfield, C.E. and Jukes, A.A. (1985) Residue studies – pesticide distribution following film-coating. *Report of the National Vegetable Research Station for 1984*, 33.

Sutton, J.C. (1982) Epidemiology of wheat blight and maize ear rot caused by *Fusarium graminearum. Canadian Journal of Plant Pathology* 4, 195–209.

Sutton, M.D. and Bell, W. (1954) The use of aureomycin as a treatment of swede seed for the control of black rot (*Xanthomonas campestris*). *Plant Disease Reporter* 38, 547–552.

Suzuki, H. (1930) Experimental studies on the possibility of primary infection of *Pyricularia oryzae* and *Ophiobolus miyabeanus* internal of rice seed. *Annals of the Phytopathological Society of Japan* 2, 245–275.

Suzuki, H. (1975) Meteorological factors in the epidemiology of rice blast. *Annual Review of Phytopathology* 13, 239–256.

Swaroop, S. (1951) The range of variation of the most probable number of organisms estimated by the dilution method. *Indian Journal of Medical Research* 39, 107–134.

Swinburne, T.R. (1963) Infection of wheat by *Tilletia caries* (DC.) Tul., the causal organism of bunt. *Transactions of the British Mycological Society* 46, 145–156.

Sykes, G. (1965) *Disinfection and Sterilization*, 2nd edn. E & FN Spon, London, UK.

Tahvonen, R. (1978) Seed-borne fungi on parsley and carrot. *Journal of the Scientific Agricultural Society of Finland* 50, 91–102.

Talavera-Williams, C.G., Pacek, A.W., Bujalski, W. and Nienow, A.W. (1991) A feasibility study of the bulk priming and drying of tomato seeds. *Transactions of the Institute of Chemical Engineering, Part C, Food and Bioproducts Processing* 69, 134–144.

Tamietti, G. and Garibaldi, A. (1984) Eradication of *Pseudomonas syringae* pv. *phaseolicola* in bean seeds in Italy. In: Quacquarelli, A. and Casano, F.J. (eds) *Phy-*

tobacteriology and Plant Bacterial Diseases of Quarantine Significance. Instituto Sperimentale per la Patologia Vegetale, Rome, Italy, pp. 69–76.

Tao, K.L., Khan, A.A., Harman, G.E. and Eckenrode, E.K. (1974) Practical significance of application of chemicals in organic solvents to dry seeds. *Journal of the American Society of Horticultural Science* 99, 217–220.

Tapke, V.F. (1931) Influence of humidity on floral infection of wheat and barley by loose smut. *Journal of Agricultural Research* 43, 503–516.

Tapke, V.F. (1948) Environment and the cereal smuts. *Botanical Review* 14, 360–402.

Tarr, S.A.J. (1972) *The Principles of Plant Pathology.* Macmillan, London, UK, 632 pp.

Taya, R.S., Tripathi, N.N. and Panwar, M.S. (1990) Influence of texture and nutritional status of soil on the efficacy of fungicides for the control of dry root-rot of chickpea (*Cicer arientinum* L.). *Indian Journal of Mycology and Plant Pathology* 20, 14–20.

Taylor, A.G. and Harman, G.E. (1990) Concepts and technologies of selected seed treatments. *Annual Review of Phytopathology* 28, 321–339.

Taylor, A.G., Klein, D.E. and Whitlow, T.H. (1988) SMP: solid matrix priming of seeds. *Scientia Horticulturae* 37, 1–11.

Taylor, A.G., Min, T.G., Harman, G.E. and Jin, X. (1991) Liquid coating formulation for the application of biological seed treatments of *Trichoderma harzianum*. *Biological Control* 1, 16–22.

Taylor, J.D. (1970a) Studies on halo-blight disease of beans caused by *Pseudomonas phaseolicola* (Burkh.) Dowson. PhD thesis, University of Bath, Bath, UK.

Taylor, J.D. (1970b) The quantitative estimation of the infection of bean seed with *Pseudomonas phaseolicola* (Burkh.) Dowson. *Annals of Applied Biology* 66, 29–36.

Taylor, J.D. (1970c) Bacteriophage and serological methods for the identification of *Pseudomonas phaseolicola* (Burkh.) Dowson. *Annals of Applied Biology* 66, 387–395.

Taylor, J.D. (1972) Field studies on halo-blight of beans (*Pseudomonas phaseolicola*) and its control by foliar sprays. *Annals of Applied Biology* 70, 191–197.

Taylor, J.D. (1986) Bacterial blight of compounding pea. *Proceedings of the 1986 British Crop Protection Conference – Pests and Diseases* 2, 733–736.

Taylor, J.D. and Dudley, C.L. (1977) Seed treatment for the control of halo-blight of beans (*Pseudomonas phaseolicola*). *Annals of Applied Biology* 85, 223–232.

Taylor, J.D., Dudley, C.L. and Presly, L. (1979a) Studies of halo-blight seed infection and disease transmission in dwarf beans. *Annals of Applied Biology* 93, 267–277.

Taylor, J.D., Phelps, K. and Dudley, C.L. (1979b) Epidemiology and strategy for the control of halo-blight of beans. *Annals of Applied Biology* 93, 167–172.

Taylor, J.D., Day, J.M. and Dudley, C.L. (1983) The effect of *Rhizobium* inoculation and nitrogen fertiliser on nitrogen fixation and seed yield of dry beans (*Phaseolus vulgaris*). *Annals of Applied Biology* 103, 419–429.

Taylor, J.D., Bevan, J.R., Crute, I.R. and Reader, S.L. (1989) Genetic relationship between races of *Pseudomonas syringae* pv. *pisi* and cultivars of *Pisum sativum*. *Plant Pathology* 38, 364–375.

Taylor, J.D., Phelps, K. and Roberts, S.J. (1993) Most probable number (mpn) method: origin and application. In: *Proceedings of the First ISTA Plant Disease Committee Symposium on Seed Health Testing.* Agriculture Canada Central Seed Laboratory Ottawa, Ontario, Canada, pp. 106–114.

Teng, P.S. (1985) A comparison of simulation approaches to epidemic modelling. *Annual Review of Phytopathology* 23, 351–379.

Tenover, F.C. (1988) Diagnostic deoxyribonucleic acid probes for infectious diseases. *Clinical Microbiological Reviews* 1, 82–101.

Teverson, D.M. (1991) Genetics of pathogenicity and resistance in the halo-blight disease of beans in Africa. PhD thesis, University of Birmingham, Birmingham, UK.

Teviotdale, B.L. and Hall, D.H. (1976) Factors affecting inoculum development and seed transmission of *Helminthosporium gramineum*. *Phytopathology* 66, 295–301.

Thakur, D.P. (1983) Epidemiology and control of ergot disease of pearl millet. *Seed Science and Technology* 11, 797–806.

Thakur, D.P. and Kanwar, Z.S. (1977) Internal seed-borne infection and heat therapy in relation to downy mildew of *Pennisetum typhoides* Staff. and Hubb. *Science and Culture* 43, 432–434.

Thomas, C.A. (1959) Control of pre-emergence damping-off and two leaf diseases of sesame by seed treatments. *Phytopathology* 49, 461–463.

Thomas, M.D. and Leary, J.V. (1980) A new race of *Pseudomonas glycinea*. *Phytopathology* 70, 310–312.

Thomas, P.L. (1974) The occurrence of loose smut of barley on commercially grown cultivars possessing genes for resistance from Jet. *Canadian Journal of Plant Science* 54, 453–455.

Thomas, P.L. and Metcalfe, D.R. (1984) Loose smut resistance in two introductions of barley from Ethiopia. *Canadian Journal of Plant Science* 64, 255–260.

Thresh, J.M. (1974) Temporal patterns of virus spread. *Annual Review of Phytopathology* 12, 111–128.

Thresh, J.M. (1976) Gradients of plant virus diseases. *Annals of Applied Biology* 82, 381–406.

Thresh, J.M. (1978) The epidemiology of plant virus diseases. In: Scott, P.R. and Bainbridge, A. (eds) *Plant Disease Epidemiology*. Blackwell Scientific Publications, Oxford, UK, pp. 79–91.

Tichelaar, G.M. (1967) Studies on the biology of *Botrytis allii* on *Allium cepa*. *Netherlands Journal of Plant Pathology* 73, 157–160.

Tillet, M. (1755) Dissertation sur la cause qui corrompt et noircit les grains de bled dans le épis, et sur moyens de prevenir ces accidens. (English translation by H. H. Humphrey in *Phytopathological Classics*, 1937, No. 5, 191 pp. American Phytopathological Society, Ithaca.)

Tomlin, C. (ed.) (1994) *The Pesticide Manual*, 10th edn. BCPC Publications, Farnham, UK.

Tomlinson, J.A. (1962) Control of lettuce mosaic virus by the use of healthy seed. *Plant Pathology* 11, 61–64.

Tomlinson, J.A. (1987) Epidemiology and control of virus diseases of vegetables. *Annals of Applied Biology* 110, 661–681.

Tomlinson, J.A. and Faithfull, E.M. (1973) Lettuce mosaic virus. *Report of the National Vegetable Research Station for 1972*, 97.

Tonkin, J.H.B. (1984) Pelleting and other pre-sowing treatments. In: Thompson, J.R. (ed.) *Advances in Research and Technology of Seeds – Part 9*. International Seed Testing Association, Wageningen, the Netherlands, pp. 94–127.

Torrance, L. (1992) Serological methods to detect plant viruses: production and use of monoclonal antibodies. In: Duncan J.M. and Torrance, L. (eds) *Techniques for the Rapid Detection of Plant Pathogens*. Blackwell Scientific Publications, Oxford, UK, pp. 7–33.

Tourte, C. and Manceau, C. (1991) Direct detection of *Pseudomonas syringae* pathovar *phaseolicola* using the polymerase chain reaction. *Proceedings of the 4th International Working Group on* Pseudomonas syringae *pathovars*, 402–403.

Trigalet, A. and Bidaud, P. (1978) Some aspects of epidemiology of bean halo-blight. *Proceedings of the Fourth International Conference on Plant Pathogenic Bacteria*, 895–902.

Trigalet, A., Samson, R. and Coleno, A. (1978) Problems related to the use of serology in phytobacteriology. *Proceedings of the Fourth International Conference on Plant Pathogenic Bacteria*, 271–288.

Trinci, A.P.J. and Ryley, J.F. (eds) (1984) *Mode of Action of Antifungal Agents*. Cambridge University Press, Cambridge, UK, 405 pp.

Triplett, L.L. and Haber, A.H. (1973) Dichloromethane and lettuce seed germination. *Science* 179, 95–96.

Trujillo, G.E. and Saettler, A.W. (1979) A combined semi-selective medium and serology test for the detection of *Xanthomonas* blight bacteria in bean seed. *Journal of Seed Science and Technology* 4, 35–41.

Tschanz, A.T., Horst, R.K. and Nelson, P.E. (1975) Ecological aspects of ascospore discharge in *Gibberella zeae*. *Phytopathology* 65, 597–599.

Tsvetkova, N.A. and Guseva, T.A. (1980) The effect of *Aphanomyces euteiches* Drechs., pathogen of root rot of pea, and Tachigaren on the chemical composition of pea plants. *Byulleten' Vsesoyuznogo Nauchno-Issledovatel'skogo Instituta Zashchity Rastenii* 47, 45–49.

Tu, J.C. (1983) Epidemiology of anthracnose caused by *Colletotrichum lindemuthianum* on white bean (*Phaseolus vulgaris*) in southern Ontario: Survival of the pathogen. *Plant Disease* 67, 402–404.

Tu, J.C. (1988) Development and evaluation of new seed treatment formulations for pea root rots. *Mededelingen van de Faculteit der Landbouwwetenschappen van de Rijksuniversiteit Gent* 40, 389–393.

Tu, J.C. (1992) *Colletotrichum lindemuthianum* on bean: population dynamics of the pathogen and breeding for resistance. In: Bailey, J.A. and Jeger, M.J. (eds) Colletotrichum *Biology, Pathology and Control*. CAB International, Wallingford, UK, pp. 203–224.

Tucker, H. and Foster, R.E. (1958) Statistical technique in indexing lettuce seed for mosaic content. *Plant Disease Reporter* 42, 1339–1341.

Tull, J. (1733) Of smuttiness. In: *The Horse-hoing Husbandry: or, an Essay on the Principles of Tillage and Vegetation*. G. Strabon, Cornhill, UK.

Tullis, E.C. (1936) Fungi isolated from discoloured rice kernals. *Technical Bulletin of the United States Department of Agriculture* 540, 1–11.

Umekawa, M. (1987) Studies on angular leaf spot of cucumber. *Bulletin of the Vegetable and Ornamental Crops Station, Iwate, Japan, Series B* 7, 82–85.

Umekawa, M. and Watanabe, Y. (1978) Dry heat and hot water treatments of cucumber seeds for control of angular leaf spot. *Bulletin of the Vegetable and Ornamental Crops Station, Iwate, Japan, Series B* 2, 55–61.

Uppal, B.N. and Malelu, J.S. (1928) A preliminary report on experiments in the control of grain smut of Jowar (*Andropgon sorghum*). *Agricultural Journal of India* 23, 471–472.

Urbain, W.M. (1986) *Food Irradiation (Food Science and Technology)*. Academic Press, New York.

Utikar, P.G., Gadre, W.A. and More, B.B. (1978) Seed treatment to control *Fusarium* wilt of pea. *Pesticides* 12, 29–30.

Utkhede, R.S. and Rahe, J.E. (1980) Biological control of onion white rot. *Soil Biology and Biochemistry* 12, 101–104.

Van der Plank, J.E. (1963) *Plant Diseases: Epidemics and Control*. Academic Press, New York, 349 pp.

Van der Plank, J.E. (1975) *Principles of Plant Infection*. Academic Press, New York, 210 pp.

Van Vuurde, J.W.L. and Maat, D.Z. (1983) Routine application of ELISA for the detection of lettuce mosaic virus in lettuce seeds. *Seed Science and Technology* 11, 505–513.

Van Vuurde, J.W.L. and Maat, D.Z. (1985) Enzyme-linked immunosorbent assay (ELISA) and disperse-dye immuno assay (DIA): comparison of simultaneous and separate incubation of sample and conjugate for the routine testing of lettuce mosaic virus and pea early-browning virus in seeds. *Netherlands Journal of Plant Pathology* 91, 3–13.

Van Vuurde, J.W.L. and Van Den Bovenkamp, G.W. (1981) Routine methods for the detection of halo-blight (*Pseudomonas phaseolicola*) in beans. In: *Report of the 17th International Workshop on Seed Pathology*. International Seed Testing Association, Zurich, Switzerland, p. 22.

Vanderwalle, R. (1942) Note sur le biologie *d'Ustilago nuda tritici* Schaf. *Bulletin de l'Institut Agronomique et des Stations de Recherches de Gembloux* 11, 103–113. (From *Review of Applied Mycology* 25, 159–160.)

Vanine, S.I. and Kotchkina, E.M. (1932) Methods of pathological investigation of the seeds of arboreal species. *Bulletin of the Leningrad Institute for Controlling Farm and Forest Pests* 2, 285–297.

Varma, B.K. (1990) Plant quarantine inspection, procedures and facilities for the import and export of seeds and vegetative propagules at the International Crops Research Institute for the Semi-Arid Tropics (ICRISAT). *Tropical Pest Management* 36, 216–219.

Vaughan, D.A., Kunwar, I.K., Sinclair, J.B. and Bernard, R.L. (1988) Routes of entry of *Alternaria* sp. into soybean seed coats. *Seed Science and Technology* 16, 725–731.

Venette, J.R. and Kennedy, B.W. (1975) Naturally produced aerosols of *Pseudomonas glycinea*. *Phytopathology* 65, 737–738.

Venis, M.A. (1969) Streptomycin inhibition of protein synthesis in peas reversed by divalent cations. *Nature* 221, 1147–1178.

Verma, V.S. (1971) Effect of heat on seed transmission of mosaic disease of cowpea (*Vigna sinensis* Savi). *Acta Microbiologica Polonica, Series B* 3, 163–165.

Vincent, J.M. (1970) *A Manual for the Practical Study of the Root-nodule Bacteria*. IBP Handbook No. 15. Blackwell Scientific Publications, Oxford, UK.

Vivian, A. (1992) Identification of plant pathogenic bacteria using nucleic acid technology. In: Duncan J.M. and Torrance, L. (eds) *Techniques for the Rapid Detection of Plant Pathogens*. Blackwell Scientific Publications, Oxford, UK, pp. 145–161.

Von der Pahlen, A. and Crnko, J. (1965) El virus del mosaico de la lechuza (*Marmor lactucae* Holmes) en Mendoza en Beunos Aires. *Revista Investigacions Agropecurias* 11, 25–31.

Vyas, S.C., Andotra, P.S. and Joshi, L.K. (1981) Effects of systemic fungicides on control of root rot on vegetables caused by *Rhizoctonia bataticola* and plant growth. *Pesticides* 15, 22–24.

Wain, R.L. and Carter, G.A. (1977) Nomenclature and definitions. In: Marsh, R.W. (ed.) *Systemic Fungicides*, 2nd edn. Longman, London, UK, pp. 1–5.

Wainwright, A., Morris, D.B. and Meredith, R.H. (1994) New seed treatments based on biteranol, tebuconazole and triazoxide for seed-borne disease control in barley. In: Martin, T.J. (ed.) *Seed Treatment: Progress and Prospects.* BCPC Monograph No. 57. BCPC Publications, Farnham, UK, pp. 103–108.

Walker, A. and Welch, S. (1990) Enhanced biodegradation of dicarboximide fungicides in soil. In: Racke, K.D. and Coats, J.R. (eds) *Enhanced Biodegradation of Pesticides in the Environment.* American Chemical Society Symposium Series No. 426. American Chemical Society, Washington DC, pp. 53–67.

Walker, A., Entwistle, A.R. and Dearnaley, N.J. (1984) Evidence for enhanced degradation of iprodione in soils treated previously with this fungicide. In: Hance, R.J. (ed.) *Soils and Crop Protection Chemicals.* Monograph No. 27. BCPC Publications, Croydon, UK, pp. 117–123.

Walker, A., Brown, P.A. and Entwistle, A.R. (1986) Enhanced degradation of iprodione and vinclozolin in soil. *Pesticide Science* 17, 183–193.

Walker, J.C. (1923) The hot water treatment of cabbage seed. *Phytopathology* 13, 251–253.

Walker, J.C. (1926) *Botrytis* neck rots of onions. *Journal of Agricultural Research* 33, 893–928.

Walker, J.C. (1934) Production of cabbage seed free from *Phoma lingam* and *Bacterium campestre. Phytopathology* 24, 158–160.

Walker, J.C. (1950) The mode of seed infection by the cabbage black-rot organism (abstract). *Phytopathology* 40, 30.

Walker, J.C. (1969) *Plant Pathology*, 3rd edn. McGraw-Hill, New York.

Walker, J.C. and Patel, P.N. (1964) Splash dispersal and wind as factors in the epidemiology of halo blight of bean. *Phytopathology* 54, 140–141.

Walkey, D.G.A. (1991) *Applied Plant Virology*, 2nd edn. Chapman & Hall, London, UK, 338 pp.

Walkey, D.G.A. and Dance, M.C. (1979) High temperature inactivation of seedborne lettuce mosaic virus. *Plant Disease Reporter* 63, 125–129.

Wallace, H.R. (1978) Dispersal in time and space: soil pathogens. In: Horsfall, J.G. and Cowling, E.B. (eds) *Plant Disease, an Advanced Treatise, Vol. II.* Academic Press, New York, pp. 181–202.

Wallen, V.R. (1953) Treatment of vegetable seed for improved emergence. *Plant Disease Reporter* 37, 620–622.

Wallen, V.R. (1964) Host–parasite relations and environmental influences in seedborne diseases. In: Smith, H. and Taylor J. (eds) *Microbial Behaviour* in vivo *and* in vitro. Cambridge University Press, Cambridge, UK, pp. 187–212.

Wallen, V.R. and Bell, W. (1956) Treatment of vegetable seed for improved emergence – 1955. *Plant Disease Reporter* 40, 129–132.

Wallen, V.R. and Hoffman, I. (1959) Fungistatic activity of captan in pea seedlings after treatment of the seeds or roots of seedlings. *Phytopathology* 49, 680–683.

Wallen, V.R. and Skolko, A.J. (1951) A comparison of seed testing methods in relation to the stem-break and browning disease of flax, caused by *Polyspora lini* Laff. *Canadian Journal of Botany* 29, 138–142.

Wallen, V.R. and Sutton, M.D. (1965) *Xanthomonas phaseoli* var. *fuscans* (Burkh.) Starr & Burkh. on field bean in Ontario. *Canadian Journal Of Botany* 43, 437–446.

Wallen, V.R., Cuddy, T.F. and Grainger, P.N. (1967) Epidemiology and control of *Ascochyta pinodes* on field peas in Canada. *Canadian Journal of Plant Science* 47, 395–403.

Wallen, V.R., Wong, S.I. and Jeun, J. (1967) Isolation, incidence and virulence of *Ascochyta* spp. of peas from soil. *Canadian Journal of Botany* 45, 2243–2247.

Walther, D. and Gindrat, D. (1987) Biological control of *Phoma* and *Pythium* damping-off of sugar-beet with *Pythium oligandrum*. *Journal of Phytopathology* 119, 167–174.

Wang, D., Woods, R.D., Cockbain, A.J., Maule, A.J. and Biddle, A.J. (1993) The susceptibility of pea cultivars to pea seed-borne mosaic virus infection and virus seed transmission in the UK. *Plant Pathology* 42, 42–47.

Warham, E.J., Prescott, J.M. and Griffiths, E. (1989) Effectiveness of chemical seed treatments in controlling Karnal bunt disease of wheat. *Plant Disease* 73, 585–588.

Warne, L.G.G. (1943) A case for club-root of swedes due to a seed-borne infection. *Nature* 152, 509.

Warwick, D.N.R., Urben, A.F., Tenente, R.C.V. and Fonseca, J.N.L. (1983) Plant quarantine activities at the National Centre of Genetic Resources (EMBRAPA) in Brazil. *Seed Science and Technology* 11, 1225–1229.

Waterworth, H.E. and White, G.A. (1982) Plant introductions and quarantine: the need for both. *Plant Disease* 66, 87–90.

Watson, D.R.W. and Dye, D.W. (1971) Detection of bacterial disease in New Zealand garden pea stocks. *Plant Disease Reporter* 55, 517–521.

Watson, R.D., Coltrin, L. and Robinson, R. (1951) The evaluation of materials for heat treatment of peas and beans. *Plant Disease Reporter* 35, 542–544.

Webster, J. (1951) Graminicolous Pyrenomycetes II. The occurrence of the perfect stage of *Helminthosporium teres* in Britain. *Transactions of the British Mycological Society* 43, 309–317.

Webster, J. (1986) *Introduction to Fungi*, 2nd edn. Cambridge University Press, Cambridge, UK.

Weller, D.M. (1988) Biological control of soilborne plant pathogens in the rhizosphere with bacteria. *Annual Review of Phytopathology* 26, 379—407.

Weller, D.M. and Saettler, A.W. (1980) Evaluation of seedborne *Xanthomonas phaseoli* and *X. phaseoli* var. *fuscans* as primary inocula in bean blights. *Phytopathology* 70, 148–152.

Wells, H.D., Burton, G.W. and Jackson, J.E. (1958) Burning of dormant dallasgrass shows promise for controlling ergot caused by *Claviceps paspali* Stev. & Hall. *Plant Disease Reporter* 42, 362–363.

Welty, R.E. (1986) Detecting viable *Acremonium* endophytes in leaf sheaths and meristems of tall fescue and perennial ryegrass. *Plant Disease* 70, 431–435

Weltzien, H.C. (1957) Untersuchungen uber den Befall von Winterweizen durch *Tilletia tritici* (Bjerk.) Winter unter besonderer Berucksichtigung der Frage der Beizmittelresistenz. *Phytopathologische Zeitschrift* 29, 121–150.

Wenzl, H. (1959) The importance of seed infection of beet seed by *Cercospora beticola* and its control. *Pflanzenschutzberichte* 23, 33–58

Whipps, J.M. and Lumsden, R.D. (1991) Biological control of *Pythium* species. *Biocontrol Science and Technology* 1, 75–90.

White, J.G. (1986) The association of *Pythium* spp. with cavity spot and root dieback of carrots. *Annals of Applied Biology* 108, 265–273.

Whitehead, R. (ed.) (1995) *The UK Pesticide Guide 1995*. CAB International, Wallingford/British Crop Protection Council, Farnham, UK.

Williams, J., Kubelik, A., Lival, K., Rafalski, J. and Tingey, S. (1990) DNA polymorphisms amplified by arbitary primers are useful as genetic markers. *Nucleic Acids Research* 18, 6531–6535.

Williams, M.J., Backman, P.A. and Crawford, M.A. (1982) Fungicidal control of a fungal endophyte in seed and established plants of tall fescue. *Phytopathology* 72, 971.

Williams, P.H. (1973) Report of the crucifer discussion group – 8 September, 1973. In: *Second International Congress of Plant Pathology, Minneapolis, Minnesota*, pp. 1–3.

Williams, P.H., Wade, E.K. and Norgren, R.L. (1973) Recommendations for minimizing the threat of blackleg and black rot of cabbage. *Extension Bulletin CPD, University of Wisconsin* 78, 1–3.

Willingale, J. and Mantle, P.G. (1987) Stigmatic constriction in pearl millet following infection by *Claviceps fusiformis*. *Physiological and Molecular Plant Pathology* 30, 247–257.

Willingale, J., Mantle, P.G. and Thakur, R.P. (1986) Postpollination stigmatic constriction, the basis of ergot resistance in selected lines of pearl millet. *Phytopathology* 76, 536–539.

Wilson, R.D. (1947) Rainfall in relation to the production of bean seed free of bacterial diseases. *Agricultural Gazette of New South Wales* 58, 15–20.

Wold, A. (1983) Opening addresses. International Symposium on Seed Pathology, Copenhagen 11–16 October 1982. *Seed Science and Technology* 11, 464–466.

Wolfe, M.S. and Dinoor, A. (1973) The problem of fungicide tolerance in the field. *Proceedings of the 7th British Insecticide and Fungicide Conference* 1, 11–19.

Wolfe, M.S., Minchin, P.N. and Slater, S.E. (1984) Dynamics of triazole sensitivity in barley mildew, nationally and locally. *Proceedings of the 1984 British Crop Protection Conference – Pests and Diseases* 2, 465–470.

Wood, P.McR. and Barbetti, M.J. (1977) The role of seed infection in the spread of blackleg of rape in Western Australia. *Australian Journal of Experimental Agriculture and Animal Husbandry* 17, 1040–1044.

Woodcock, D. (1978) Microbial degradation of fungicides, fumigants and nematicides. In: Hill, I.R. and Wright, S.J.L. (eds) *Pesticide Microbiology*. Academic Press, New York, pp. 731–797.

Worthing, C.R. and Hance, R.J. (eds) (1991) *The Pesticide Manual*, 9th edn. BCPC Publications, Farnham, UK.

Wu, W.S. and Lu, J.H. (1984) Seed treatment with antagonists and chemicals to control *Alternaria brassicicola*. *Seed Science and Technology* 12, 851–862.

Wurster, D.E. (1959) Air-suspension technique of coating drug particles. *Journal of the American Pharmaceutical Association* 48, 451–454.

Wylie, S., Wilson, C.R., Jones, R.A.C. and Jones, M.G.K. (1993) A polymerase chain reaction assay for cucumber mosaic virus in lupin seeds. *Australian Journal of Agricultural Research* 44, 41–51.

Yao, C.L., Fredriksen, R.A. and Magill, C.W. (1990) Seed transmission of sorghum downy mildew: detection by DNA hybridisation. *Seed Science and Technology* 18, 201–207.

Yarham, D.J. (1988) The contribution and value of cultural practices to control arable crop diseases. In: Clifford, B.C. and Lester, E. (eds) *Control of Plant Diseases: Costs and Benefits*. Blackwell Scientific Publications, London, UK, pp. 135–154.

Yarham, D.J. and Jones, D.R. (1992) The forgotten diseases: why should we remember them. *Proceedings of the Brighton Crop Protection Conference – Pests and Diseases* 3, 1117–1126.

Yarham, D.J. and Norton, J. (1981) Effects of cultivation methods on disease. In: Jenkyn, J.F. and Plumb, R.T. (eds) *Strategies for the Control of Cereal Diseases*. Blackwell Scientific Publications, London, UK, pp. 157–166.

Yarwood, C.E. (1938) *Botrytis* infection of onion leaves and seed stalks. *Plant Disease Reporter* 22, 428–429.

Ye, H.Z. (1980) On the biology of the perfect stage of *Fusarium graminearum* Sch. *Acta Phytophylacica Sinica* 7, 35–42.

Zadoks, J.C. and Schein, R.D. (1979) *Epidemiology and Plant Disease Management*. Oxford University Press, Oxford, UK, 427 pp.

Zaumeyer, W.J. (1929) Seed infection by *Bacterium phaseoli*. *Phytopathology* 19, 96.

Zaumeyer, W.J. (1932) Comparative pathological histology of three bacterial diseases of bean. *Journal of Agricultural Research* 44, 605–632.

Zazzerini, A., Capelli, C. and Panattoni, L. (1985) Use of hot-water treatment as a means of controlling *Alternaria* spp. on safflower seeds. *Plant Disease* 69, 350–351.

Zeigler, R.S. and Alvarez, E. (1987) Bacterial sheath brown rot of rice caused by *Pseudomonas fuscovaginae* in Latin America. *Plant Disease* 71, 592–597.

Zeigler, R.S. and Alvarez, E. (1989) Grain discoloration of rice caused by *Pseudomonas glumae* in Latin America. *Plant Disease* 73, 368.

Zimmer, R.C., McKeen, W.E. and Campbell, C.G. (1990) Development of *Peronospora ducometi* in buckwheat. *Canadian Journal of Plant Pathology* 12, 247–254.

Zink, F.W., Grogan, R.G. and Welch, J.E. (1956) The effect of the percentage of seed transmission upon subsequent spread of lettuce mosaic virus. *Phytopathology* 46, 662–664.

Zinnen, T.M. and Sinclair, J.B. (1982) Thermotherapy of soybean seeds to control seedborne fungi. *Phytopathology* 72, 831–834.

Index

The scientific and common names and the crop hosts of the organisms, most of which are seedborne pathogens, are listed in Tables 1–4 of this book. The names of organisms applied to seeds for biological control purposes are given in this index.